MAINTAINABILITY OF FACILITIES

For Building Professionals

MAINTAINABILITY OF FACILITIES

For Building Professionals

CHEW Yit Lin, Michael

Department of Building, National University of Singapore

World Scientific

NEW JERSEY • LONDON • SINGAPORE • BEIJING • SHANGHAI • HONG KONG • TAIPEI • CHENNAI

Published by

World Scientific Publishing Co. Pte. Ltd.

5 Toh Tuck Link, Singapore 596224

USA office: 27 Warren Street, Suite 401-402, Hackensack, NJ 07601

UK office: 57 Shelton Street, Covent Garden, London WC2H 9HE

British Library Cataloguing-in-Publication Data
A catalogue record for this book is available from the British Library.

MAINTAINABILITY OF FACILITIES: FOR BUILDING PROFESSIONALS

ISBN-13 978-981-4291-75-0
ISBN-10 981-4291-75-7

Typeset by Stallion Press
Email: enquiries@stallionpress.com

Printed in Singapore by Mainland Press Pte Ltd.

ACKNOWLEDGMENTS

I would like to thank the research team in NUS for their invaluable contribution to the work. A special thanks to Das Sutapa for her assistance in conducting the research for the past four years, with such passion and devotion. A big thanks to DeSilva Nayanthara, who helped to lead the initial research team and her very capable teammates Tan Phay Ping, Tan Shan Shan, Elizabeth Elsa Soemara, Petikirige Ranga Yashoman Edirisinghe, Kang Kok Hin and Indira Siagian.

I would like to acknowledge the valuable contribution from colleagues and friends from the industry:

Mr Chin Chi Leong — BCA
Ms Yvonne Soh — BCA
Mr Tay Kwee Hian — HDB
Dr Liu Tai Ker — RSP Architects Planners and Engineers Pte Ltd
Mr Michael Wong — Hitchins
Mr James Lim — Uniseal Pte Ltd
Francis Chan — Vision Waterproofing Pte Ltd
Ms Wee-Lim Eng Hon — Diethelm Industries Pte Ltd
Mr Jean Edouard Perrine (Eddy) — Permatsteelisa Pacific Holdings Ltd
Mr Michael Lim — Centre for Cleaning Technology Pte Ltd
Mr Sunny Teo — Tractel Singapore Pte Ltd
Mr Stefan Gisin — Alcan Alucobond (Far East) Pte Ltd
Mr David Lim — Integral Components Equipment Pte Ltd
Mr Wong Chung Wan — Setsco Services Pte Ltd
P.P. Choo and Goh Joo Wee — MHE-Dematic (S) Pte Ltd
Eddie Loh and Cedric Yip — Campaign Cleaning Services (Pte) Ltd
Steven Tan — Shaw Organisation

Tan Teck Bin — JTC Corporation
Dr D. Martin-Vosshage — Fraunhofer Representative Office Singapore

As well as colleagues and students in NUS:

Lam Tsz Shan, Chan Hui Chuan, Hing Kai Sin, Ho Kar Woon, Chan Huang Teng, Francis Linda Lourdes, Kwok Hon Meng Andy, Lye Mei Sian, Ng Chew Peng Camellia, Quek Yuan Bing Lilin, Tan Joice, Tan Teck Sheng Adrian, Toh Chim Wei, Wang Tung Shu, Yeo Yang Song, Yu Wing Sze, Lai Tsui Fen, Lim Chee Phoon Samuel, Lim Keng Hwee, Neo Chee Siong, Ng Kah Kiat Jefrfrey, Samshade Mosaheb, Teo Ding Luen, Teo Whye Hong Darren, Chan Lai Chin, Seetoh Yi Tu, Foon Fong Yee, Yang Wen Tirng, Fu Ai Lian, Wee Sean Han Adrian, Sulaiman Nur Hafizah Bte., Yeong Hoe Kiat, Chong Ya Hui Adeline, Ho Shin Siew Tanny, Wee Chorng Yee Sam and Qi Feng.

PREFACE

This book focuses on spearheading the integration of maintainability right from the design stage. To enable such integration, improving the knowledge of maintainability and setting maintainability benchmarks are two key issues set out. These objectives are fulfilled with the development of a comprehensive defect library, a material manual and a maintainability scoring system. These tools serve to define acceptable standards in design and construction practices which thus enhance long-term maintainability of facilities.

The book aims to improve the standard and quality of design, construction and maintenance practices to produce efficient facilities that require minimum maintenance. The text covers technical issues related to maintainability of major components of a facility for wet area, façade, basement, roof, M&E in Chapters 2 to 6. The implications and selection of materials for high maintainability is discussed in Chapter 7. Chapter 8 discusses the common diagnostic techniques and Chapter 9 maintainability grading system.

The target readers are practitioners and students in the related professions including architecture, engineering, building, real estate, construction, project management, facilities management, quantity and building surveying.

This book is best read with its supplementary website (http://www.hpbc.bdg.nus.edu.sg). The website requires a 'username' and 'password' to log in. As the owner of this book, you will be given access to the website. Please email me (bdgchewm@nus.edu.sg) stating the special code for this book at the bottom of this page. You will be informed once your exclusive account is activated. Please note that each special code can be activated only once.

Thank you.

Chew Yit Lin, Michael
Professor, Department of Building
School of Design and Environment
National University of Singapore

84P8S3

CONTENTS

CHAPTER 1
INTRODUCTION

1.1. General

Maintainability is defined in this book as "the ability to achieve the optimum *performance* throughout the *lifespan* of a facility within the minimum *life cycle cost* (LCC)" (Fig. 1.1).

It embraces all stages of the delivery process, from planning, design, detailing, construction, operation, to maintenance and disposal of a facility. It requires a clear understanding of issues associated with durability, serviceability and sustainability for the proper selection of materials, components, systems, as well as the effect of and impact to, the environment.

Initial capital costs and schedules are often given the top priority in most projects, ignoring the fact that the main bulk of the total LCC is attributed to the operation and maintenance costs. These high operation and maintenance costs are best minimized right from the planning and design stage.

In the face of current shrinking maintenance budget, increasing customer expectation and demand for better total building performance, maintainability management has become a significant concern to building

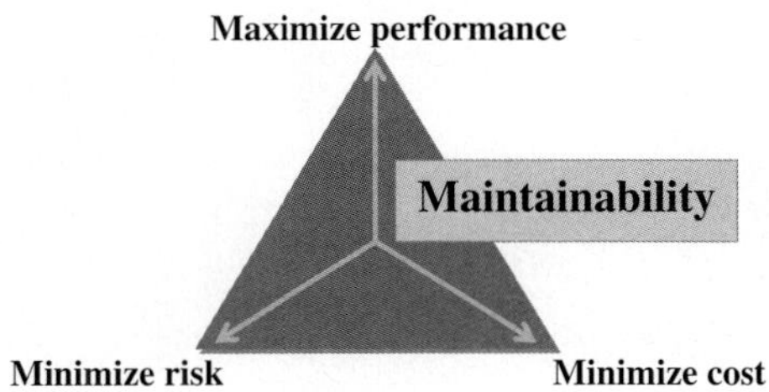

Fig. 1.1. Definition of maintainability.

owners, developers and facility managers. One of the more impending tasks of the delivery process of a facility is to minimize the rising total facility related cost-in-use [1] and increasing maintenance costs associated with defects arising from design [2–5], inferior construction workmanship [6, 7], maintenance practices [8–12] and materials' response to the environment [13, 14].

The issue of maintenance management of facilities is a universal one and its consideration at the planning and design stage is of paramount importance on the future performance attributes and its LCC. As modern facilities are designed to meet higher standards, the effect of the decisions made during the planning and design stage would have a far reaching effect and significant impact on the future maintainability of a facility. Over the years, the expectation and sophistication of facility owners/users have escalated. Facility management has shifted from corrective approach to preventive in the 1980s and to predictive approach of today (Fig. 1.2).

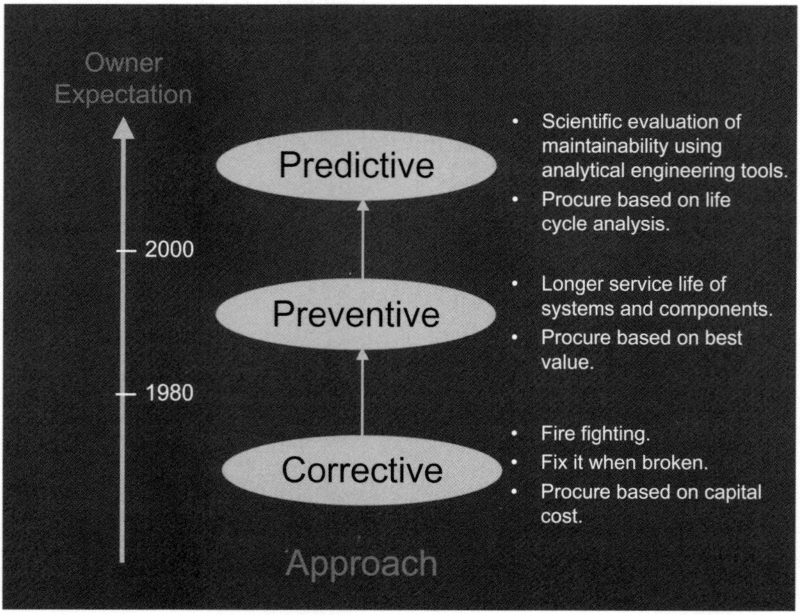

Fig. 1.2. A shift of owners' expectation.

The continuous development and enhancement of sophisticated facilities have increased the demand for higher quality maintenance strategies. One of the principal issues at the heart of a quality maintenance program is the availability and reliability of an analytical engineering model, designed to predict the life cycle and deterioration patterns of the subsystems, components and assembly of components of a facility [15]. An improved initial design, good construction workmanship to translate and deliver the design details as originally intended and a comprehensive maintenance program to be put in place, are relevant and critical to effectively extend the lifespan of a facility and for the enhancement of maintainability [16].

1.2. Maintainability Analysis

An analysis of maintainability of a facility includes:

- *Strategic analysis* — To derive the most efficient strategy aimed at reducing the whole life costs of a facility while achieving the desired service quality, by optimization of economic, technological, managerial and human comfort throughout the lifetime of the facility.
- *Functional consideration* — To assess functions for which the facility is designed, constructed and maintained. Table 1.1 shows example of a checklist, for various components of a facility, with a matrix mapping the desired performances at various stages of the delivery process. Identification of the significance of the desired performance for each component throughout the delivery process right from the planning and design stage is crucial.
- *Cost-benefit ratio optimization* — To provide maximum benefit at minimum cost of resource allocation throughout the product delivery process over the lifetime of a facility.
- *Selection of optimum building systems and technology* — To select the optimum combination of alternative system and technology with consideration of factors including management, economics, architecture, law, engineering, technology, aesthetics and user satisfaction.

Table 1.1. A checklist mapping the desired performances at various stages of the delivery process.

Component: Façade/Wet Area/Roof/Wall/Floor/ Doors/Windows/M&E/External Works **Component Detail:**	**Integrity**	**Acoustics**	**Lighting**	**Thermal**	**IAQ**	**Visual**
Design and Construct						
1. *Design*						
a. Buildability, e.g. BDAS [17,18]						
b. Upgradability						
c. Retrofitability						
d. Accessibility						
2. *Detailing*						
a. Interface — Between materials/components/ elements						
b. Joints — Width/spacing/continuity						
c. Drainage — Channels/drains/slope/flow rate/flow pattern						
d. Openings — e.g. open balcony leading to water ingress						
e. Penetration — Ducts/aircon/pipes						
f. Fixtures — Protrusion/intrusion/louves						
3. *Materials*						
a. Durabilty — Organic/inorganic						
b. Life cycle — Organic/inorganic						
c. Repairability — Type/subtype of components						
d. Replaceability — Affect neighboring components						
e. Sustainability — Disposal						
4. *Construction*						
a. Prefabrication — Jointing/waterproofing						
b. Workmanship — e.g. CONQUAS [19]						
c. Warranty — Suppliers/fabricator/contractor						
d. Commissioning — Tests conducted						
Operational						
1. *Cleaning*						
a. Material — Ease of cleaning/detergents/tools						
b. Labor — Skill level/supervision						

(*Continued*)

Table 1.1. (*Continued*)

Component: Façade/Wet Area/Roof/Wall/Floor/ Doors/Windows/M&E/External Works **Component Detail:**	**Integrity**	**Acoustics**	**Lighting**	**Thermal**	**IAQ**	**Visual**
c. Access — Gondola/falseworks/recess/extrusion d. Down time — Obstruction/blockage/corporate image e. Loss in productivity — Disturbance to tenants f. Frequency 2. *Energy* 3. *Inspection* a. Frequency **Maintenance** — Check defect databank according to component details 1. *Repair* a. Diagnostic — Access/instrumentation b. Materials — Type/subtype of components c. Labor — Experties/manhour d. Down time — Obstruction/blockage/corporate image e. Loss in productivity — Disturbance to tenants f. Frequency 2. *Replace* a. Displacement — Neighboring components b. Materials — Type/subtype of components c. Labor — Experties/manhour d. Down time — Obstruction/blockage/corporate image e. Loss in productivity — Disturbance to tenants f. Frequency						

- *Simulation/modeling of lifetime process of a facility* — To verify selection by conducting scientific simulation or mathematical modeling of the whole life performance of a facility. The financial impact over the life time of the facility due to variation in (a) design details,

(b) construction and maintenance practices, (c) climatic conditions, (d) risk factors, (e) infrastructure, (f) initial project costs and (g) customer needs, are mathematically expressed to enable accurate decision making.

Integrating the above elements requires solid technical knowledge of the various maintainability issues and setting of maintainability benchmarks. To address this, a website — http://www.hpbc.bdg.nus.edu.sg/ — comprising of a comprehensive defect library, a material manual and a maintainability scoring system has been developed. These tools serve to define acceptable standards in design and construction practices which in turn enhance the long-term maintainability of a facility (Sec. 1.3).

Please see 'Perface' to gain access to website.

1.3. Maintainability Website (http://www.hpbc.bdg.nus.edu.sg/)

The website focuses on two main tasks:

(a) Improving the knowledge on maintainability

- Project teams at the planning and design stage are often not well-focused to achieve maintainable designs and construction, owing to poor awareness of the maintainability problems during the service life.
- Facility managers are facing new challenges with the increasing evolvement of new anomalies arising from innovative design and construction, beyond the existing knowledge and tools available in their organizations could address.

(b) Setting the maintainability benchmarks

- It is of compelling logic to improve the level of maintainability right from the planning and design stage.
- The rationale of the benchmark is to enable the decision maker to choose a more maintainable design option. Yardsticks like

scoring systems would assist in the selection of highly maintainable design and construction.

The website seeks to create an awareness of the current obstacles to good maintainability for facilities. It attempts to achieve this by a compilation of a "*Defect Library*" with information about types of defects and their causes, maintenance and diagnostic methods, good practices and guidelines which can be applied during the design, construction and operation stage of a facility to control repetitive problems. In addition, information on the performances and durability of materials is compiled in a "*Material Manuals*". A "*Maintainability Scoring System*" is developed to facilitate the selection of maintainable alternatives for projects. This system seeks to appraise the various factors that could affect long-term maintainability and provide optimum design solutions for individual components (Fig. 1.3).

1.3.1. *Defect Library*

Common defects occurring in different elements in a facility are compiled in the first section of the database called "Defect Library". Figure 1.4 shows the database interface of the website.

Under the heading of defect library, the user could select any one of the intended building elements including wet area, façade, basement, roof and M&E.

An example using an element of façade is discussed in this section.

Under the façade section, defects are classified according to the various materials used (Fig. 1.5).

The navigation menu (Fig. 1.6) allows users to explore the various defects relating to the available façade materials. In the first two sections, the problem type and the possible causes are explained and illustrated with photographs, animations and diagrams. Good practices are compiled into the next section with visual illustrations to provide guidelines to prevent the occurrence of repetitive errors and omissions. In the fourth section, maintenance and diagnostic techniques are illustrated giving insight of the

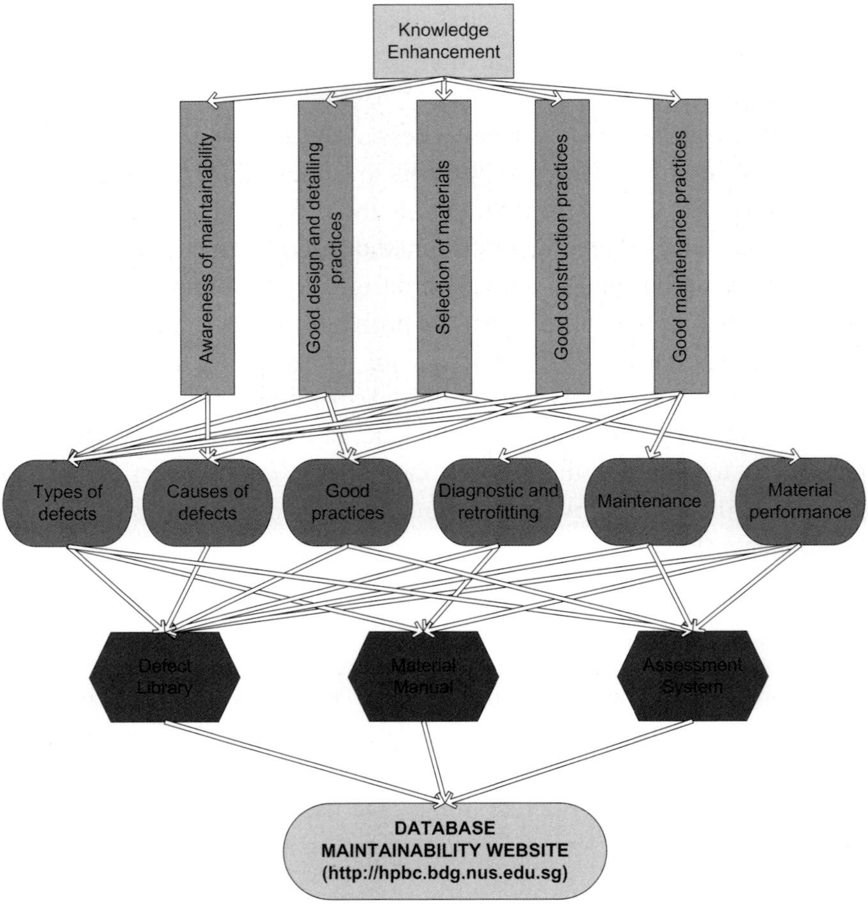

Fig. 1.3. Framework of improving the knowledge on maintainability.

preventive measures and maintenance actions that could be considered. Possible remedial techniques are illustrated in the fifth section. A point to highlight in this defect library is the section of "similar cases", where it serves as a "databank" in which defect records could be updated.

1.3.2. *Material Manual*

The Material Manual seeks to provide information on the various performances including durability, sustainability and cleanability of

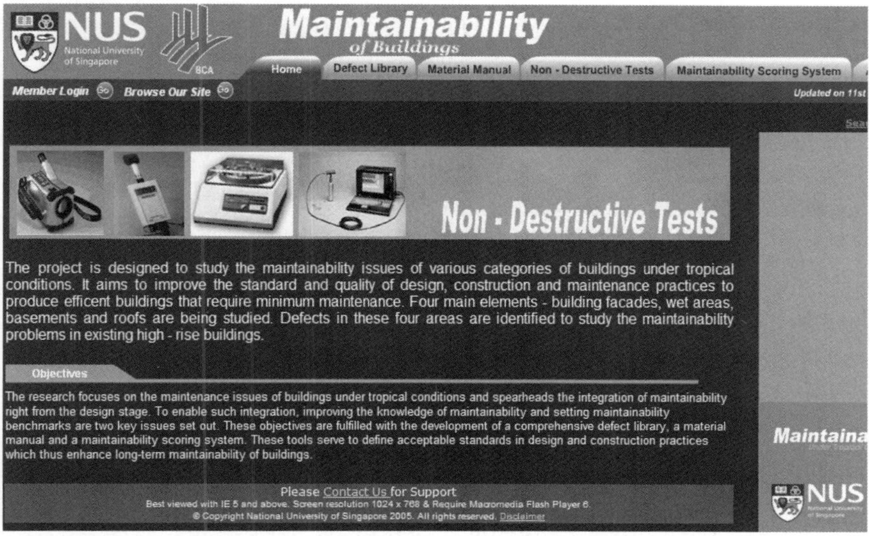

Fig. 1.4. Interface of database.

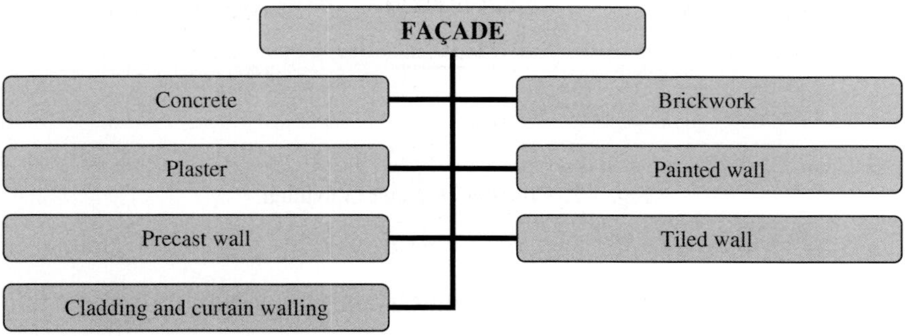

Fig. 1.5. Classification of façade materials and systems.

materials. It may be used as a reference tool in the selection of appropriate materials that contribute to long-term performance of a facility throughout its lifetime.

Components covered in the Material Manual are shown in Fig. 1.7.

Each link leads the user to the various sub-components of interest, which describes quantitatively in terms of the material types and durability,

Fig. 1.6. Navigation menu for façade.

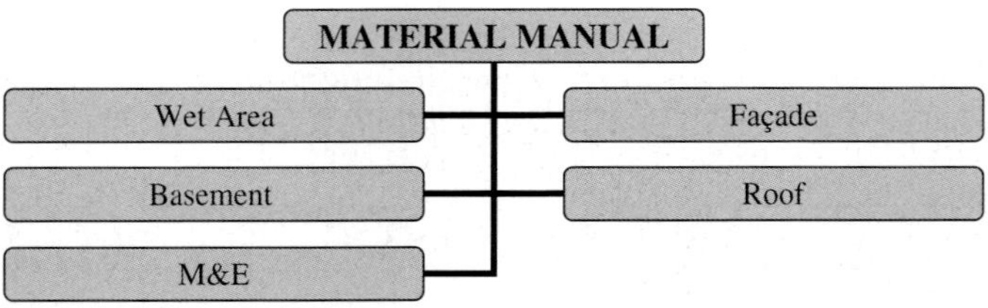

Fig. 1.7. Contents of material manual.

properties, maintenance/cleaning methods and the potential types of failures and repair methods. Figure 1.8 shows an example illustrating aluminum as infill panels under the heading of "Curtain Walling". Technical knowledge could be found in the various links in the side menu. References are also provided for further reading.

1.3.3. *Maintainability Grading System*

The grading system takes on the approach of modeling the optimum maintainability upon minimum LCC. Figure 1.9 shows an example on how façade may be graded.

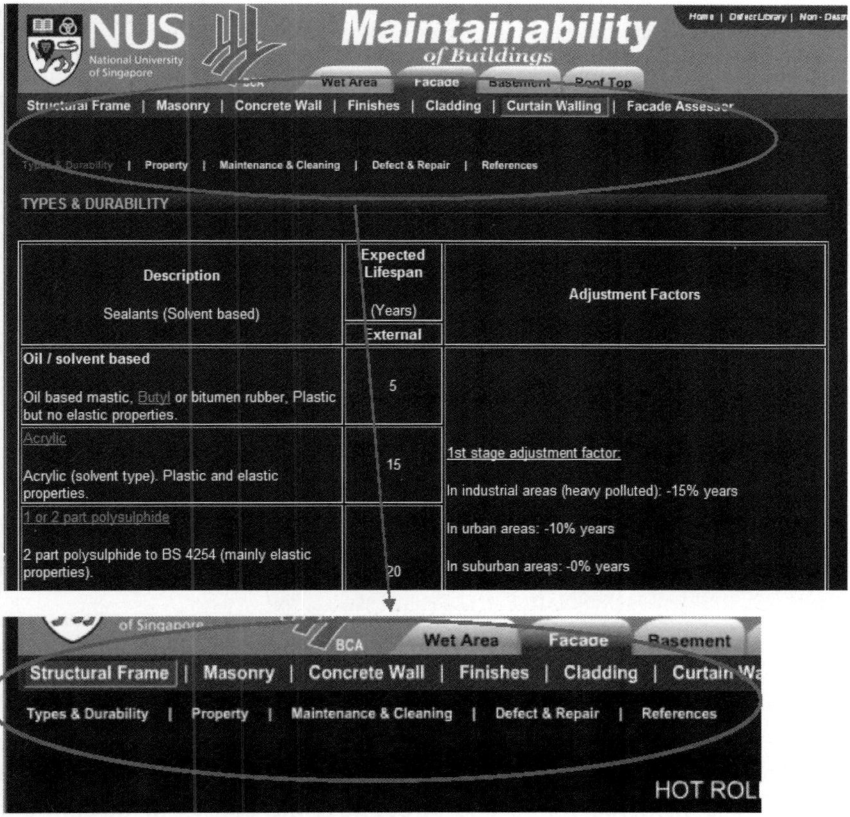

Fig. 1.8. Typical case of aluminum for material manual.

The principle of risk management is applied here, where the risk is the high level of maintainability [20, 21]. Risk factors to control the level of maintainability considered include:

(1) design (e.g. water-tightness, stress accommodation, water flow control, accessibility, downtime and material performance),
(2) construction (quality),
(3) maintenance (quality),
(4) environment and usage,
(5) building's profile.

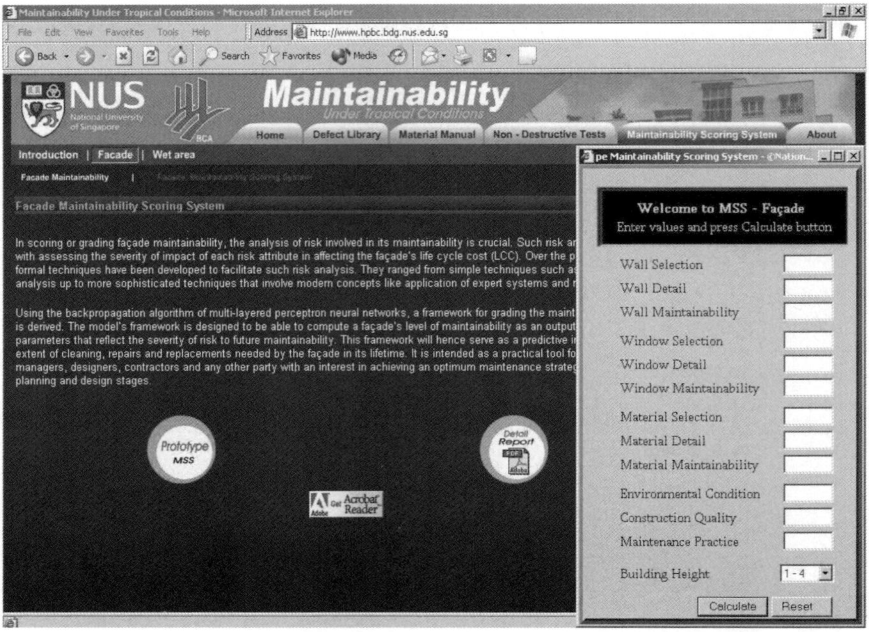

Fig. 1.9. Example of façade maintainability grading system.

Figure 1.10 illustrates an example of the preliminary risk factors that should be considered for façade throughout the phases of design, construction and maintenance.

In scoring maintainability, principles of risk management are applied through a series of decision making. It involves the best possible selection of building elements that require minimum cleaning, repair, replacement and downtime from a wide range of available materials, variety of systems and designs. This has become rather multifaceted with variability in the performance and difference in the latent risk of failures associated with these different systems and components (Fig. 1.11). One of the ways to model maintainability may be in the form of LCC, in which case, the optimum maintainability is the optimum LCC which can be set at the minimum level of risk.

The LCC associated with maintenance cost, which includes routine cleaning, repair and replacement, is used to work out the level of maintainability. With the total number of defects at a given base period, the

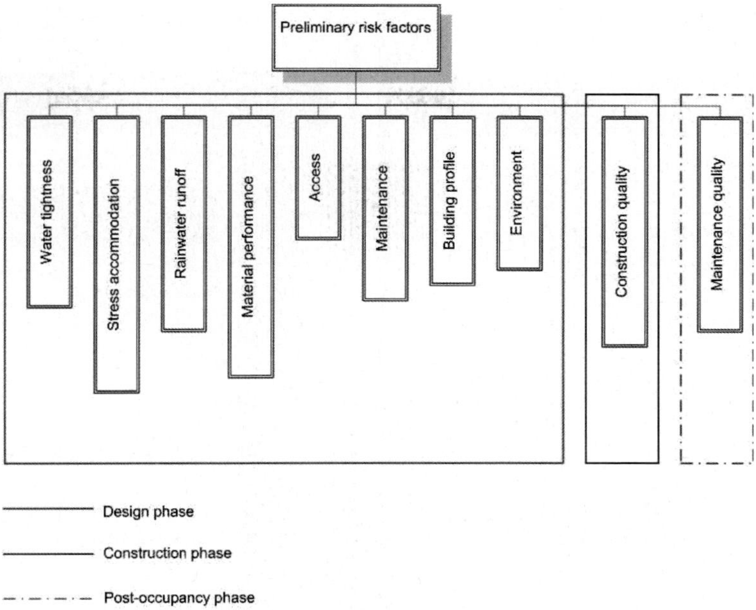

Fig. 1.10. An example of maintainability risk factors for façade.

following normalization can then be set to assign the grading of the maintainability (Maintainability Score, MS),

$$\mathrm{MS} = 1 - \left[\frac{\sum_{i=1}^{d} \mathrm{RRDC}_i + \mathrm{CC}}{\sum_{i=1}^{D} \mathrm{RRDC}_i + \mathrm{BCC}} \right],$$

where d = the current number of defects in a building,
D = total number of defects in the building at the base period,
RRDC_i = the respective repair replacement and downtime cost of ith defect,
CC = actual cleaning cost,
BCC = the basic cleaning cost derived from required cleaning cycles.

Details on maintainability grading system is discussed in Chap. 9.

Fig. 1.11. Interface of façade and wet area maintainability scoring system.

1.4. Integrated Approach to Maintainability

As we learn from experience, new problems evolve.

1.4.1. *Performance Mandates*

They could be related to a single or a combination of the following performance mandates (Fig. 1.12):

- integrity,
- spatial,
- thermal,
- lighting,
- acoustical and
- environment and energy.

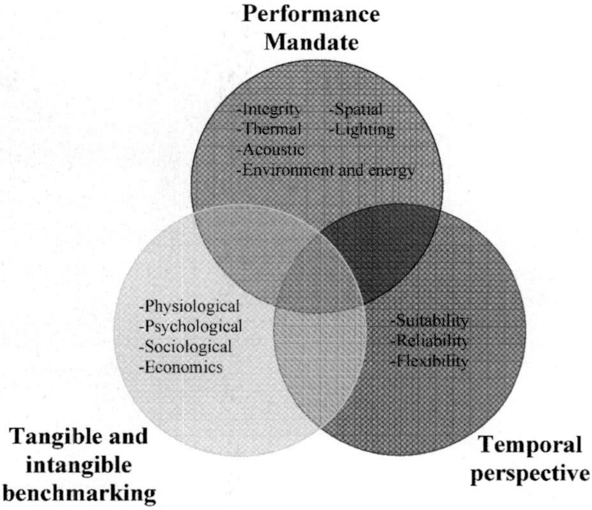

Fig. 1.12. Integrated approach to maintainability.

1.4.2. *Performance Benchmarks*

Some performance mandates are intangible, they cannot be measured scientifically and may be difficult to benchmark. They could be deficient in a single or a combination of the following:

– physiological,
– psychological,
– sociological, and
– economics.

1.4.3. *Temporal Perspective*

They may be old problems which were not considered as significant in the past but are crucial today, e.g. environmental and energy issues (Fig. 1.13). They could be new problems evolving from untested innovation, lost of knowledge and craftsmanship, negligence and the like in any or a combination of the following:

- design,
- detailing,

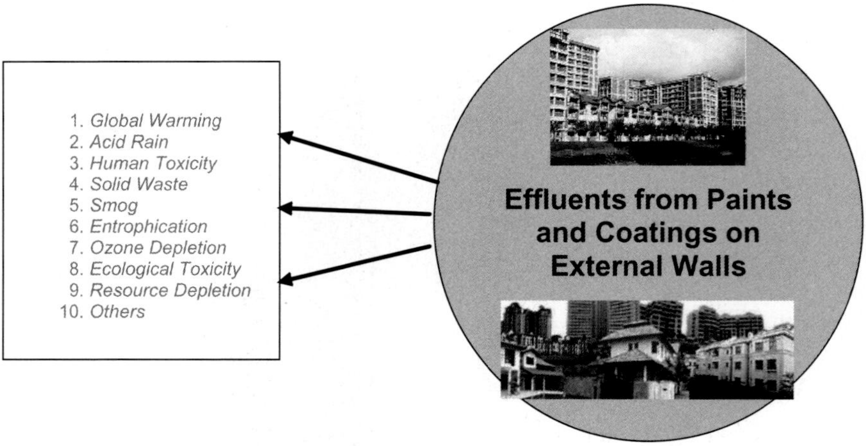

Fig. 1.13. Environmental issues related to paints in high-density high-rise housing.

- environment,
- materials,
- construction,
- operation,
- maintenance, and
- disposal.

Some of these problems may be temporal but disastrous. An example is falling windows in cities with high-density high-rise residential buildings (Figs. 1.14 and 1.15). The same problem source applies to almost every residential unit with public safety in question.

An integrated approach is hence needed right from the planning and design stage on the maintainability of a facility. Maintainability requires the management of risk:

- study existing buildings and their conditions before failure,
- explore risk attributes of building maintenance,
- minimize defects through best possible mix of technical attributes.

Accessibility to every crucial part of a facility must be considered right from the planning and design stage, particularly for the maintenance, repair and

Fig. 1.14. Falling windows of high-density, high-rise residential buildings in densely populated cities.

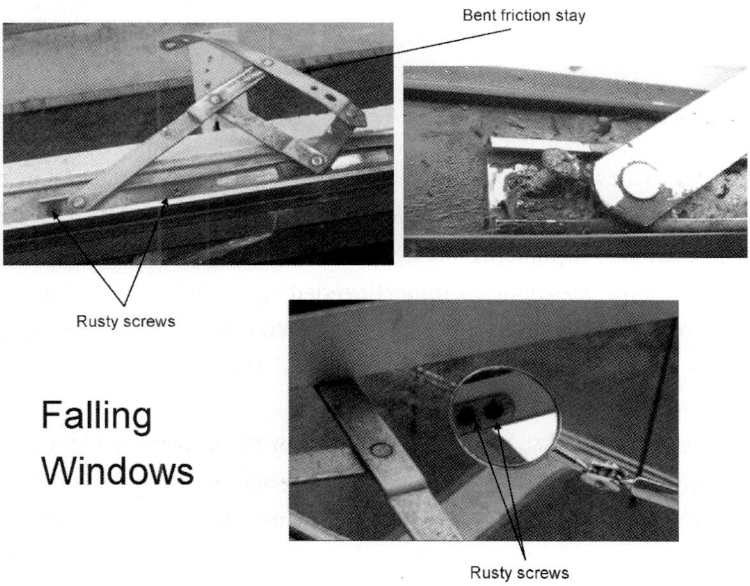

Fig. 1.15. Falling windows due to rusty aluminum screws and inadequate friction stay.

Fig. 1.16. Temporary access to the façade of buildings with irregular shapes.

replacement of the façade of tall buildings, which is not only expensive but detrimental to the image of the property (Fig. 1.16).

References

[1] M. Y. L. Chew, S. S. Tan and K. H. Kang, "A technical evaluation index for curtain wall cladding façades", *Structural Survey*, **22**(4), 210–227, 2004.

[2] C. Briffett, "Balancing design and maintenance issues", *Building Maintenance and Modernisation Worldwide*, **2**, 1097–1105, 1990.

[3] P. C. Mselle, "Management of maintenance records and their use in planning", *Proc. 5th Int. Conf. Inspection, Repairs and Maintenance of Buildings and Structures*, 211–215, 1997.

[4] A. Tesfaye, "A holistic approach towards low maintenance buildings: A case study of sample esidential buildings in Botswana", *Proc. 5th Int. Conf. Inspection, Repairs and Maintenance of Buildings and Structures*, 357–379, 1997.

[5] D. Arditi and M. Nawakorawit, "Designing buildings for maintenance: Designers' perspective", *Journal of Management in Engineering*, **5**(4), 107–115, 1999.

[6] S. Assaf, A. Al-Hammad and M. Al-Shihah, "The effect of faulty design on building maintenance", *Journal of Quality in Maintenance*, **3**(1), 29–39, 1995.

[7] P. S. Dunston and C. E. Williamson, "Incorporating maintainability in constructability review process", *Journal of Management in Engineering*, **15**(5), 56–60, 1999.

[8] T. Kondo, S. Hatakeyama and O. Isobata, "Simplified inspection method for long term maintenance of existing buildings", *Building Maintenance and Modernisation Worldwide*, **2**, 650–658, 1990.

[9] M. Y. L. Chew, "Efficient maintenance: Overcoming building defects and ensuring durability", *Conf. Building Safety, The Asia Business Forum*, Kuala Lumpur, 1994.

[10] B. S. Blanchard, D. Verma and E. L. Peterson, *Maintainability: A Key to Effective Serviceability and Maintenance Management*, Wiley, New York, 1995.

[11] Q. P. Shen, K. K. Lo and Q. Wang, "Priority setting in maintenance management of public buildings: A modified multi-attribute approach using analytic hierarchy process", *Construction Management and Economics*, **16**, 693–702, 1998.

[12] K. C. Lam, "Planning and execution of building services maintenance beyond year 2000", *Southeast Asia Facility Management*, March/April, 2000.

[13] M. Y. L. Chew, C. W. Wong and L. H. Kang, *Building Façades — A Guide to Common Defects in Tropical Conditions*, World Scientific Publishing, Singapore, 1999.

[14] M. Y. L. Chew and X. Zhou, "Enhanced resistance of polyurethane sealants against cohesive failure under prolonged combination of water and heat", *Polymer Testing*, UK, 2001.

[15] I. M. Shohet, "Building evaluation methodology for setting maintenance priorities in hospital buildings", *Construction Management and Economics*, **21**(7), 681–692, 2003.

[16] J. Christian and A. Pandeya, "Cost predictions of facilities", *Journal of Management in Engineering*, **13**(5), 52–61, 1997.

[17] BCA, *Guide to the Buidable Design Appraisal System*, Technology Development Division of the Building and Construction Authority, Singapore, 2001.

[18] BCA, *Code of Practive on Buildable Design*, Building Construction Authority, Singapore, 2005.

[19] BCA, *CONQUAS 21: The BCA Construction Quality Assessment System*, 6th Edition, Building Construction Authority, Singapore, 2005.

[20] M. Y. L. Chew and D. Sutapa, "Building grading systems: A review of the state-of-the-art", *Architectural Science Review*, **50**(1), 2–12, 2008.

[21] R. Flanagan and G. Norman, *Risk Management and Construction*, Blackwell Scientific, Oxford, England, 1993.

CHAPTER 2
WET AREA

2.1. General

One of the building elements that demands high attention is internal wet areas. Internal wet areas are areas subjected to constant damp conditions with alternating drying and wetting cycles. Although wet areas may only occupy less than 10% of the gross floor area, the annual maintenance cost for wet area can range from 35% to 50% of the total maintenance cost depending on the type of facility [1].

Various systems and components such as waterproofing membrane, screed, architectural finishes, service pipes and sanitary fittings are used in wet areas. *In-situ* construction and pre-cast technology have been adopted to construct these units. These systems and components are made up of numerous sub-components whose performances are depended on for the overall structural, functional and aesthetic performance of the whole system. The failure of any component may result in subsequent defects surfacing in other components and ultimately bringing about the total failure of the entire wet area. It is hence crucial for the design, construction and maintenance of all components to be carried out soundly and with due considerations to the likely impacts they would have on each other and on the overall performance of the system [2].

2.2. Systems and Components Selection

The selection of primary components and systems such as waterproofing system, screed, tile and tile bedding, grout, paint and service pipes are mainly based on their serviceability so that they may continue to perform their intended functions through their design life. Durability of these components should be checked under the real exposure conditions in wet areas. Compatibility among components is crucial.

Details of material durability, property etc. can be found in the "Material Manual" of the Maintainability Website (http://www.hpbc.bdg.nus.edu.sg/).

2.2.1. *Waterproofing System*

Waterproofing systems are applied over the substrates of wet area floors and walls to improve the water-tightness of the wet area units. Water-tightness of wet areas relies mainly on the adequacy of waterproofing over penetrations, projections and joints and how these discontinuities of the floor and wall elements can achieve a "monolithic" structure. The decision to select a specific waterproofing system should take into consideration the design of the substrate structure and area of coverage.

Generally, waterproofing systems can be classified into three types [3]:

(1) preformed membrane,
(2) liquid-applied membrane, and
(3) integral systems

Figure 2.1 shows the various classifications of waterproofing systems of preformed, liquid-applied and integral systems. Typical examples are given in Table 2.1.

Preformed membrane is often used for large areas such as roofs and basements. Liquid-applied membrane is popular for small areas such as wet areas because:

(1) The membrane is seamless (no physical joints).
(2) The membrane has good elongation and recovery properties to accommodate substrate movements and to bridge cracks.
(3) Of its unique built-in primer, it provides positive adhesion to both base substrate and top wearing (screed) surface (i.e. screed can be applied directly on membrane without debonding).
(4) Of its even adhesion to every part of the substrate.
(5) It offers the flexibility of applying over the substrate.

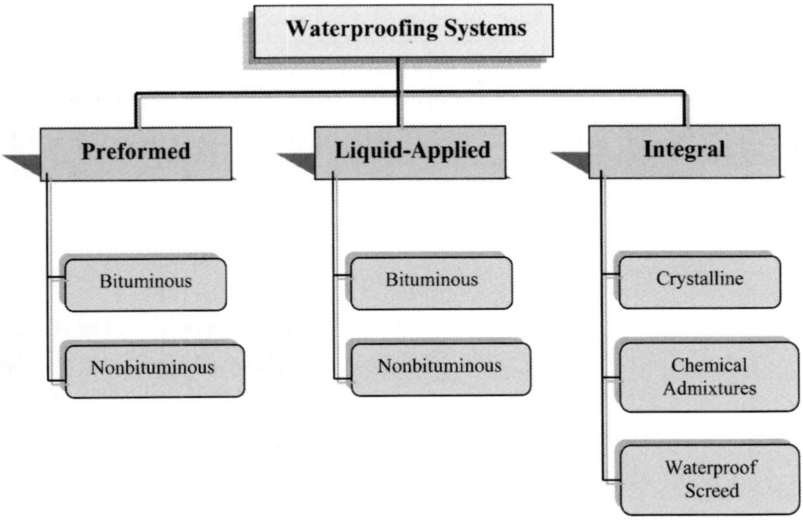

Fig. 2.1. Classification of waterproofing systems.

(6) Of its continuity between (a) horizontal and vertical planes, (b) at abutments of joints and (c) around penetrations and projections offering homogeneity to the whole system.

Integral systems are also becoming popular in these areas due to thier, easy application. They reduce the permeability of the concrete by introducing water-reducing agents, pozzolanic products, organic binders or pore blocking additives.

Selection criteria of waterproofing system should include:

(1) Compatibility with other sub-components such as screed and tile bedding.
(2) Ease of application, especially at pipe penetration areas.
(3) Elasticity and ability to bridge over cold joints and differing materials.
(4) Good adhesion and cohesion strengths.
(5) Ability to retain its adhesion and remain stable under extreme fluctuations of temperature and moisture.
(6) Ability to resist mechanical damage such as peeling, cracking and rupture prior to screed finish.

Table 2.1. Examples of different waterproofing systems [4].

Types of Systems	Nature	Examples
Preformed membrane	Bituminous	Asphalt Bitumen Modified bitumens
	Nonbituminous	Ethylene-copolymer bitumen (ECB) Polyvinyl chloride (PVC) Polyethylene (PE) Chlorinated polyethylene (CPE) Butyl rubber Ethylene propylene diene (EPDM) Neoprene Polyisobutylene (PIB) Bentonite clays
Liquid-applied membrane	Bituminous	Bituminous modified
	Nonbituminous	Acrylic Epoxy Polyurethane (PU) Polyisorene/rubber Cementitious Silicon Hypalon Neoprene
Integral systems		Crystalline waterproofing Chemical admixtures Waterproof screeds

2.2.2. *Screed*

A screed is a layer of well-compacted material applied *in situ* to a structural base or some other substrate and finished to a designated level [5, 6]. Its functions include:

(1) to provide a smooth, flat surface to a correct level for bedding of tiles,
(2) to serve as a protection for the waterproofing coating,
(3) to provide falls to drain water off,
(4) to accommodate service pipes.

Table 2.2. Various types of wet systems.

Types	Definitions
Monolithic screed (on green concrete)	Screed which is laid on *in-situ* concrete base before it has set (within 3 h of placing) is known as monolithic screed.
Bonded screed (on hardened concrete/ precast concrete)	Also known as separate construction and bonding agent, it can be used to ensure maximum bond with the base.
Unbonded screed	Unbonded construction is adopted when the screed is laid on the damp proof membrane or when the concrete contains water repellent admixture.
Floating screed	This type of screed is unbonded to concrete base and is designed as floating screed to accommodate heating elements or sound insulation.
Proprietary floor screed	Screeds that are basically denser and modified by inclusion of materials in emulsion form are known as proprietary floor screeds.

Wet systems are often used in the construction of wet areas including monolithic, bonded, unbonded, floating and proprietary screeds (Table 2.2) [7].

2.2.3. *Tile*

Man-made and natural tiles are popular finishing materials due to their tested performance such as durability, cleanability and good appearance (Fig. 2.2) [8, 9].

2.2.3.1. *Man-Made Tiles*

There are typically two types of man-made tiles: ceramic and homogeneous tiles.

Ceramic tiles can be classified into extruded tiles and dust pressed tiles. Extruded tiles are shaped in the plastic state in an extruder and the resulting column cut into tiles of predetermined lengths. Dust-pressed tiles are formed from powder or small grains, shaped in moulds under high pressure before firing. The surfaces of these tiles are either glazed or unglazed

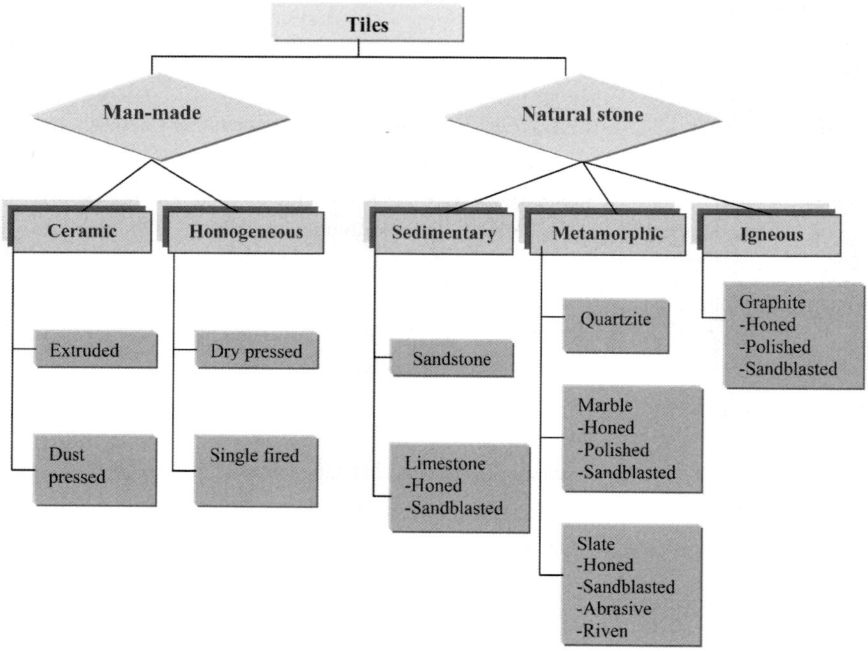

Fig. 2.2. Classification of tiles.

(Figs. 2.3a and 2.3b). Glazed tiles undergo an additional phase in the firing process with a layer of liquid glass. A range of colors are achievable by mixing certain mineral elements, e.g. gold, silver, copper, cobalt, zinc. The nonporous layer of liquid glass makes the tile surface more stain resistant but less slip and scratch resistant. For flooring of wet areas, unglazed tiles are more suitable due to its better slip resistant.

Homogeneous tiles are obtained from very low water absorption ceramic tiles that are dry pressed and produced by single firing. They are superior to ordinary ceramic tiles as they are made of a single compact body mass and are virtually porous free with very low water absorption (1–3%). Unlike ceramic tiles, homogeneous tiles are uniform in color throughout the tiles and so when a homogenous tile cracks or is chipped, it will not expose any indifferent color and is hence less noticeable. The surfaces of these tiles can be polished, matt finished, rock finished or structured finished.

(a)

(b)

Fig. 2.3. **(a)** Glazed tiles and **(b)** unglazed tiles.

Table 2.3 tabulates the basic properties of ceramic versus homogeneous tiles.

2.2.3.2. *Natural Stone Tiles*

Granites and marbles are the more common types of natural stones used in wet areas. Compared to marble, granite offers more regularity in color tone and texture. Its unique speckled and crystalline appearance distinguishes granite from other natural stones (Fig. 2.4a). Marble is usually adorned for its "natural beauty look" due to its irregularities in veining and color tone (Fig. 2.4b). Table 2.4 tabulates the basic properties of granite versus marble.

2.2.4. *Tile Bedding*

Bedding materials provide the bonding between tiles and substrates. Ordinary cement mortar with additives is commonly used in wet areas. These materials are normally characterized by their adhesive strength, open time (allowable time for application) and deformability. The selection of the correct type of bedding material for tiling needs to be compatible with the type of substrate, tile and the designed bedding thickness, to minimize consequences such as debonding, cracking and efflorescence [10].

Table 2.3. Properties of ceramic tiles and homogeneous tiles.

Properties	Ceramic	Homogeneous
Size/dimension (mm)	200 × 200/300 × 300, with tolerance in deviation 1–1.5%.	Commonly available size: 150 × 150, 200 × 200, 300 × 300, 600 × 600, 2–5% allowance, thickness is 7–9 mm for floor and wall.
Constituents	Clay about 50–60%, flint about 30–35% and feldspar about 10–15%.	Clay, feldspar and silica sand are sieved to a particle size of only 45 microns.
Weight	Ranged from 10 to 15 kg/m^2 for wall tiles and 19 to 24 kg/m^2 for floor tiles.	
Color fastness	Colors in ceramic tiles are unaffected and would not change or fade on exposure to atmospheric conditions. 	Various colors and textures. The color and brightness of homogeneous tile is unaffected when exposed to UV light. 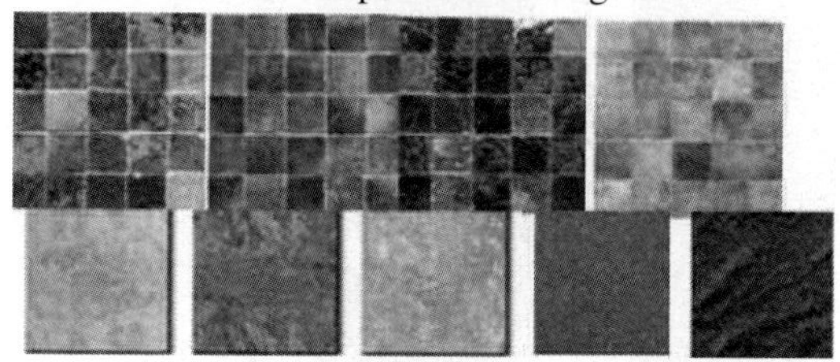
Dirt resistance	Both do not retain dust or residues. Routine maintenance, usually no more than wiping tiles with a damp sponge.	
Fire resistance	Both are completely fire-proof at any temperature. They will not burn or feed a fire. Their surface will not give off any toxic gasses or fumes during a fire. In addition to being fire-proof, tiles have been found to provide excellent protection for structural surfaces during fire.	

(a)

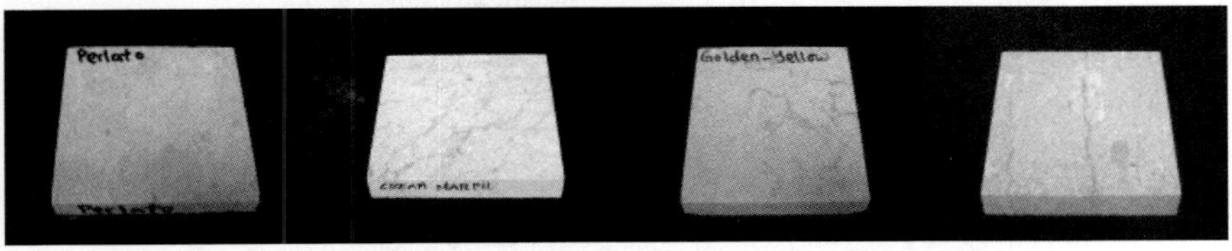

(b)

Fig. 2.4. **(a)** Granite and **(b)** marble.

Table 2.5 shows the properties of different tile beddings. Recommended compatibilities of types of bases, tiles and tile beddings are shown in Tables 2.6 and 2.7.

2.2.5. *Grout*

Grout serves as a sealer for tile joints to prevent water seepage. Grouting materials should have good working characteristics, low shrinkage and good adhesion properties to seal up tile joints. Commonly available grouts include cementitious, organic polymer-based compositions and epoxide resins.

Important performance parameters include:

(1) impermeability,
(2) resistance to water, heat, cleaning agents, chemicals,
(3) resistance to mold growth and bacteria,
(4) resilience and compressibility,
(5) resistance to high-pressure water cleaning,
(6) compatibility of substrate and grouts.

Minimum standards that grouts should achieve for sustainability in wet areas are illustrated in Table 2.8.

Table 2.4. Properties of granite and marble tiles.

Properties	Granite	Marble
Group	Igneous	Metamorphic
Mineral group	Siliceous	Calcareous
Weight (kg/m^3)	2611–2707	2651–2731
Composition	Feldspar, quartz and small amounts mica are the primary minerals as below	Calcite, dolomite or a combination of both and accessory minerals acting as coloring agents. Veining minerals range from calcite, quartz and a variety of other minerals
Colors	Granite is found in more than a hundred distinct colors with varying patterns	Every color is possible due to the variability of accessory minerals
Finishes	Polished, honed, flamed, bush hammer	Polished, honed
Thickness	10–15 mm for internal wall; up to 20 mm for floor	
Treatment	It is important to seal the stone to ensure it can last to its maximum lifespan by resisting scratch, water penetration, stains, etc. This is usually done before installation on the five sides (front and four sides) of the individual cut-panel, or even six sides (includes the back side). When necessary, some treatments should be involved, such as color enhancer, anti slip resin, etc.	
Hardness	Hard	Soft/medium/hard

(Continued)

Table 2.4. *(Continued)*

Properties	Granite	Marble
Modulus rupture (MN/m^2)	10.27–21.10	9.2–18
Density (g/cm^3)	2.5–2.7	2.4–2.8
MOHS' Scale	5–7	3–5.5
Porosity (%)	0.24–2.3	0.6–2.3
Water absorption (%)	0.05–0.5	0.1–0.6
Coefficient of linear expansion ($/^{o}C \times 10^{-6}$)	3.7–6.0	2.7–5.1
Abrasion resistance (mm)	0.68–4.1	0.25–2.9
Compressive strength (MN/m^2)	107.56–212.37	62.46–122.64
Flexural strength (MPa)	11–26	6.15–21.4
Tendencies	Absorbs oils and other liquids. Generally not affected by acidic foods and drinks, however, polished granites can be etched by hydrofluoric acid and concentrated acids which are found in some household cleaning products	Absorbs oils and other liquids. Easily scratched. Acid sensitive

Table 2.5. Properties of different tile beddings.

Tile Bedding	Wet Density (kg/m^3)	Water Absorption (%)	Compression Strength (MPa)	Shear/ Tensile Bond (MPa)	Thermal Shock Resistance[a] (MPa)	Shear Strength 28 Days (MPa)
Cementitious adhesive	±1300–1650	4% max	7.3–20.7	1.0–3.4	NA	1.6–2.3
Organic adhesive	±1700	4% max	10.0–27.6	1.5–4.1	NA	NA
Epoxy-based adhesive	±1300–1700	NA	24.0–55.2	12.4–13.8	>8.3	5.0–8.6

[a] High temperatures remove the water in Portland cement mortars very rapidly and prevent the Portland cement from hydrating, curing completely and gaining full strength. In addition, the rapid drying often causes mortar to crack, crumble or lose bond to the substrate, tile or stone.

Table 2.6. Compatibility of materials with various types of base [13].

Background Material	Cementitious Adhesive (with Liquid Latex)	Organic Adhesive/ Mastic	Epoxy-Based Adhesive
Concrete	S	S	S
Masonry	S	U	S
Cementitious backer board/plaster	S	S	S
Existing ceramic, stone, terrazzo	S	C	S

S = Suitable; U = Unsuitable; C = Consult supplier.

Table 2.7. Compatibility of adhesives with different types of tiles [13].

Tiles	Cementitious Adhesive (with Liquid Latex)	Organic Adhesive/ Mastic	Epoxy-Based Adhesive
Ceramic	S	S	S
Homogeneous	S	S	S
Granite	U	S	S
Marble	U	S	S

S = Suitable; U = Unsuitable.

Table 2.8. Minimum standards of grout [11].

Properties	Test Method	Passing Criteria
Linear shrinkage	ANSI A 118.6: 1992	≤ 0.2% at 7 days
Water absorption	ANSI A 118.6: 1992	≤ 10% immersion to dry
Compressive strength	ANSI A 118.6: 1992	≥ 15N/mm^2 at 28 days
Tensile strength	ANSI A 118.6: 1992	≥ 2.4N/mm^2 at 28 days
Flexural strength	ANSI A 118.6: 1992	≥ 5.0N/mm^2 at 10 days (3 days before demoulding + 7 days air cure)
Color fastness	SS 483: 2000	No color change after 28 days or UV exposure

2.2.6. *Paint*

Paints serve as decorative and protective coatings over ceilings and portions of the un-tiled walls in wet areas. A typical paint system is composed of primer, sealer, undercoat and topcoat (Table 2.9). Appropriate selection of paint is important as the paint film is subjected to constant moist conditions in wet areas. Common defects include peeling, flaking, blistering, biological attack and efflorescence.

For long-term performance, the selection should hence consider:

(1) type of substrate,
(2) type of environment,
(3) application method,
(4) surface preparation,
(5) overcoating interval.

2.2.7. *Service Pipe*

Cast iron and plastic pipes have traditionally been used in internal wet areas of high-rise buildings. Due to the poor performance of cast iron under high humidity conditions, with corrosion and subsequent water leakage as common maintenance problems [12, 13], they are not used in new constructions of wet areas. Plastic materials are therefore recommended for long-term performance of the plumbing system. Chlorinated polyvinyl-chloride (cPVC) and polyvinyl chloride (PVC) are the main

Table 2.9. Components of paint system.

Component	Functions	Comments
Primer	• To reduce porosity of receiving surface so as to prevent undue absorption of resin from subsequent coats of paints	The primers should be used to reduce the porosity of the surface and have adequate moisture resistant capability
Sealer	• To resist alkali attack from cement-based material • To stabilize loose surface particles • To reduce porosity of receiving surface so as to prevent undue absorption of resin from subsequent coats of paints	Sealers should be used to reduce the porosity of the surface and have adequate moisture resistant capability
Undercoat	• To level out minor irregularities and provide uniform texture • To improve inter-coating adhesion	Choice of undercoat should be compatible with the primer/sealer and the finishing coats
Topcoat	• To provide the required functions of paint	The finishing coat of a paint system

categories of plastics pipes used in wet areas (Fig. 2.5) [14, 15]. Major reasons for the growth in use for plastic pipes are longer durability and cost saving in installation, labor and equipment due its lightweight.

2.3. Common Defects in Wet Areas

Common defects often found in wet areas, their possible cause and the role of the parties involved are shown in Fig. 2.6 and Table 2.10.

2.3.1. *Waterproofing*

2.3.1.1. *Design Aspects*

Water-tightness of wet areas relies mainly on the adequacy of waterproofing over penetrations, projections and joints and how these discontinuities of the floor and wall elements can achieve a "monolithic"

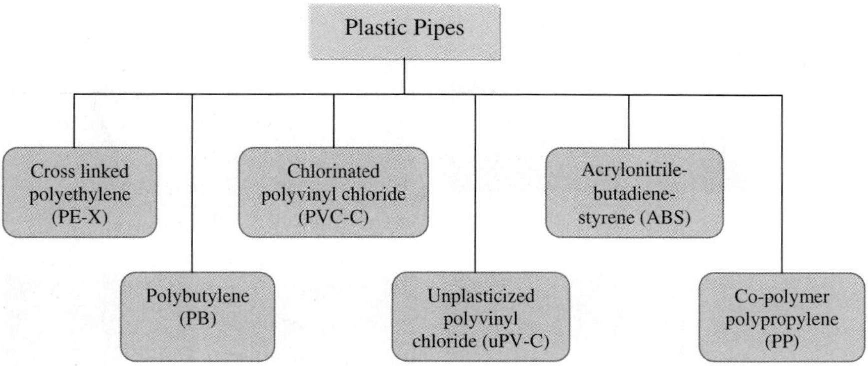

Fig. 2.5. Classifications of plastic pipes.

structure. Water-tightness at interfaces between different components could be enhanced by the following design features [16–18]:

(1) *Floor wall joints* — Use of fiberglass mesh and an angle fillet could help to reinforce waterproofing membrane at joints and to accommodate any stresses attributed to the structural movement at the abutment of wall and floor (Fig. 2.7). A bond breaker or a flexible sealant can also be used (Figs. 2.8a and 2.8b).
(2) *Pipe penetrations and floor traps* — The waterproofing membrane should be dressed up at pipe penetrations and down into the floor outlets. Provision of upstands of the membrane together with fiberglass mesh would maintain the imperviousness at these locations. The details are shown in the Figs. 2.9a and 2.9b.
(3) *Shower/bath area* — Walls adjacent to the shower and bath area should be covered with two coats of waterproofing membrane of more than 1500 mm in width and 1800 mm in height. For bathtubs that are designed only for bathing, an upstand of more than 1500 mm of membrane at adjacent wall should be provided.

2.3.1.2 *Construction Aspects*

Substrate preparation

The substrate receiving the waterproofing membrane should be flat, even, dry and free from debris.

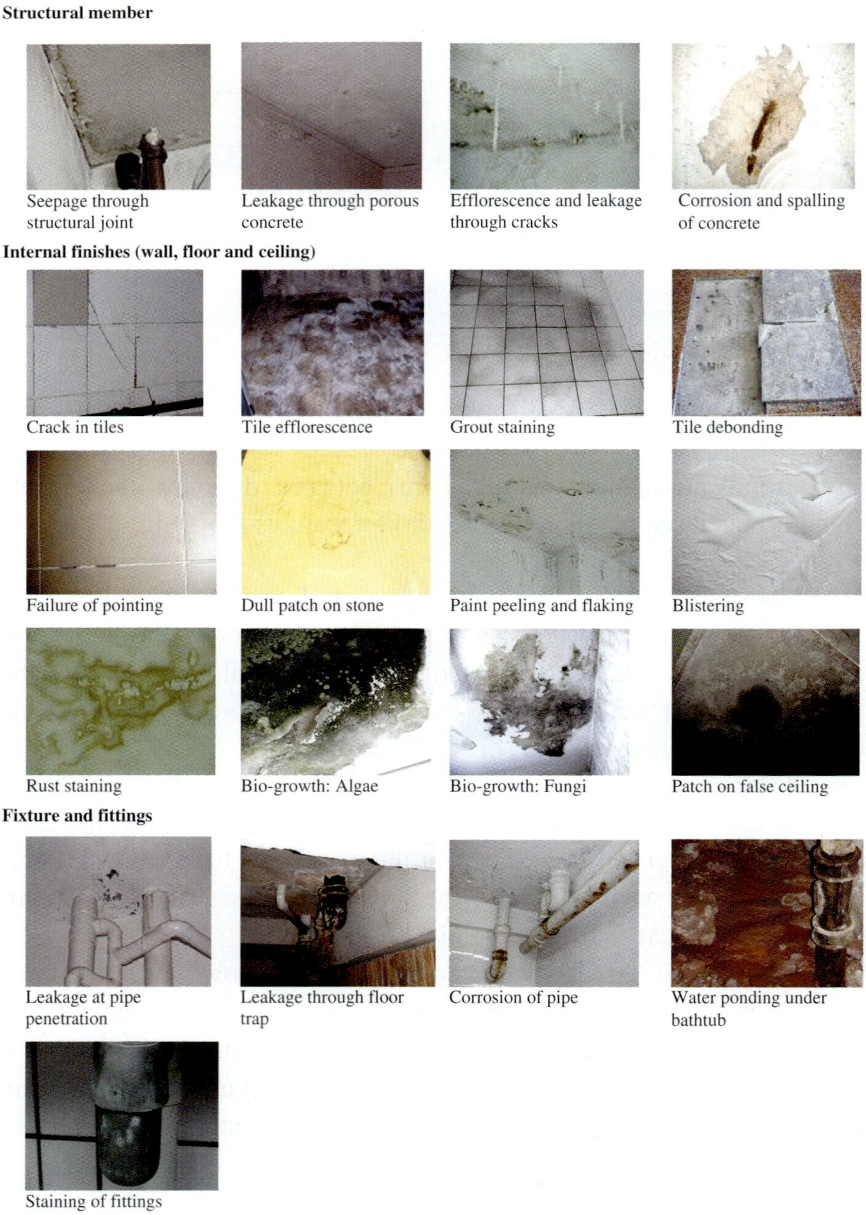

Fig. 2.6. Common defects in wet area.

Table 2.10. Common defects in wet area.

Defects and Coding	Possible Causes of Defects	Cause gr. D*	C*	M*	E*
Structural elements					
Seepage through structural joints (C1)	Premature failure of the poor quality or incompatible membrane or insufficient water-tightness due to poor application	▲	▲		
	Deterioration of the waterproofing membrane with time. Need replacement generally after 10–15 years				▲
	Ineffective accommodation for movement between two elements (slab and wall) can damage the membrane. Fibreglass mesh at joints is recommended	▲			
	Dry zone (usually without waterproofing) is not segregated from wet zone or water ponding	▲			
	Main water source remains undetected and unattended			▲	
Leakage through wall or floor (C2)	Poor design of waterproofing adjacent to fixtures to prevent seepage through wall and slab. More the number of wet walls, higher is the chance of such defect	▲			
	No segregation of dry and wet zone	▲			
	Aging of waterproofing. Also influenced by type, quality, durability, feasibility of application and level of usage	▲	▲		▲
	Porous concrete (improper mix design or inadequate mixing) may retain water and remain damp for long. Together with faulty waterproofing, water can leak through the surface	▲	▲		
	Type and mounting of fixture adds to water seepage	▲			
	Dry zone is not segregated from wet zone and water ponding	▲			

(*Continued*)

Table 2.10. *(Continued)*

Defects and Coding	Possible Causes of Defects	Cause gr. D*	Cause gr. C*	Cause gr. M*	Cause gr. E*
Efflorescence and leakage through cracks (C3)	Leaching of soluble salt or unhydrated lime from cement — A result of poor quality control for cement and improper curing		▲		
	Water seepage due to cracks resulting from excessive shrinkage of the concrete (very low water cement ratio or inadequate curing)	▲	▲		
Corrosion and spalling of concrete (C4)	Permeability of concrete is high due to very low water cement ratio, no use of admixture, poor compaction	▲	▲		
	Poor specification of rebars, e.g. epoxy coating, steel bars, use of corrosion inhibitors, etc.	▲			
	Use of bars already with rust and scaled		▲		
	Seepage through floor trap and other penetrations keeps concrete wet	▲	▲	▲	▲
	Ageing of membrane. Also influenced by its type, quality and feasibility of application	▲	▲		▲
	Volume increase of corrosion products causes spalling. Such stage can reach if the initial signs of rusting are neglected			▲	
Finishes					
Tile cracks (C5)	Lack of proper designing or installation of movement joints to accommodate differential movements of the structure	▲	▲		
	Excessive shrinkage cracks due to lack of curing of concrete backing		▲		

(Continued)

Table 2.10. *(Continued)*

Defects and Coding	Possible Causes of Defects	Cause gr. D*	C*	M*	E*
	Tiles are porous and if not soaked properly draws up water from substrate and cause cracks in mortar	▲	▲		
	Tiles not tapped in place, i.e. hollow backing or damaged by other construction activities		▲		
Efflorescence on tiles (C6)	The presence of loosely bonded sand, pre-hydrated cement and un-hydrated cement mix due to poor mixing during the construction may leach out soluble salts. Depends on material of screed also	▲	▲		
	Cracks in tile (C5) or grout (C9) allows water to seep in. It may lead to leaching of soluble salts in concrete backing	▲	▲		
Staining or dull patch on tiles (C7)	Tiles are not stain resistant or selected as per color consistency	▲	▲		
	Grout material is not durable. With contact of water, joints erode and black stains may form	▲			
	Inadequate cleaning especially for inaccessible parts or cleaning with harmful material or methods. Cleanability is poor	▲		▲	
	Careless use of wet area, vandalism				▲
Tile debonding or adhesive failure (C8)	Inadequate provision or wrong detailing of expansion joints	▲	▲		
	Poor bonding between incompatible tile and bedding	▲			
	Excessive drying shrinkage of the substrate due to wrong mix design or insufficient curing	▲	▲		

(Continued)

Table 2.10. *(Continued)*

Defects and Coding	Possible Causes of Defects	Cause gr. D*	C*	M*	E*
	Absorption of water by inadequately soaked porous tile may lead to a weakening of bond		▲		
	Contamination of the contact surface between tiles and substrate by dust and dirt. It prevents proper adhesion between the surfaces		▲		
	Incorrect tiling, e.g. insufficient tapping of tiles into position or loss of stickiness of adhesive due to delay		▲		
	Poor bonding by inadequate mechanical keys at the tile back	▲	▲		
Failure of pointing (C9)	Poor quality of grout — Inferior standard or improperly mixed	▲	▲		
	Dirt on the tile edge may cause poor bonding		▲		
	Inconsistency in size, thickness or alignment		▲		
	The tiled surface is not properly cleaned after tile installation		▲		
Paint peeling and flaking (C10)	Poor preparation of the substrate may cause peeling due to improper bonding between the paint film and substrate		▲		
	Use of a paint of poor grade or one with expired shelf life	▲	▲		
Blistering (C11)	Painting done on a surface which is warm or the moisture content is > 6% Its result of inadequate curing		▲		
	Substrate remains wet due to seepage from upper floor at pipe penetration (C15), floor trap (C16), joints (C1) or floor (C2)	▲	▲	▲	▲
	The paint is not breathable type	▲			

(Continued)

Table 2.10. *(Continued)*

Defects and Coding	Possible Causes of Defects	Cause gr. D*	C*	M*	E*
Rust stain (C12)	Migration and leaching of corrosion products to the surface. May be by corrosion of concrete rebars (C4), piping (C17) or fittings (C19)	▲	▲	▲	▲
	Painting on substrate already having rust stains. In such cases a primer coat should be applied		▲		
Biological/other stain on floor/walls (C13)	Main source of water remains unattended and untreated. Prolonged dampness helps in abundant growth of algae and fungi. This is prevalent in inaccessible areas behind or between fixtures or with frequent splashing of water from basin or bath. Problem increases with more number of wet walls	▲		▲	
	Poor ventilation maintains a damp humid condition	▲			
	Paint is not fungi or algae resistant or absorbs water (nonwashable)	▲			
	Unsuitable or irregular cleaning, careless usage, vandalism			▲	▲
	First sign of algae/fungi remain unattended and not scrubbed off. Many times poor accessibility is the reason	▲		▲	
Biological stain on ceiling (C14)	Water leakage of pipes/HVAC ducts above the ceiling. Corrosion resistant plastic pipes or proper joining required	▲	▲		
	Excessive condensation on HVAC ducts	▲			
	The plaster board material retains moisture that attracts the growth of micro-organisms (esp. poorly ventilated areas)	▲			
	No periodic checking of pipes and HVAC ducts			▲	

(Continued)

Table 2.10. *(Continued)*

Defects and Coding	Possible Causes of Defects	Cause gr. D*	C*	M*	E*
Sanitary plumbing					
Leakage at pipe penetration and joints (C15)	Porous infill of the cold joint at pipe penetration. Unplanned piping is accommodated by gap left in slab or hacking	▲	▲		
	No/defective water proofing membrane around the pipe	▲	▲		
	Membrane is damaged during careless installation of fixture/finishes		▲		
	Joints are fitted badly or poor accessibility for fitting or repair	▲	▲		
	Main water source remains undetected and unattended			▲	
Leakage through floor trap (C16)	Improper laying of floor trap		▲		
	Short discharge pipe or gully pipe due to improper laying		▲		
	Porous infill of the cold joint at the penetration (usually unplanned)	▲	▲		
	Poor detailing or application of membrane around pipe	▲	▲		
	Deterioration of the membrane with time may result in lost of water-tightness. Needs replacement				▲
Corrosion of pipes (C17)	Water carrying acidic iron salts may lead to the corrosion of cast iron pipes. PVC pipe is recommended	▲			▲
	Rust forming at joints if not attended in time may lead to water leak due to volumetric expansion of the joint. A difficult access hinders regular maintenance	▲		▲	

(Continued)

Table 2.10. (*Continued*)

Defects and Coding	Possible Causes of Defects	Cause gr.			
		D*	C*	M*	E*
Water ponding under bathtub or on floor (C18)	No/wrongly placed floor trap beneath the bathtub to handle water leakage from the drainage pipe	▲			
	If the bath is not enclosed, there is high amount of splash on floor	▲			
	Poor grading of floor	▲	▲		
	The joint between tub and wall if not watertight, allows seepage through to the underneath of the tub. Custom made tubs may leak	▲	▲		
Staining of fixtures and fittings (C19)	Presence of high content of sulfide and chloride ions in the environment (usually industrial area) may cause pitting. Selection of corrosion resistant material (e.g. PVC) recommended	▲			▲
	High chloride content in water can destroy the passive film to form corrosion products. Suitable material should be selected	▲			▲
	Inadequate cleaning especially for inaccessible parts or cleaning with harmful material or methods	▲		▲	
	Inferior material quality, poor cleanability or not cleaned regularly	▲		▲	
	Careless use of wet area, vandalism				▲

D* — Design; C* — Construction; M* — Maintenance; E* — Environment.

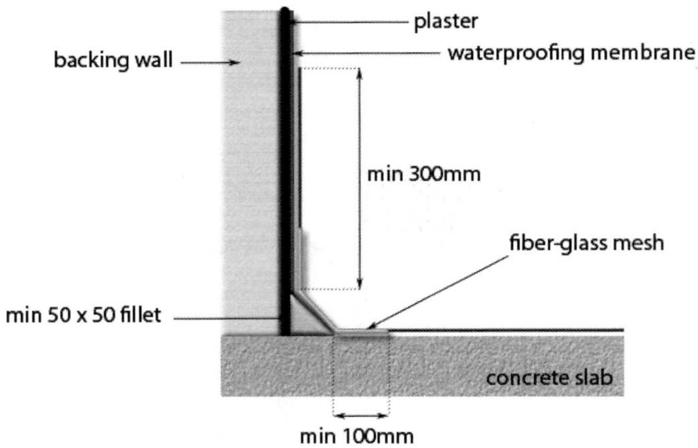

Fig. 2.7. Waterproofing at joints.

(**a**)

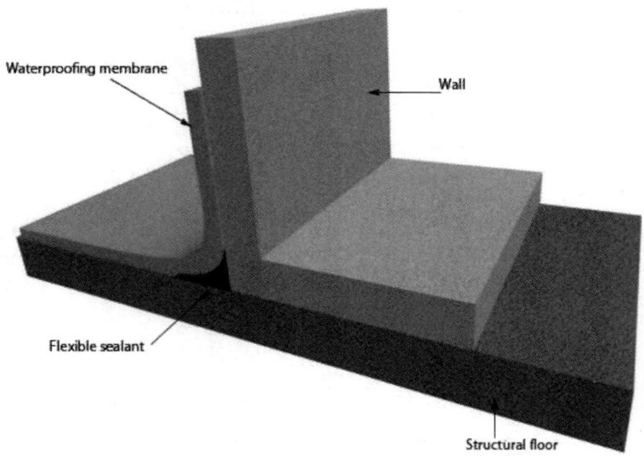

(**b**)

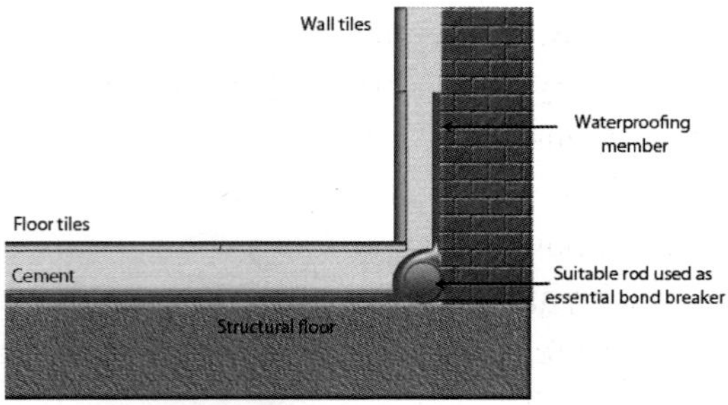

Fig. 2.8. (**a**) Use of flexible sealant at floor and wall joint. (**b**) Use of bond breaker at floor and wall joint.

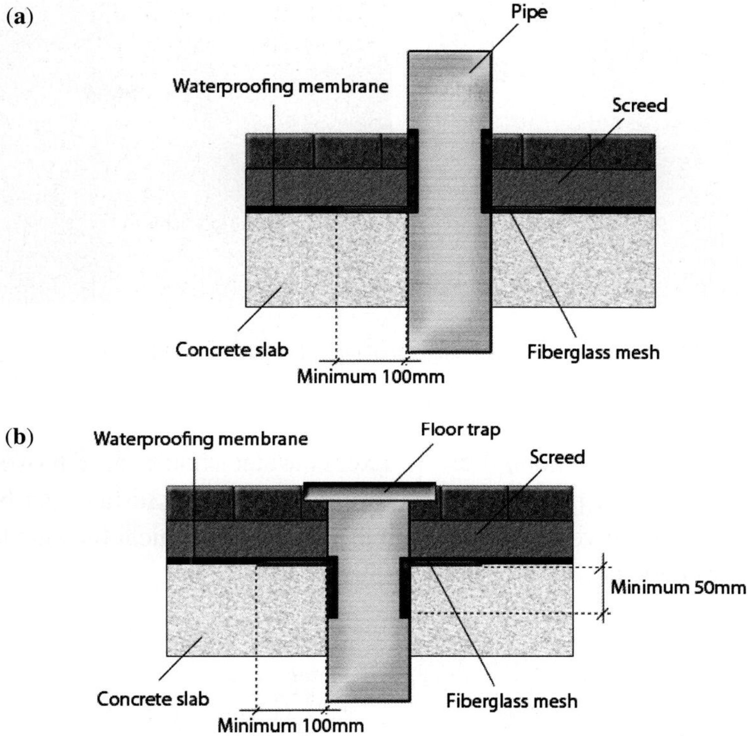

Fig. 2.9. Waterproofing detail at (**a**) pipe penetrations and (**b**) floor traps.

(1) *Removing imperfections* — When the surface has imperfections, it should be repaired or removed down to sound concrete and patched up with suitable mortar such as shrinkage compensated grout, polymer modified mortar, epoxy mortar, etc. (Fig. 2.10) [19]. Table 2.11 shows the common types of surface defects and their repairs.

(2) *Cleaning of substrate* — Cleaning of the substrate is essential to remove loose and foreign materials from the surface before the application of waterproofing membrane to ensure excellent bonding between the substrate and membrane. Some cleaning methods include high-pressure water jet, cleaning with vacuum cleaner and with brooms (Figs. 2.11a–2.11c). Both chemical and mechanical cleaning is recommended for surfaces which contain greasy or oily contaminants.

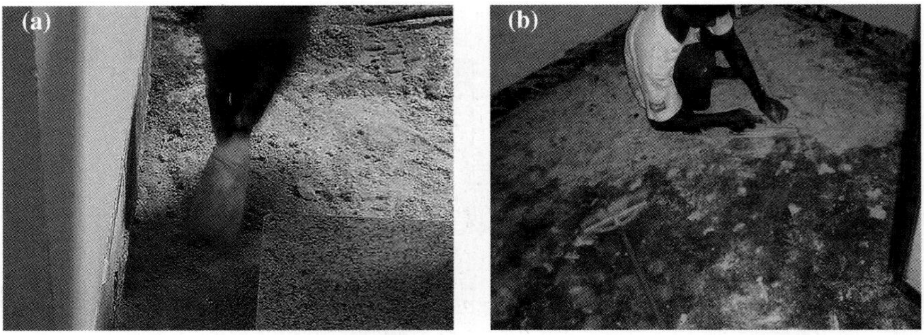

Fig. 2.10. (**a**) Removing a defective concrete. (**b**) Patching with mortar.

(3) *Removing surface dampness* — Excess water should be removed from the surface prior to waterproofing application. Surface can be dried by natural ventilation or with the use of mechanical blowers to accelerate the drying process (Fig. 2.12).

Membrane application

Considerations should include:

(1) *Size of the team* — In internal wet areas, the area to be waterproofed is relatively small thus the size of the team should not be too large given the space constraints.
(2) *Size of the mixes* — Mixing of waterproofing products should be done in a controlled environment according to the manufacturer's specifications. All waterproofing products should be mixed mechanically to ensure even mixing. The proportion of a two-part system if not pre-packed, should be measured out as accurately as possible and mixed within the specified time.
(3) *Tools* — Tools and equipment used for application should be cleaned before and immediately after application. Table 2.12 shows the different tools used for different types of systems.

The application of membrane should begin at a corner: diagonal to the entrance and upturns at wall floor junctions, to avoid stepping on applied area. Rollers or brushes can be used for application of membrane in most

Table 2.11. Common surface imperfections and repair methods [16].

Defect	Method	Tool
Cracks	Hack out a V-groove to a minimum depth of 25 mm. Remove all loose debris from affected area. Plaster back to original profile using nonshrink grout or specified patch repair plaster	Hack out a V-groove using a hammer and chisel or a mechanical hacker. Mix patching compounds using drills and mixing attachment and mixing bucket
Holes/voids	Clean out all dust, dirt and loose particles. Plaster back to original profile using nonshrink grout or a sand/cement/polymer mix	Clean using brush, broom or vacuum cleaner
Rough surfaces	Grind back to a smooth profile. Apply sand/cement screed over affected area. Apply recommended layer of plaster topping over affected area	Grind using electrical grinder or concrete grinder. Apply screed using a steel float and finish off with a wood float
Honeycomb cracks	Hack back to sound concrete. Plaster back to original profile using nonshrink grout or recommended repair plaster	Hack using a hammer and chisel or a mechanical hacker. Mix patching compounds using drills, mixing attachment and bucket
Internal corners and around penetrations	Apply a sand/ cement fillet or as specified by manufacturer, to all internal corners coving trowels	Mix small quantities of sand/cement plaster, patching compounds, etc. use diamond point trowels. For constructing coves and fillets, use margin and
Sharp protrusions and corners	Grind off to smooth level	Use electrical grinder
Dirt/oil grease loose plaster etc.	Remove with a high-pressure water jet	

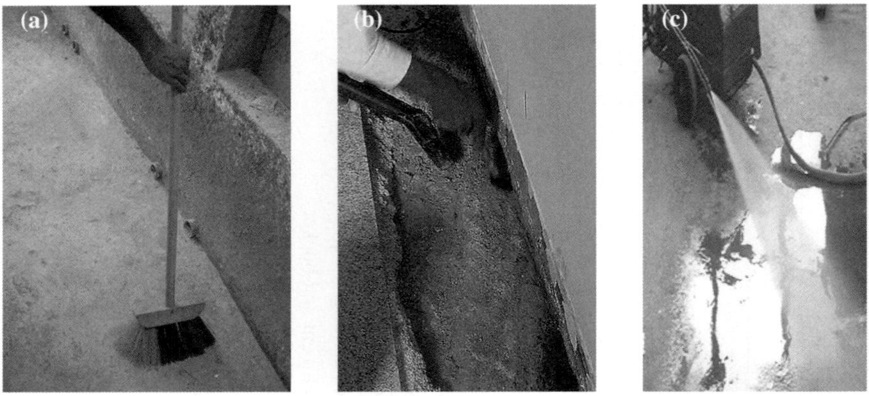

Fig. 2.11. Cleaning of substrate: (**a**) cleaning with a broom, (**b**) cleaning with a vacuum cleaner and (**c**) cleaning with high pressure water jet.

Fig. 2.12. Mechanical blowers to dry the surface faster.

Table 2.12. Different tools used for application of different membranes.

Type of System	Tools for Application
Polyurethane-based	Usually by brush, squeegee or broom. Vertical grade by brush or trowel
Rubber-based	Brush, rollers or airless spray
Acrylic-based	Usually applied by rollers to work material into fiberglass reinforcement
Cementitious-based	Usually by brush or spray followed by troweling

locations. Rollers having the same width as the upturn could achieve an even and uniform application [20]. However, critical locations where pipe penetrations and wall joints exist should be waterproofed using a brush to ensure good coverage. Figure 2.13 illustrates the application procedure for wall and floor.

Curing

It is vital to allow the waterproofing membrane to cure properly before performing any tests so as to prevent any premature failure of the waterproofing system. Table 2.13 highlights the common curing method for different systems.

Protection

Protecting the waterproofing membrane is the last step in the waterproofing process. The waterproofing contractor should ensure that works are carried out under controlled conditions and protection should be provided to the membrane immediately after application. The area waterproofed should be cordoned off.

Quality assurance

To uphold the final performance of membrane, it is important to carry out intermittent inspections during the application. Table 2.14 illustrates basic checks that can be performed before and after the application of waterproofing membrane.

2.3.2. *Screed*

2.3.2.1. *Design Aspects*

Design of screeds requires careful considerations of fitting layout in wet areas to provide required falls and accommodate services for plumbing. These include:

(1) Gradient of falls specified should be adequate to allow for efficient drainage of runoff. The directions of slopes should be clearly

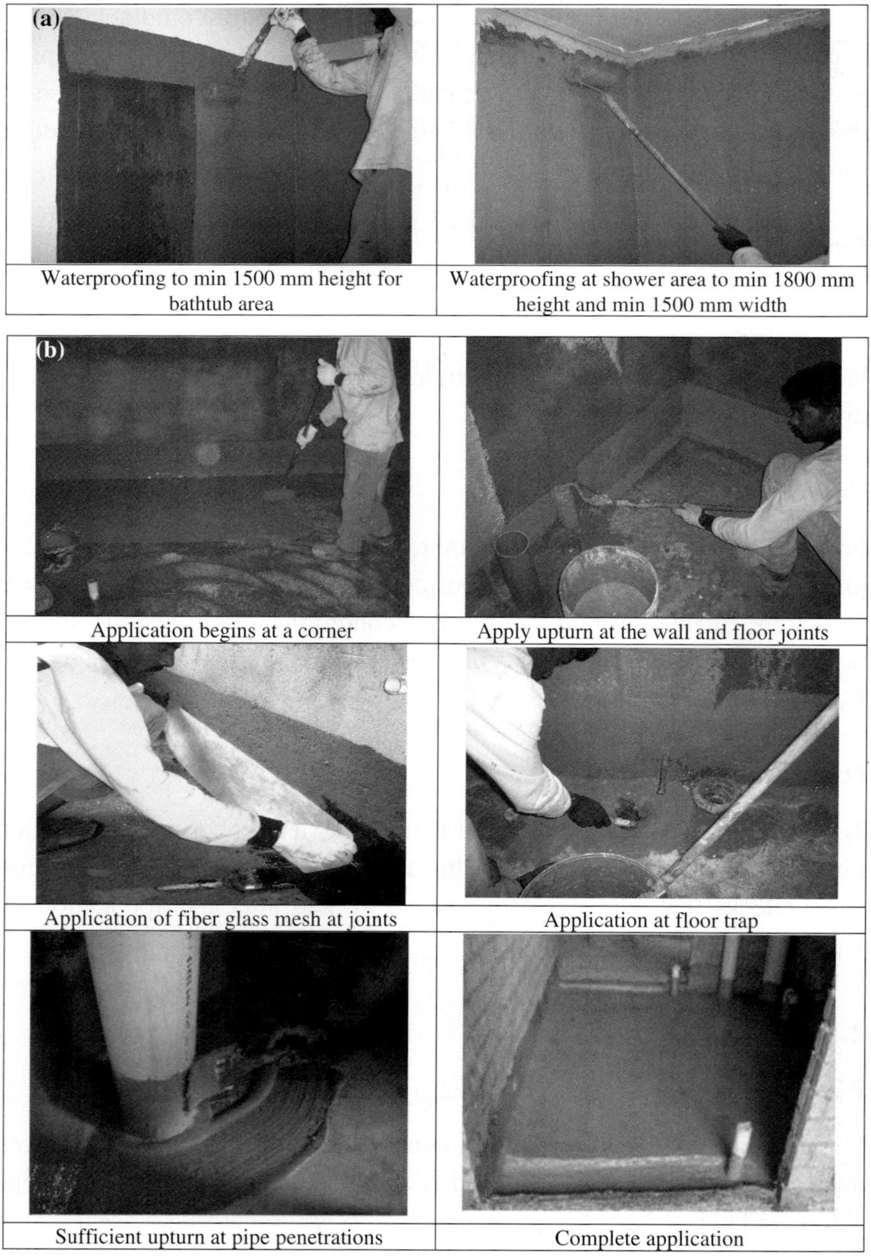

Fig. 2.13. (**a**) Waterproofing application on wall. (**b**) Waterproofing application process of floor.

Table 2.13. Curing methods and curing time.

Type of System	Curing
Polyurethane-based	Moisture cured Recommended 24 h curing time Contains solvent thus requires ventilation To be cured at least 72 h before flood test
Polyisoprene/ rubber-based	Air-drying About 1 h in exposed condition, 3–4 h in enclosed areas Recommended 72 h before flood test
Acrylic-based	Air drying Within 1 h, but requires minimum four to five coats due to reinforcement Total system therefore requires longer drying time
Cementitious-based	Moisture cured Normally 1–2 h Flood test within 24 h to assist in curing

Table 2.14. Inspection during the application of waterproofing membrane.

Stage of Work	Check	Criteria for Rejection
Checking for surface preparation	Dust	If excessive white powder on the cloth when wiping with a dark cloth
	Oil	If standing droplets appears while sprinkling the water
	Laitance	If excessive lose powdered materials are observed when the surface is scraped with a knife
Checking for application 	Water leakage	Waterponding test: Temporarily seal all outlets and flood the whole area to a minimum depth of 25 mm Ensure that the water level is not above the finished level of the membrane upturn at wall at the end of the test Flooding time is 24 h. Inspect deck below for leakage

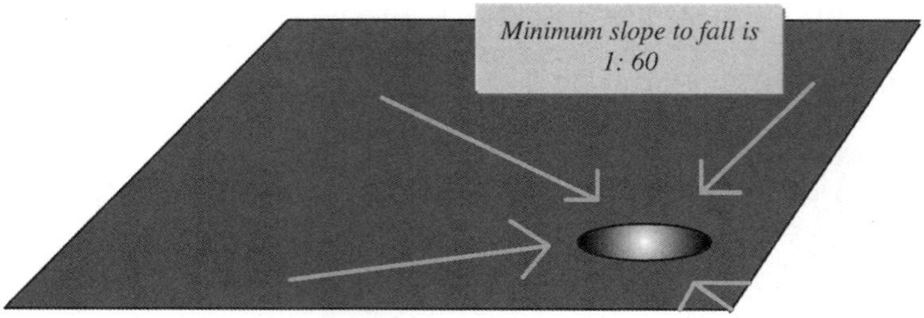

Fig. 2.14. Minimum slope to fall to achieve efficient drainage of water towards the drainage point.

indicated in the drawings (Fig. 2.14). The minimum gradient of the fall at shower areas or wash areas should be maintained at 1:60 towards the floor trap.

(2) The thickness of screed may be determined by the size of the services that are installed within screed, as the overall depth of the screed is to be increased by the depth of the particular service accommodated.

2.3.2.2. *Construction Aspects*

Substrate preparation

The surface of the substrate should be roughened to form mechanical key to enhance good adhesion of screed. For reinforced concrete wall, the substrate to receive render should be spatter-dashed with cement/sand (ratio 1:2 by volume) or latex adhesive. The protrusion thickness of spatter-dash should be about 3–5 mm [8].

Screed application

Proper application can be achieved through the following:

Planning — Prior to application, the entire wet area surface should be planned to accommodate required levels and falls according to the drawings. Level packs can be used to mark the required level of the screed layer and to ensure the correct falls to the floor outlets (Fig. 2.15).

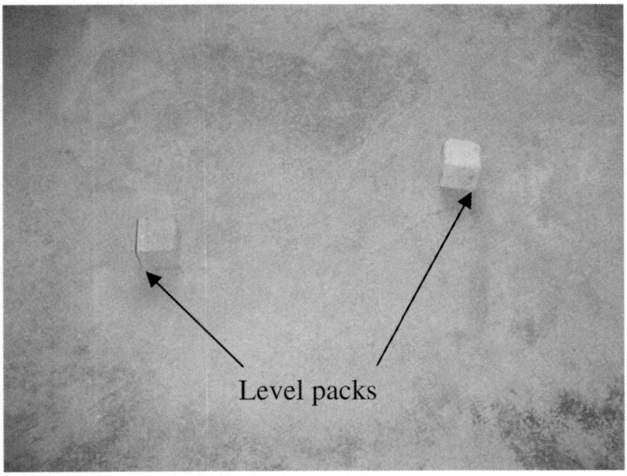

Fig. 2.15. Use of level packs.

Mixing — To achieve consistency, pre-packed mortar mix is recommended over conventional site mixing of cement sand and mortar. Pre-packed mortar is a dry product of cement, sand and performance-enhancing additive. Proper mixing of correct proportion of water and dry mixture should be carried out using mechanical mixers prior to screeding (Fig. 2.16).

Placing — Laying of screed should be carried out by spreading 10–20 mm surcharge over the base surface. The total acceptable thickness should not be less than 50 mm for unbonded screed [21, 22]. Thinner (less than 50 mm) layers may be acceptable with high-quality screeds and sound workmanship. Where thickness is greater than 50 mm, it should be laid in two layers and each layer should be well-compacted to remove voids [23, 24]. The following practices can be followed during the laying of screed:

- A wooden float can be used to roughen the surface to receive the next coat. The recommended curing time for the first coat is 24 h.
- Subsequently, the next coat could be applied in the same manner until the required thickness is achieved (Fig. 2.17).

Fig. 2.16. Use of a mechanical mixture.

Fig. 2.17. Leveling the surface.

This may be achieved either by setting the timber battens accurately to the required level or by the use of level packs.

- To ensure consistency in the surface flatness, the width of the screed bay should be limited to a maximum of 4–5 m.

  - The layout of the screed bays and joints should be consistent with the movement joints in the structural base or with rigid tile joints.
  - A plumb or level should be used to check the levelness of the surface.

Curing

The screed should be air-cured for at least seven days to achieve maximum strength. Water may be sprayed during the first two days to facilitate the hydration of cement.

Quality assurance

The following are tolerances for finished surface regularity [24–29]:

- Floor:
  - 3 mm for tiles to be fixed with adhesive
  - 5 mm for tiles to be fixed with cement sand and mortar
  - 10 mm for tiles to be fixed with cement and sand semi-dry mix
- Wall:
  - 3 mm for tiles to be fixed with adhesive
  - 6 mm for tiles to be fixed with cement sand mortar

2.3.3. *Tiling*

2.3.3.1. *Design Aspects*

Good detailing of tiling would enhance prolonged durability and better performance of the system. Movement joints, tile sizes, grouting materials and pointings are important issues to be considered during the design stage to minimize defects occurrence during construction and post occupancy stages.

Movement joint

Movement joints extending through the tiling and the substrate should be installed to accommodate stresses due to shrinkage, deflection and moisture. The recommended maximum spacing and width of movement joints [23] are:

- *Internal wall*: Horizontal and vertical directions: 5–6 m
- *Wall joint width*: Minimum 3 mm, preferred 5 mm
- *Internal floor*: 5–7 m in all directions
- *Floor joint width*: Minimum 5 mm

The spacing and width of joints should be consistent with the movement joints provided in the substrate (Fig. 2.18). Stress relief joints designed in the tile layer should coincide with joints in screed to ensure consistent differential movements throughout the system (Fig. 2.19).

Pointing width

Pointing accommodates movement of tiles due to temperature change, dimensional instability or moisture absorption. Tables 2.15–2.17 show the recommended widths for various types of tiles and grouts.

2.3.3.2. *Construction Aspects*

Substrate preparation

The substrate surface should be flat and even to receive the tiles. Dirt and foreign particles should be removed by cleaning prior to tiling works (Fig. 2.20).

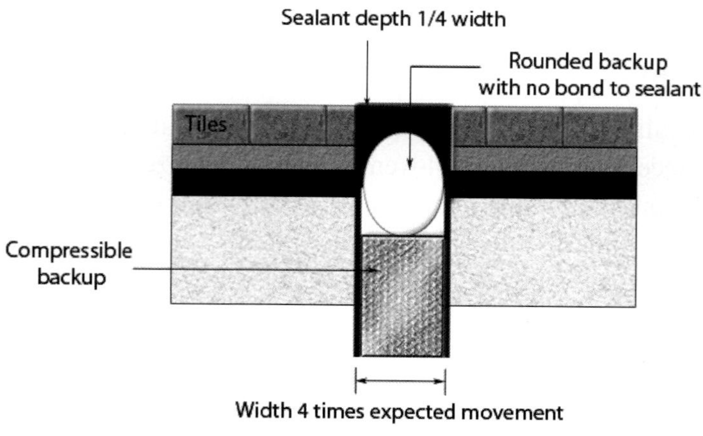

Fig. 2.18. Details of a typical movement joint [23].

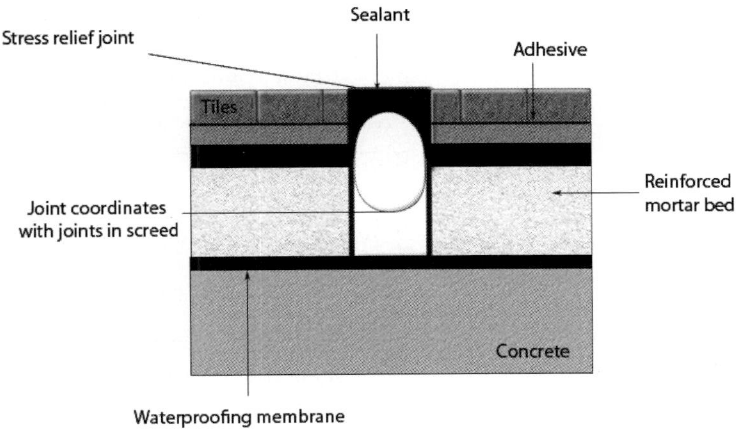

Fig. 2.19. Typical details of a stress relief joint [23].

Table 2.15. Recommended pointing widths for polymer modified grout.

	Joint Width (mm)			
Tile Size (mm)	**1.5**	**3**	**6**	**9**
25 × 25 × 6	4.8	2.6	1.6	1.3
50 × 50 × 6	8.8	4.8	2.6	1.9
106 × 106 × 7.5	14.9	7.7	4.0	2.8
100 × 200 × 12	11.7	5.9	3.1	2.1
100 × 200 × 30	4.8	2.5	1.3	0.9
150 × 150 × 9	17.5	8.9	4.6	3.2
150 × 150 × 12	13.1	6.7	3.5	2.4
200 × 200 × 9	23.0	11.8	6.0	4.0
300 × 300 × 9	34.7	17.5	8.9	6.0
325 × 325 × 9	—	18.9	9.6	6.5
400 × 400 × 9	—	23	11.8	8
600 × 600 × 9	—	34.6	17	11.8

Approximate m^2 coverage. *Source:* www.laticrete.com (*Courtesy*: Laticrete International Inc).

Planning

(1) *Selection of tile* — Inherently, tiles may have some dimensional variations due to the production process. To ensure consistent width of tile joints, the dimension of tiles to be laid in each area should be

Table 2.16. Recommended joint width for cementitious grout.

	Joint Width (mm)			
Tile Size (mm)	**3**	**6**	**9**	**12**
25 × 25 × 6	5.8	—	—	—
50 × 50 × 6	10.4	—	—	—
100 × 100 × 12	9.8	5.2	3.7	2.7
150 × 150 × 9	19.2	10.0	7.0	5.5
150 × 150 × 9	14.4	7.4	5.2	4.0
200 × 200 × 9	25.4	13.0	9.0	7.0
300 × 300 × 9	37.7	19.0	13.0	9.8
325 × 325 × 9	40.7	20.7	14.0	10.8
400 × 400 × 9	50.0	25.4	17.0	13.0
600 × 600 × 9	74.3	37.0	25.4	19.2

Approximate m^2 coverage. *Source*: www.laticrete.com (*Courtesy*: Laticrete International Inc).

Table 2.17. Recommended joint width for unsanded grout.

	Joint Width (mm)			
Tile Size (mm)	**1.5**	**3**	**6**	**9**
106 × 106 × 7.5	42	26	—	—
150 × 150 × 7.5	58	30	—	—
200 × 200 × 9	65	33	—	—
300 × 300 × 9	96	49	—	—

Approximate m^2 coverage. *Source*: www.laticrete.com (*Courtesy*: Laticrete International Inc).

Fig. 2.20. Surface cleaning.

Fig. 2.21. Sorting tiles in a "V" box (*Courtesy*: Housing Development Board).

controlled within a 1 mm variance. This can be done by sorting in the factory and providing a marking of the size range on the tile box for identification. Sorting of tiles should be carried out on site with the use of a V-box as shown in Fig. 2.21 or other devices. During the sorting process, tilers should also look out for other blemishes, such as cracks, warp, chip and nonsquareness.

(2) *Cutting and Termination* — Tiles laid around doorframes and sanitary fittings should be cut to maintain consistent tiling work. The cutting positions should be planned, marked (Fig. 2.22a) and cut (Fig. 2.22b). A grinder should be used to smoothen the edge of the cut tiles to prevent the cut tiles from being either jagged or sharp (Fig. 2.22c).

(3) *Marking* — To achieve alignment and consistent joint width, adequate reference lines should be marked on the screed to facilitate tile installation. A string can be used to imprint a colored dye on screed as the reference line for alignment of tiles (Fig. 2.23).

Installation

Man-made tile

Manmade tiles are usually fixed with modified cement mortar:

- Tiles should be dampened for at least 12 h to saturated-surface-dry (SSD) condition to prevent excessive absorption of water from the

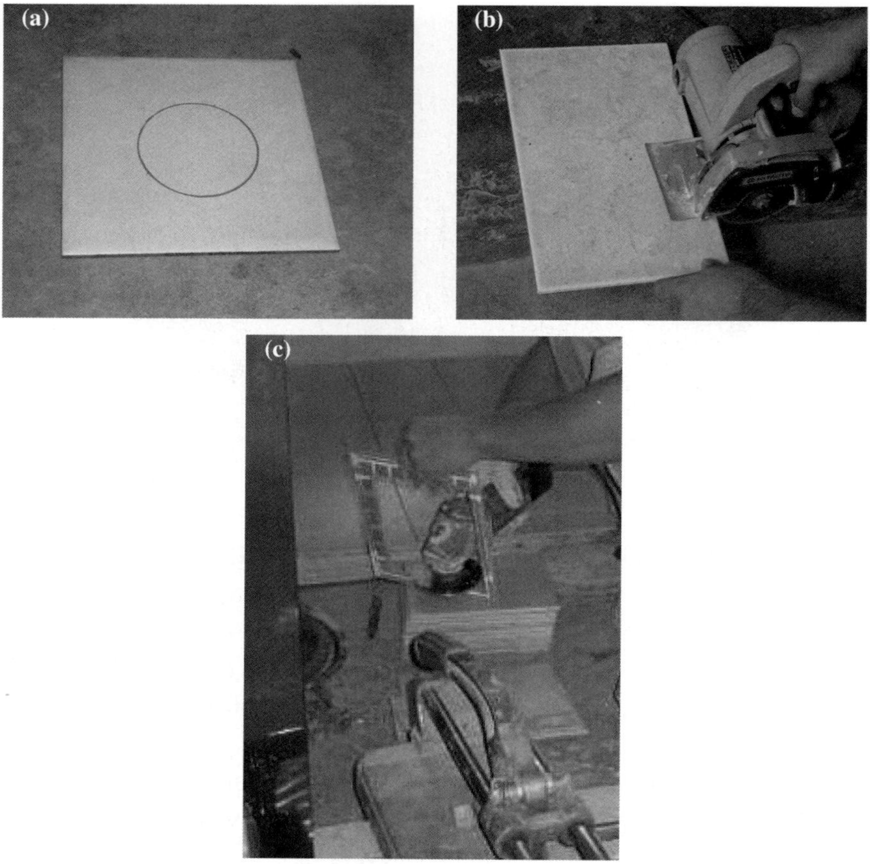

Fig. 2.22. Cutting and termination of tiles. (**a**) Mark the recess. (**b**) Cutting along the outline. (**c**) A grinder is used to smoothen the cut edges.

screed (Fig. 2.24a) and they should be stacked vertically to drain off the water (Fig. 2.24b). At the point of laying, the surface of the tiles should be dry. Tests should be carried out on glazing tiles to confirm that soaking would not lead to crazing.

- Mortar should be prepared in accordance with the manufacturer's instructions (Figs. 2.25a–2.25d).
- Mortar should be applied to a consistent thickness of not more than 10 mm and should not be spread over a coverage that could not be covered with tiles in time as the surface of adhesive may form a skin which prevents bonding. The maximum spread should be limited to

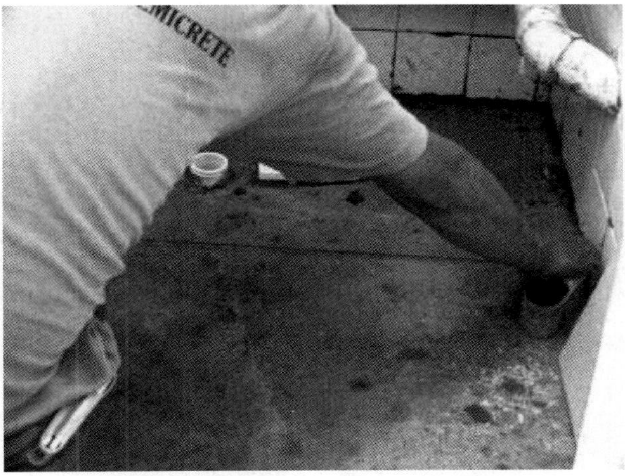

Fig. 2.23. Imprinting lines for alignment of tiles.

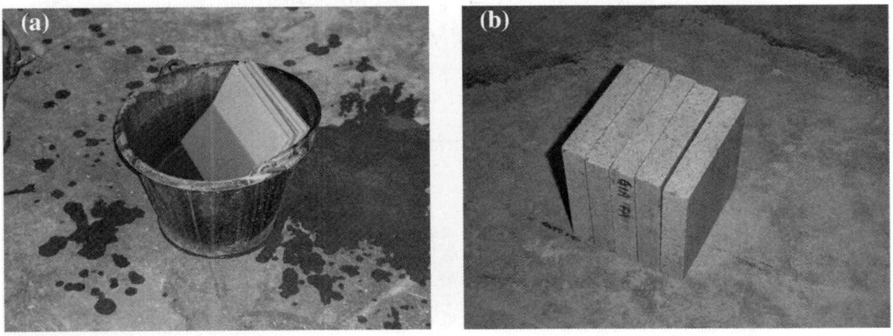

Fig. 2.24. (**a**) Soaking of tiles. (**b**) Drying of tiles.

about 1 m^2 (Fig. 2.26a). However, for adhesives, "open time" of the adhesive film should be considered in this regard.

- The mortar bed shall be allowed to stiffen slightly.
- The background of the tiles should be cleaned thoroughly (Fig. 2.26b).
- Mortar should be spread on the cleaned background of the tile. The keys and frogs on the tile back should be completely filled up with mortar, with a thickness of not less than 2 mm but not more than 6 mm exceeding the depth of keys or frogs (Fig. 2.27).
- The tiles should be then placed firmly with a twisting motion (do not slide tiles in), ensuring regular and straight joints to a recommended

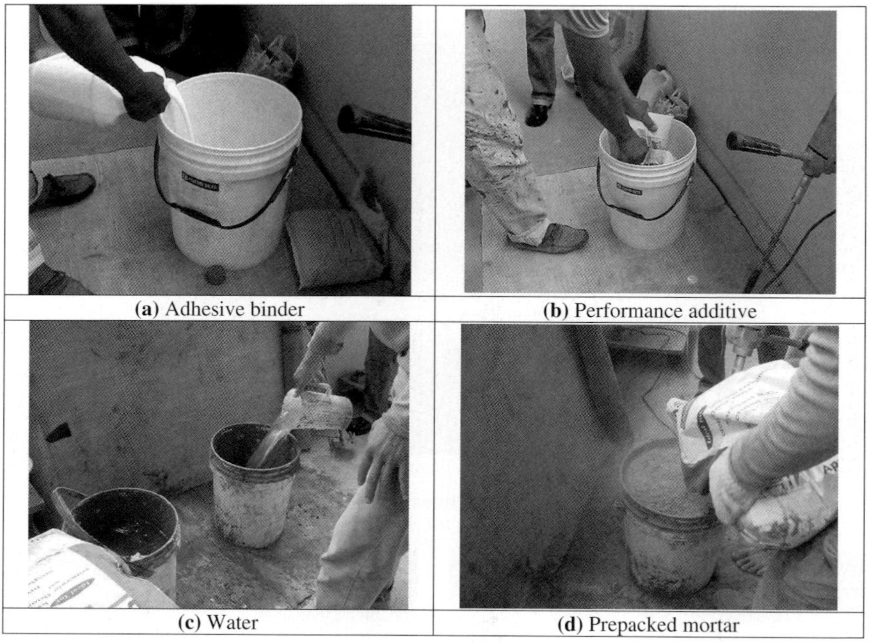

Fig. 2.25. Mortar preparation.

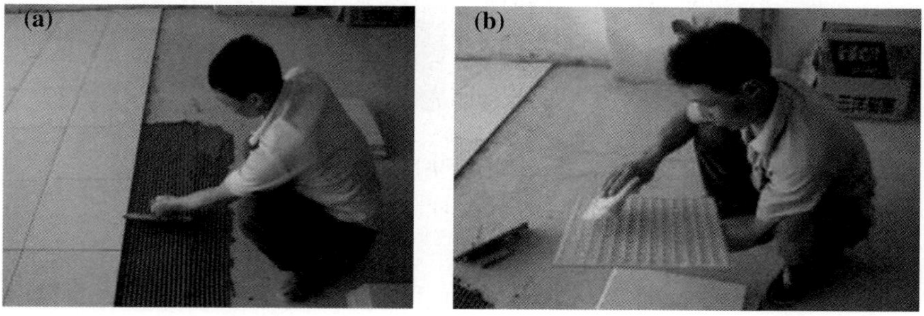

Fig. 2.26. (**a**) Spreading of the adhesive. (**b**) Cleaning of the tile background.

width. Once they have been put in position, they should be uniformly tapped using a hammer into place to achieve good contact. A block (Fig. 2.28) should be used to prevent over-hammering causing the tiles to chip or crack.

Fig. 2.27. Application of mortar to tiles.

Fig. 2.28. Fixing of the tile firmly into the bedding.

- Spirit levels should be used to check surface levelness and evenness during tile installation and after fixing (Fig. 2.29). For bigger size tiles, a further check of levelness can be carried out about 2–3 h after laying. This is to prevent unevenness of tiles caused by tile settlement due to self-weight.

Fig. 2.29. Checking surface levelness.

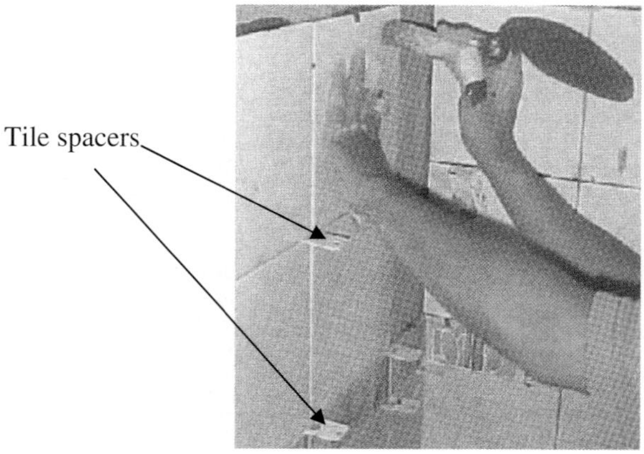

Fig. 2.30. Tile spacers for consistent spacing between tiles.

- Tile spacers can be used to achieve neat and consistent joint widths (Fig. 2.30). These spacers should be easily removable and not cause staining to the tiles.

Natural stone tile

During installation, preparation of natural stone tiles is very important to prevent defects such as, efflorescence, cracking, debonding and tonality.

Fig. 2.31. Application of impregnator to natural stones.

(1) *Pre-treatment* — Pre-treat the edges and the front surface with a water repellent impregnator by spreading it evenly as natural stone tiles have high water affiliation (Fig. 2.31). However, caution should be exercised when attempting to coat the back face of the tiles with a sealer as it may act as a bond breaker preventing the bedding material from adhering properly.
(2) *Curing* — The treated tile should be allowed to cure properly before laying.
(3) *Setting natural pattern* — The floor tiling should be set out in line with the natural pattern of the tiles starting from the center of the room and in straight lines parallel to the axes of the room. However, for irregularly shaped rooms where cutting of tiles are unavoidable, cut tiles should be placed at the least noticeable locations.
(4) *Cleaning* — The back of the tiles should be wiped with a damp cloth or brush to remove dust (Fig. 2.32).

Fixing of tiles should be carried out with suitable adhesive that should be in accordance with manufacturer's instructions.

- A skim-coat of the bedding mix should be spread on the receiving surface using the flat side of a notched trowel so as to cover it completely. A V-notched adhesive trowel should be used, holding it at

Fig. 2.32. Cleaning the back of the stone.

a 45°–75° angle (Fig. 2.33). Trowel's angle should be kept consistent and straight to create uniform ridges in the adhesive so as to obtain a level setting surface.

- A layer of adhesive should be applied over the backside of the tile to cover the surface entirely (Fig. 2.34).
- The tile should be laid onto the bedding with a firm pressure and adjusted within a limited time specified by the bedding material's manufacturer.
- Once the tiles are put in position, tap uniformly on the tiles to achieve good surface contact.
- A spirit level ensures that the tiles laid are even and level.
- Appropriate plastic spacers or spacing rod (for wider joints) should be used to control the uniformity of the joint widths. Normally, joints for smooth tiles are at least 1 mm wide while for textured tiles are between 6 and 10 mm (Fig. 2.35). Widths should comply with the standard requirements (Tables 2.15 to 2.17).

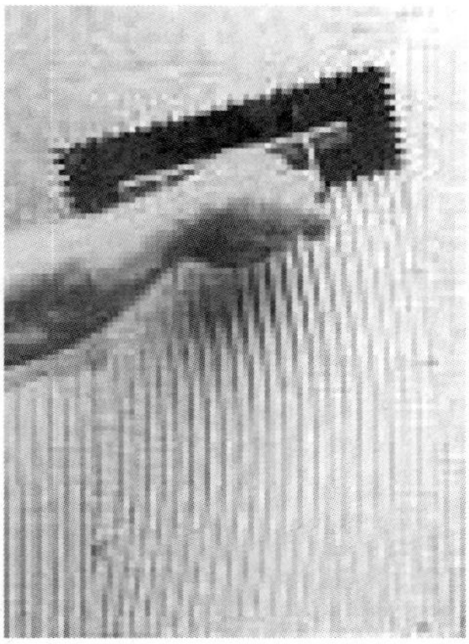

Fig. 2.33. Troweling at 45°.

Fig. 2.34. Application of adhesive to tile.

- To check for complete adhesion behind large tiles, random removal of tiles could be carried out as work proceeds (Fig. 2.36). In general, adhesive must cover 90% of the back of any tile which is randomly removed for checking. If this condition is satisfied, the adhesive is cautiously removed from the tile and the substrate and re-applied with fresh adhesive. If this condition is not satisfied, other tiles in the vicinity of the original tile should be removed to check if the fault is general.

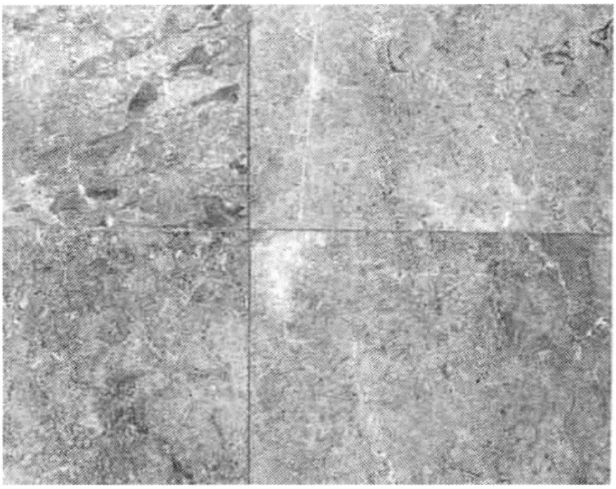

Fig. 2.35. Consistent joints.

Fig. 2.36. Removal of tiles to check for adhesion.

Grouting

Grouting of tile joints has a significant impact on the overall appearance of the finishes. White cement is commonly used due to its low costs.

Polymer modified grouts which are usually formulated with good working characteristic, low shrinkage, good adhesion to the sides of the joints and higher durability are gaining popularity.

- Prior to grouting, the joints should be brushed clean to remove excess mortar.
- Saturate the tile before grouting to prevent migration of grout pigments into the pores of tiles. Joints that are to receive resin-based or polymer-modified cementitious grouts should be dry.
- The mixing of the material should be done in accordance with the manufacturer's instructions.
- Spread a grout over the entire surface of the tile. The mortar should be pressed sufficiently with a rubber float so that it flows along the joint and occupies all voids (Fig. 2.37a). Any excess grout on the tile surface should be cleaned away immediately with a sponge so as to avoid staining to the tile surface (Fig. 2.37b).

Fig. 2.37. (**a**) Grouting. (**b**) Removing excess grout.

Fig. 2.38. Protection to completed tile works.

Cleaning and protection

All loose material and residual cement should be brushed off and the surface washed with clean water to remove dirt. Protection to the tiling works should be provided immediately after its completion. This can be done by covering the tiled area with polyethylene sheets, cardboards or wooden boards (for heavy traffic) until the entire system has cured (Fig. 2.38). The protective material should be sufficiently thick to withstand human traffic, abrasion and impact, impermeable to water and would not stain under wet and dry conditions.

Quality assurance

Table 2.18 illustrates a check list to assure the standard of the finishes in wet areas.

2.3.4. *Ventilation and Air Circulation*

In addition to removing odor, adequate air circulation is vital to extract dampness which is responsible for defects such as biological growth on floors, ceilings and walls. Proper air circulation can be achieved by natural or mechanical ventilation or a combination of both. Exhaust fans

Table 2.18. A check list for finishes in wet areas.

Checks	Wall Tiling	Floor Tiling
1. Tolerances for surface unevenness ○ Surface unevenness over 1 m length ≤ 1.5 mm ○ Unevenness between two tiles over the joint ≤ 1 mm	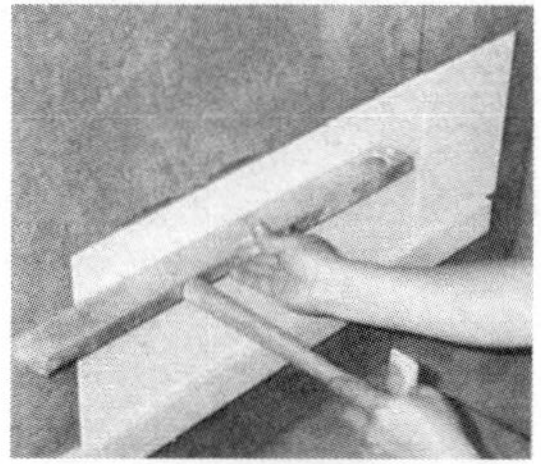	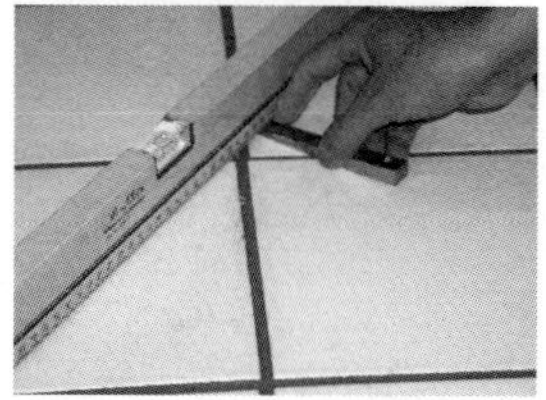
2. Interface with other elements ○ Tile interface with ceiling, window frames, door frames, pipes, etc. are aligned and consistent	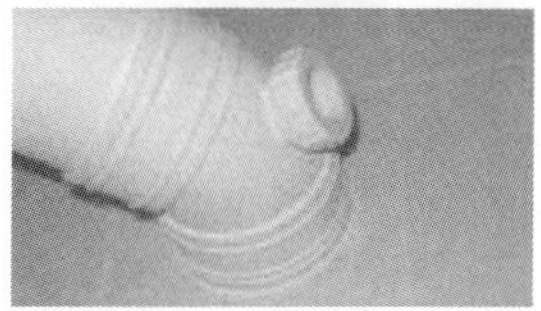	

(*Continued*)

Table 2.18. *(Continued)*

Checks	Wall Tiling	Floor Tiling
3. Joints, tonal variations ○ Joints are aligned and with consistent size between 1 and 2 mm ○ Joints are pointed neatly with no voids within them ○ The joint width shall be less than 5 mm ○ Joints with no excess or uneven grout		
4. Hollowness, chip offs, sharp edges and crack Surface is free of chipped, cracked and hollow tiles		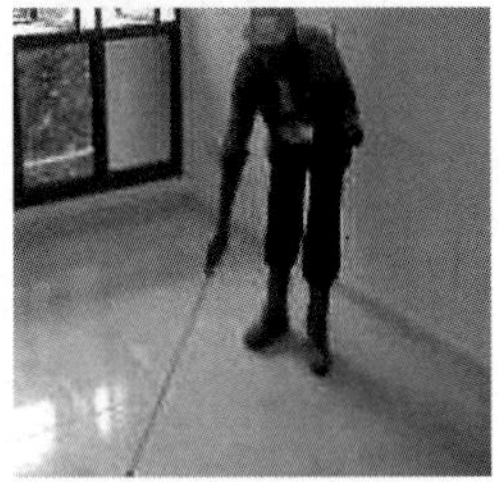

Fig. 2.39. Exhaust intake grill of mechanical ventilation.

and ventilation ducts should be located within 3 m of the fan inlet or intake grill [30]. According to SS CP 13:1999 [31], the minimum required outdoor air supply is 10 air exchange per hour (ACH). For areas where they are entirely mechanically ventilated, minimum air exchange rate should be 15 ACH.

Intake grills should be provided at low levels near to W.C.s and urinals. This would enable foul-air to be extracted quickly (Fig. 2.39). The exhaust air should be discharged at least 2 m away from the pavement level and at least 5 m away from any window.

2.3.4.1. *Plumbing*

Layout

An efficient plumbing layout reduces the need for penetrations through waterproofing membrane. Important considerations include:

(1) Detailed drawings of the layout of service pipes that are to be cast with the floor slab should be provided. These detailed layouts should include details such as gradients of horizontal pipes, joints and connections (Fig. 2.40) [32].

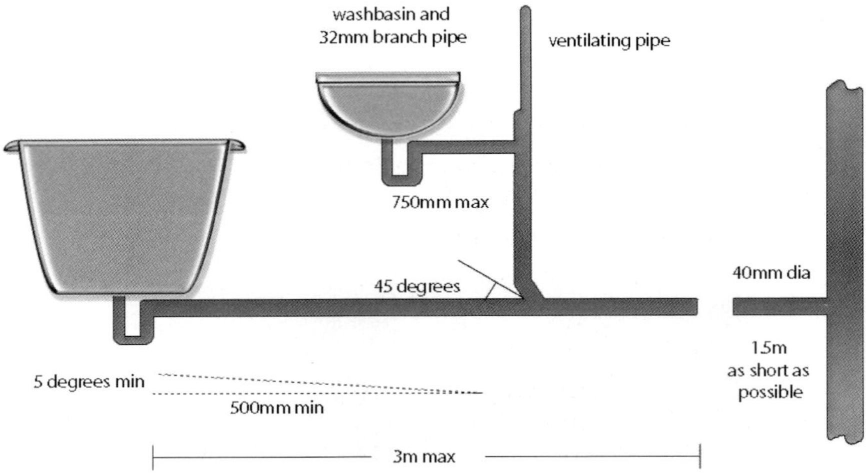

Fig. 2.40. A typical detail of plumbing system [28].

(2) The number of wet walls (Table 2.19a) and pipe penetrations (Table 2.19b) through wall and floor should be minimized. Single wet wall design with a common discharge stack can reduce the number of penetrations (Fig. 2.41).

(3) Raised floors provided around the pipe could minimize contact between water and pipes (Fig. 2.42) and reduce the chances of water leakages through pipe penetrations or corrosion of the pipe (cast iron).

Accessibility

The accessibility for repair and replacement of service pipes is one of the major concerns that should be addressed during the design stage. Openable covers and walk-in pipe ducts can be provided to facilitate access for maintenance (Figs. 2.43–2.45). In addition, designers should take into consideration the positions of service pipes with regards to the accessibility of the entire floor or wall area for cleaning. The pipes should not inhibit the ease of cleaning of floor and wall surfaces.

2.3.4.2. *Sanitary Fittings*

Sanitary fittings should be installed such that to provide convenience to the user and to facilitate maintenance work.

Table 2.19(a). Complexity of plumbing details for different numbers of wet walls.

Criteria
When the number of wet walls is **1 (minimum)**, the plumbing details are simple • The discontinuities may be minimum, reducing the risk of water leakages • Maintenance of the system may be easy due to the simple layout
When the number of wet walls is **2 (moderate)**, the plumbing details are relatively simple • The discontinuities may be relatively small, reducing the risk of water leakages • Maintenance of the system may be relatively easy due to the simple layout
When the numbers of wet walls is **>2 (large)**, the plumbing details are more • The discontinuities may be relatively high, increasing the risk of water leakages • Maintenance of the system may be relatively difficult due to the complex layout

Table 2.19(b). Recommended number of penetrations for different sanitary fittings.

Fittings	**Recommended Number of Penetrations**
Urinal/WC	1 trap shall be provided to serve a maximum of 10 urinals. For more than 10 urinals, more than one trap shall be provided
Washbasin	1 wash basin trap shall be provided to serve a maximum of 10 wash basins. For more than 10 wash basins, more than one trap shall be provided
Wash/shower/bath	1 floor trap for every 3 WC cubicles provided that the separating wall/partition does not touch the floor level and the other 2 WC cubicles are graded to fall towards the floor trap in the third WC cubicle

Source: Ministry of Environment.

(1) *Shower/bath or wash area* — Separation of shower/bath or wash areas from the common toilet areas is a good practice to minimize water from affecting the entire floor area. Enclosed areas or keeping level differences not less than 75 mm between shower/wash areas and the other areas or use of "kerbs" can control the water flow (Figs. 2.46a and 2.46b).

(2) *Lavatory* — Washbasins can be installed into the preformed countertop or vanity top with a splashback of minimum height of 100 mm and an apron edge to prevent overflowing of water from the

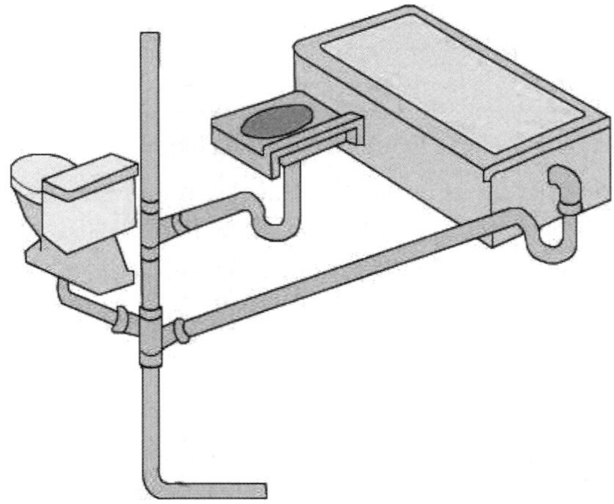

Fig. 2.41. Common stack.

Fig. 2.42. Raised levels.

Fig. 2.43. Access panel.

basin [33] (Fig. 2.47). If the wash basin is not fixed to a vanity top (i.e. free standing or wall hung type), a minimum size of 550 mm in length and 400 mm in width should be provided to prevent water from splashing onto the floor (Fig. 2.48). In addition, it can be fixed away from the wall to minimize water seepage through adjacent wall (Fig. 2.49).

(3) *Water Closet (WC)/urinals* — Wall-hung toilet bowls may be selected to facilitate the ease of cleaning the floor (Fig. 2.50) [13].

Fig. 2.44. Wall mounted pipes.

Fig. 2.45. Walk in pipe ducts.

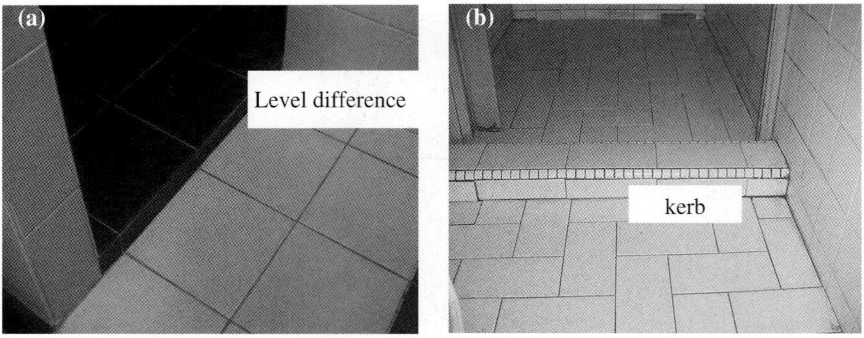

Fig. 2.46. (**a**) Level differences at a shower. (**b**) Kerb.

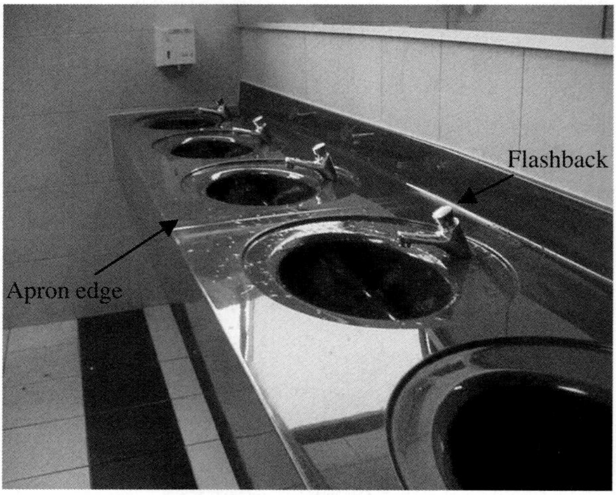

Fig. 2.47. Preformed vanity top with apron edge.

Spatial design

The location of fixtures affects greatly the ease of floor cleaning. The following guidelines may enhance efficient maintenance:

(1) between two wash basins — Minimum distance of 800 mm (Fig. 2.51),
(2) between two WCs — Minimum distance of 750 mm (Fig. 2.52),
(3) between two urinals — Minimum distance of 750 mm (Fig. 2.53).

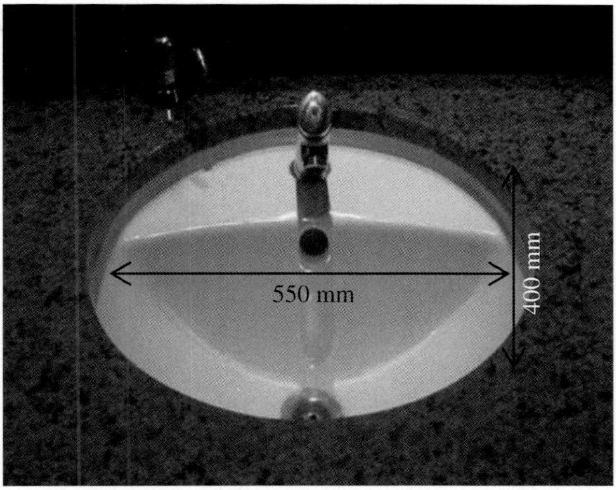

Fig. 2.48. Recommended dimensions of basins.

Fig. 2.49. Free-standing washbasin fixed way from a wall.

Fig. 2.50. Wall hung toilet bowl.

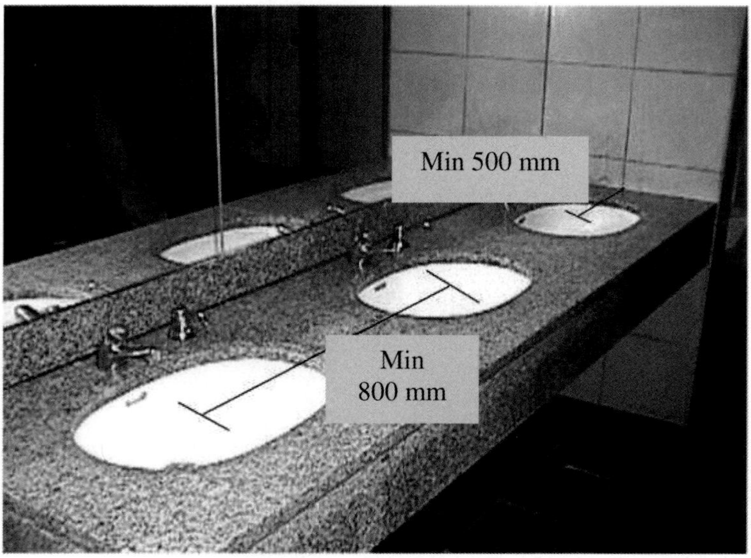

Fig. 2.51. Minimum distance between washbasins.

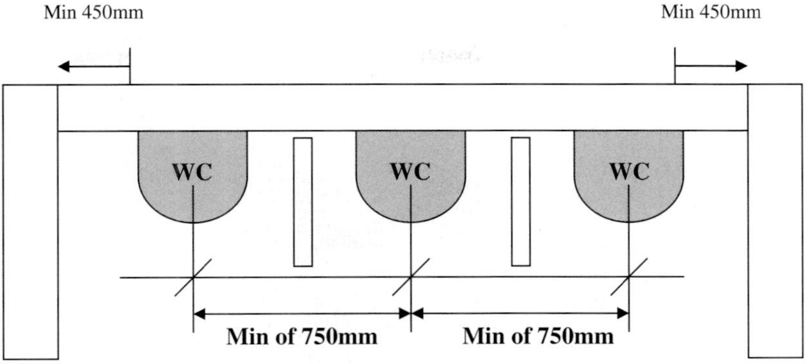

Fig. 2.52. Minimum distance between WCs.

Fig. 2.53. Minimum distance between urinals.

Fixing of sanitary fixtures

Sanitary fixtures — If a sanitary fixture abuts the wall or within 75 mm of the wall (e.g. bench top), the waterproofing membrane should be dressed up to a minimum height of 150 mm from the fixture surface (Fig. 2.54).

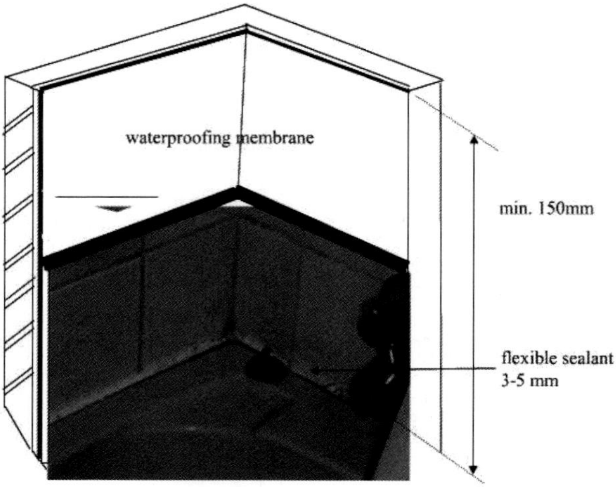

Fig. 2.54. Sanitary fitting abutting to the wall.

Figure 2.55 illustrates different applicable details of sealed joints between sanitary fittings and wall to prevent water seepage. A supporting wall can also be erected away from the backing wall to fix the fitting. This would minimize the chances of water penetration through the wall (Fig. 2.56). On the other hand, additional floor traps should be provided to facilitate efficient drainage to prevent water ponding (Fig. 2.57). For instance, areas underneath the bathtub and shower tray are critical locations where waterponding may arise.

2.4. Maintenance

All building elements start to deteriorate from the moment they are completed. Maintenance which includes cleaning, repair and replacement, are essential to regulate the pace of building deterioration.

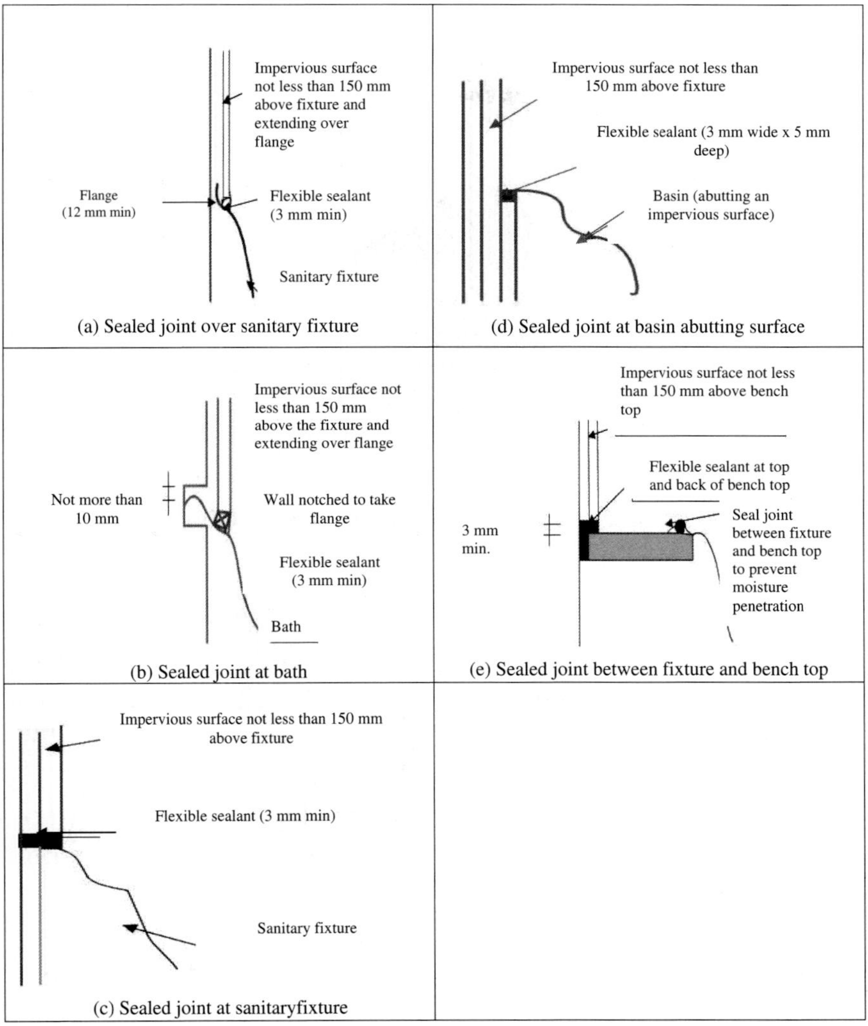

(a) Sealed joint over sanitary fixture

(d) Sealed joint at basin abutting surface

(b) Sealed joint at bath

(e) Sealed joint between fixture and bench top

(c) Sealed joint at sanitaryfixture

Fig. 2.55. Sealed joint between sanitary fittings and wall.

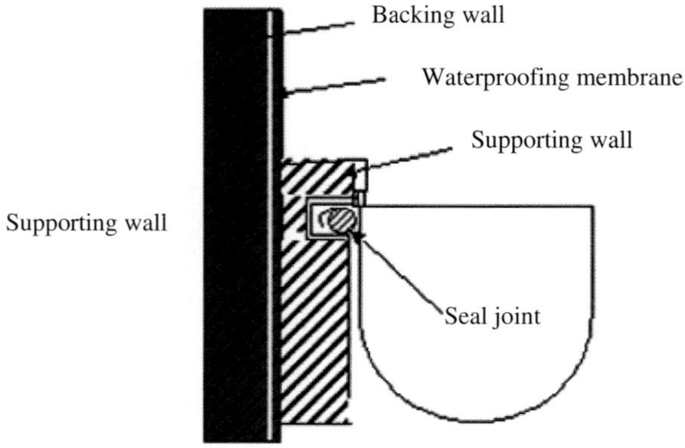

Fig. 2.56. Supporting wall.

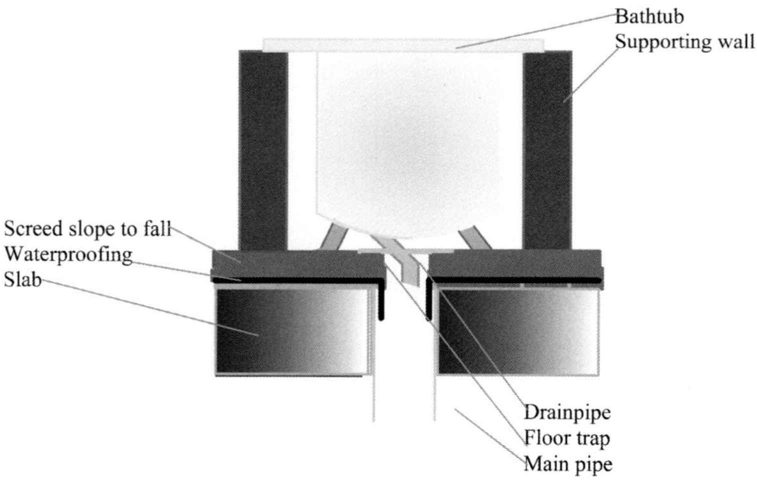

Fig. 2.57. An outlet to provide drainage from beneath the bathtub.

Cleaning method of wet area varies depending on the type of finishes used. Table 2.20 outlines certain practices for normal cleaning to keep wet areas maintainable.

Special cleanings are necessary for unexpected stains (Table 2.21). These stains can be due to various activities of humans and prolonged exposure of the surfaces for certain solutions.

Table 2.20. Cleaning methods.

Type of Finishes	Ceramic/Homogeneous Tile	Granite	Marble	Paint
Specifications	Unglazed ceramic tiles or quarry tiles to SS CP 68: 1997 or BS 6431 Part 1, 1983, with test certification to ISO 10545: 1995 Glazed ceramic tiles to BS 6431 Part 1, 1983 or to SS CP 68:1997, with test certification to ISO 10545, 1995	Natural granite and other igneous rocks specified and selected to BS 5390: 1976	—	Paint to SS CP 22
Maintenance	Regular mopping and buffing	Regular mopping and buffing	Regular mopping and buffing	Cleaning
Method	A solution of 1/4-cup low-suds detergent (solutions produce less bubbles) or 1–2 tablespoons of either washing soda or tri-sodium phosphate or commercial floor cleaning powder shall be mixed	Natural cleaner with high-foaming jet (without pressure) is suitable. It is recommended to avoid acidic, high alkaline and chlorinated products	As a rule of thumb, cleaning solutions that contain any form of acids, phosphorus, chlorine or scouring powders should not be used on polished marble	Washing with clean water is sufficient on surfaces that have minor stains and dirt

(Continued)

Table 2.20. *(Continued)*

Type of Finishes	Ceramic/Homogeneous Tile	Granite	Marble	Paint
	well in 1 gallon water to clean the floor 	However, dry cleaning methods are recommended over frequent wet cleaning methods in wet area	This would make marble looks dull or suffer from a peeling effect known as "orange peel" unless adhering to the cautions and/or formulas provided. In addition, any type of abrasive product should also be avoided, in order to prevent scratches and loss of shine	If necessary, mild detergent may be used for heavy stains
	Plain water can be used to remove dirt. However, for heavier soiling, a mild detergent solution is recommended. The solution shall		For polished finishing small areas, use of warm water would be sufficient for periodic cleaning	

(Continued)

Table 2.20. *(Continued)*

Type of Finishes	Ceramic/Homogeneous Tile	Granite	Marble	Paint
	be well-mixed before using. The surface should then be kept dry for more shine 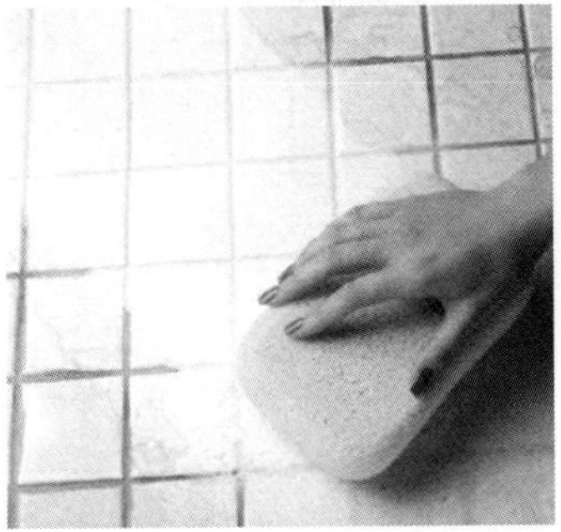			
Cleaning frequency	*Residential*: Twice a week *Commercial and institutes*: Three times a day *Offices and industrial*: Twice a day	*Residential*: Twice a week *Commercial and institutes*: Three times a day *Offices and industrial*: Twice a day	*Residential*: Twice a week *Commercial and institutes*: Three times a day *Offices and industrial*: Twice a day	No frequent cycles

Table 2.21. Typical examples of requiring special cleaning.

Types of Soiling	Example	Typical Sources	Cleaning Methods
Soap scum	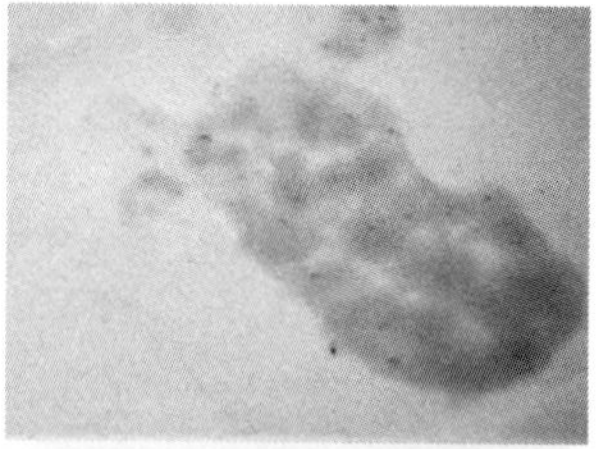	Soap	• Sponging with a mixture of 1/2 cup packaged water softener, 2 tablespoons rottenstone and 1 cup hot water • With a solution of 1–2 tablespoons trisodium phosphate in 1 gallon hot water
Mildew	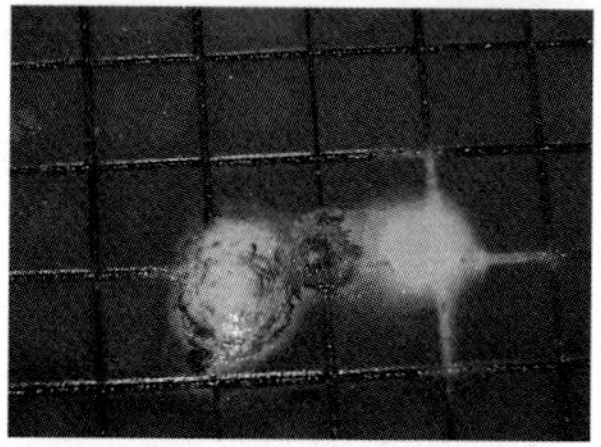	Dampness in toilet	• Clean with a dilute solution of chlorine bleach in water • Using a mildew-retardant household spray • The directions on the label of the solutions should be strictly followed
Rough white coating		Mineral built up	• It may be dissolved with a commercial tile cleaner

(*Continued*)

Table 2.21. *(Continued)*

Types of Soiling	Example	Typical Sources	Cleaning Methods
Dark stains		Fungi growth on moisture film	• They may be removed by covering them with full-strength liquid laundry detergent for a couple of hours and following by sponging with water • If the stains do not come off, they should be scrubbed with a brush
Organic stains	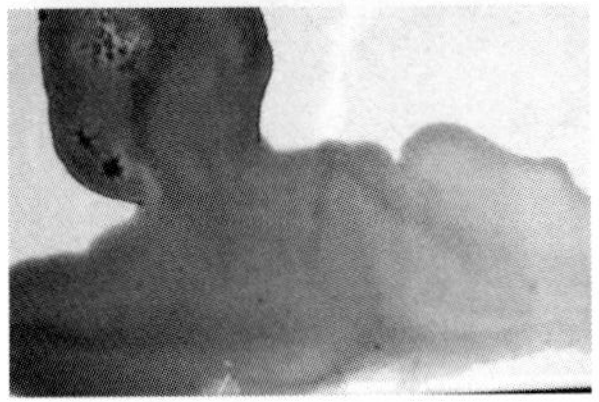	Tea, coffee	• A solution of 20% peroxide (hair bleaching strength) and a few drops of ammonia can be used to remove the stains
Oil stains		Butter, hand cream or lotion	• These can be easily removed by scrubbing with hot and detergent solution or wiping with ammonia-dampened cloth • If these alkaline solutions fail in removing all the oil, a stiff brush can be used

(Continued)

Table 2.21. *(Continued)*

Types of Soiling	Example	Typical Sources	Cleaning Methods
Rust stains		Corrosion of metallic fittings	• A commercial rust remover should be used to remove the stain. The directions for usage given on the label should be followed exactly and the solution should not be left on the surface longer than the specified duration. This is to prevent the acid in many rust removers from etching the tile surface
Acid stains		Acidic cleaning agents	• Any acidic substances should be wiped off immediately with a wet cloth • If the surface is etched, polishing may be required

References

[1] M. Y. L. Chew and N. De Siva, "Maintainability problems of wet areas in high-rise residential buildings", *Building Research and Information*, **31**(1), 60–69, 2003.

[2] M. Y. L. Chew, D. S. Nayanthara and S. S. Tan, "Handbook for wet area", http://www.hpbc.bdg.nus.edu.sg/, under the segment "About Us" and "Publication".

[3] M. Y. L. Chew, *Construction Technology for Tall Buildings*, 3rd Edition, World Scientific, Singapore, 2009.

[4] Building and Construction Authority, *Good Industry Practices — Waterproofing for Internal Wet Areas*, Singapore, 2001.

[5] British Standards, *BS 8204-Part 1: Screeds, Bases and in-situ Floorings. Concrete Bases and Cement Sand Levelling Screeds to Receive Floorings. Code of Practice*, British Standards Institution, London, 2002.

[6] British Standards, *BS 8204-Part 2: Screeds, Bases and in-situ Floorings. Concrete Wearing Surfaces. Code of Practice*, British Standards Institution, London, 2002.

[7] M. J. Gatfield, *Screeds, Flooring and Finishes: Selection, Construction and Maintenance*, Construction Industry Research and Information Association, London, 1998.

[8] Building and Construction Authority, *Good Industry Practices — Ceramic Tiling*, Singapore, 2000.

[9] Building and Construction Authority, *Good Industry Practices — Marble and Granite Tiling*, Singapore, 2000.

[10] British Standards, *BS EN 12004 Adhesives for Tiles. Definitions and Specifications*, British Standards Institution, London, 2001.

[11] Housing Development Board, HDB Standard Specifications, Singapore, 2003.

[12] M. Y. L. Chew and N. De Silva, "Factors affecting water-tightness in wet area of high-rise residential buildings", *Architectural Science Review*, **45**(4), 375–383, 2002.

[13] B. A. Richardson, *Defects and Deterioration in Buildings*, E&FN Spon, New York, 2001.

[14] N. L. Mohinder, *Piping Handbook*, 7th Edition, McGraw Hill, New York, 2000.

[15] D. A. Willoughby, R. D. Woodson and R. Sutherland, *Plastic Piping Handbook*, McGraw Hill, New York, 2002.

[16] Building and Construction Authority, *Good Industry Practices — Waterproofing for Internal Wet Areas*, Singapore, 2001.

[17] Standards Australia, *AS 3740: Waterproofing Wet Areas within Residential Buildings*, Standards Australia International Ltd, Sydney, 1994.

[18] M. T. Kubal, *Construction Waterproofing Handbook*, McGraw Hill, New York, 2000.

[19] Singapore Standards, *CP 82 — Code of Practice for Waterproofing of Reinforced Concrete Buildings*, Singapore Standards, Productivity and Innovation Board, Singapore, 1999.

[20] American Concrete Institute, "Guide to the use of waterproofing, damp-proofing, protective and decorative barrier systems for concrete", reported by ACI Committee 515, Detroit, 1988.

[21] P. W. Pye and H. W. Harrison, *Floor and Flooring: Performance, Diagnosis, Maintenance, Repair and the Avoidance of Defects*, Building Research Establishment, London, 1997.

[22] M. J. Gatfield, *Screeds, Flooring and Finishes: Selection, Construction and Maintenance*, Construction Industry Research and Information Association, London, 1998.

[23] Singapore Standards, *CP 68 — Code of Practice for Ceramic Wall and Floor Tiling*, Singapore Standards, Productivity and Innovation Board, Singapore, 1997.

[24] British Standards, *BS 8000: Part 9, Code of Practice for Workmanship on Building Site: Cementitious Leveling Screeds and Wearing Screeds*, British Standards Institution, London, 1989.

[25] British Standards, *BS 8000 Part 11.1, Workmanship on Building Sites: Code of Practice for Wall and Floor Tiling; Ceramic Tiles, Terrazzo Tiles and Mosaics*, British Standards Institution, London, 1989.

[26] British Standards, *BS 5385: Part 1, Wall and Floor Tiling. Code of Practice for the Design and Installation of Internal Ceramic and Natural Stone Wall Tiling and Mosaics in Normal Conditions*, British Standards Institution, London, 1995.

[27] British Standards, *BS 5385: Part 3, Wall and Floor Tiling. Code of Practice for the Design and Installation of Ceramic Floor Tiles and Mosaics*, British Standards Institution, London, 1985.

[28] British Standards, *BS 5385: Part 4, Wall and Floor Tiling. Code of Practice for Tiling and Mosaics in Specific Conditions*, British Standards Institution, London, 1992.

[29] British Standards, *BS 5385: Part 5, Wall and Floor Tiling. Code of Practice for the Design and Installation of Terrazzo Tile and Slab, Natural Stone and Composition Block Floorings*, British Standards Institution, London, 1994.

[30] Ministry of Environment, *Guidebook for Better Public Toilet-Design and Maintenance*, Singapore, 1999.

[31] Singapore Standards, *CP 13 — Code of Practice for Mechanical Ventilation Air-Conditioning in Buildings*, Singapore Standards, Productivity and Innovation Board, Singapore, 1999.

[32] *British Standards, BS-EN 12056-2: Gravity Drainage Systems Inside Buildings Part 2: Sanitary Pipe Work, Layout and Calculation*, British Standards Institution, London, 2000.

[33] British Standards, *BS 6340: Part 3, Shower Units. Specification for Prefabricated Shower Enclosures and Shower Cabinets*, British Standards Institution, London, 1985.

CHAPTER 3
FAÇADE

3.1. General

A building envelope serves the functions of weather and pollution exclusion, thermal and sound insulation. It also provides adequate strength, stability, durability, fire resistance, aesthetics appeal, etc. (Fig. 3.1).

The external walls of traditional buildings are mainly made of masonry and/or reinforced concrete. They are usually finished with cement render and painted or finished with various cladding materials. Cladding materials range from small tiles/mosaics, fixed to the substrate by adhesion, to large metal/stone panels by anchorage methods. They form the covering or dressing for the substrate/wall behind them. Tiles of various modular sizes are either hand-laid or incorporated into the surface during the prefabricating process.

With the advancement of prefabrication, large panel curtain walls have become popular especially for tall commercial buildings. A curtain wall may be defined as a thin wall, supported by the structural steel or concrete frame of the building, independent of the wall below and is designed to envelop the building and act as a barrier to elements of nature [1].

Considerations to be given to the design, construction and maintenance of external walls include:

Visual

- panel shape and size,
- joint locations and joint sizes,

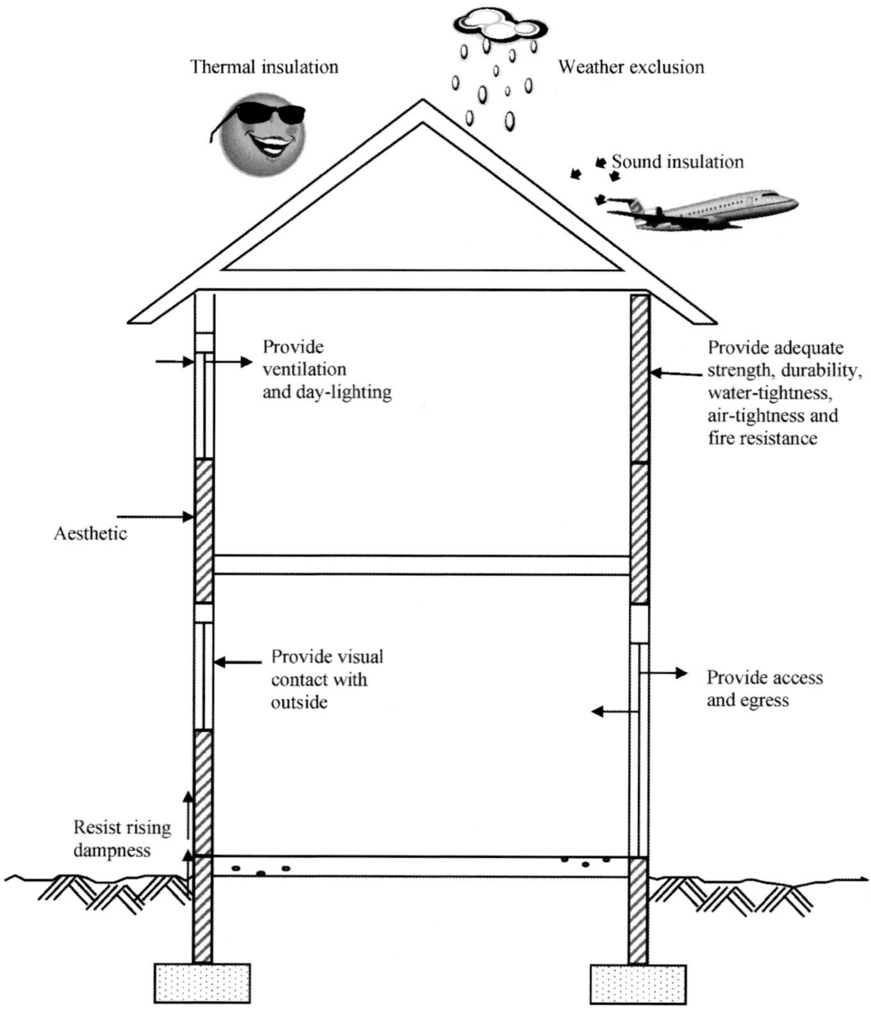

Fig. 3.1. Functions of the building envelope.

- daylighting and nightlighting,
- blinds and shades,
- materials, colors, and finishes,
- integration with interior design, e.g. cabling behind partitions.

Integrity

- air and water-tightness — Sealing, drainage, indoor air quality,
- loading — Static, dynamic, fatigue,
- movements — Load, thermal, moisture,
- exceptional loads — Blast, intrusion, impact,
- fire — Resistance, reaction, spread vertically and horizontally.

Physics/environment/comfort

- heat transfer,
- lighting,
- sound transmission — Noise from street, next room,
- ventilation,
- moisture — Rainwater, humidity, condensation, degradation, mold growth.

Buildability

- tolerance,
- pre-assembly — Stick, unitized, panelized,
- quality — QA, factory work, site work.

Maintenance

- access — Cleaning, inspection, repair, replacement,
- life cycle — Component life, inspection cycle, repair cycle,
- serviceability — Cleaning, repairability, replaceability.

3.2. Environmental Aspects

Before the project team embarks on the design as well as materials and system selection for façade, it should first consider the atmospheric conditions the building is exposed to including (a) rain (b) wind (c) sunlight (d) pollution.

With a good understanding of the environment and its microclimate, a risk analysis can be made right from the planning stage to enable the project team to identify areas that requires special attention [2]. Materials and design features that are suitable for these situations can then be selected.

3.2.1. *Rain*

Rainwater plays a central role in many of the destructive mechanisms that a building experiences, such as dimensional change, corrosion, leaching, efflorescence and water penetration, which may lead to the deterioration of internal finishes. In the staining of façades, rainwater is the initiator of the process. Movement of water over the façade of the building will result in either washing or deposition of dirt, causing the disfigurement of the building's appearance if such effects are concentrated at different locations. Water is also a major factor in the leaching and efflorescence process, as well as promoting biological growth, all of which will result in the staining of the building façade.

3.2.2. *Wind*

Most rainfalls are accompanied by wind, which makes it important to consider a building's exposure to the prevailing wind direction. One fundamental effect of wind during rainfall is that it will change the direction of fall of raindrops so that they impinge on vertical surfaces instead of falling parallel to the surfaces [3]. Wind force does not only bring about driving rain but may also alter the pattern of runoff flow on the façade. Since the direction from which the wind blows changes throughout the year, the direction from which rain will impinge on the building façade will vary accordingly.

The dominant wind direction affects how rain will strike the façade and how the resultant rain runoff will flow and redistribute the dirt on the façade. When wind moves across a building, it changes its direction especially around the edges of a building. Raindrops that are driven by the wind would not be able to follow this directional change and would strike the façade. Hence, the wetting pattern for a tall building would be heavier at windward corners, projections and parapets. The washing effect at such areas would thus be relatively stronger [4]. Once the raindrops impinge on the façade and generate enough runoff to run over the face of the wall, its flow would be rather vertical but this is also subjected to the force of lateral winds that has the ability to alter the pattern of flow. This wind force increases as the building height increases.

3.2.3. *Sunlight*

In addition to UV radiation which is detrimental to many organic materials, e.g. polymer-based waterproofing and sealants, sunlight plays an important role in affecting the formation of stains on a façade. Such stains result from the biological growth of bacteria, fungi and algae on the façade and are visible in the form of green, red, brown or red cushions that spread outwards. Staining due to biological growth usually occurs on lower sections of façades that are in close proximity to vegetation and are sheltered from sunlight by adjacent buildings or trees. A façade's exposure to sunlight may be affected by its orientation since the north and south facings generally receive less sunlight compared to the east and west facings. This may result in a slower drying process after a façade has been wetted, leaving the façade damp for longer periods and therefore promoting biological staining.

3.2.4. *Pollution*

Damaging air pollutants include sulfur dioxide which can lead to sulfation of natural stones, carbon dioxide which can lead to carbonation of concrete and other pollutants that may cause materials to decay, e.g. nitrogen oxides, chlorides and ozone.

These pollutants may be carried by wind and deposited directly on the façade or they can be deposited in solution by rainwater. Pollution sources that emit pollutants into atmosphere include:

- industrial plants and presence of pollutants emitting processes,
- vehicular traffic on heavily utilized roads,
- airborne dust from neighboring countries.

Pollutants from these sources can vary in size and they range from molecular to particulate. Table 3.1 shows some common pollutants in air.

Table 3.1. Common pollutants in air.

Pollutants	Examples	Sizes	Sources	Possible Effects
Aerosols	Sulfur dioxide (SO_2)	Gaseous	Burning of fuels	Formation of acid rain, reaction with water and other materials to form sulfates and chlorides to cause destruction to façade materials
Soot or black smoke	Smoke from tobacco, coal, fuel oil and metallurgical industry	<1 μm	Imperfect combustion	Transported by air and deposited on façades to cause stains
Coarse particulate matter	Ash, dust and rock debris, and mineral dust	>1 μm	Unburnt fuel and dust from roads and industries	Transported by air for short distances and deposited more easily on horizontal or sloping surfaces to form stains

Molecular pollutants such as sulfur dioxide are present in the air and they usually dissolve in rainwater to form acidic rain that corrodes and attacks façade materials, which causes localized losses of material. Particulate matters in the air generally comprise of dust, smoke and suspended particles. Dust particles, being large and heavy, settle quickly and will not affect large areas. Smoke and suspended particles, being smaller and lighter, remain in the atmosphere for a longer time and affect larger areas.

Staining of façades could be due to biological staining agents which can be traced to a number of diverse plant groups: algae, fungi, mosses, ferns and strangling figs.

Algae can be distinguished from fungi by their characteristic green color due to the presence of chlorophyll. They may also be orange or blue green in color depending on the amount of other pigments within it. On the other hand, fungi is colorless but may become blackish or colored as they become reproductive [5–8]. Algae thrive best where there is sun, moisture and nutrients. Dirt that is blown off the road and retained on the façade becomes a source of nutrients. Common as such, algae are found mainly outdoors on external façades. Common species of algae that may be found on the façades are Trentepohlia Odorata, Chlorococcum and an Algae comprising of Scytonema, Schizothrix and Anacystics (Table 3.2) [5–7, 9].

On surfaces shaded from the direct rays of the sun and where humidity level is high, fungi rather than algae are the organisms that are likely to proliferate [10]. Unlike algae, fungi are unable to synthesize own food and thus have to rely on the medium on which they grow on to provide the nutrients necessary for their growth. Fungi begin to grow when the relative humidity of the surface exceeds 70% and for these reasons, they are usually found on internal surfaces of buildings such as bathrooms, kitchen, ceiling boards, wallpapers and glass panels. Other than algae and fungi, biological growth on building façades may also be in the form of mosses, ferns or small plants (Table 3.2). Mosses are small plants that form mat-like patches on walls of old buildings. Ferns and other small plants growing on walls are much more visible forms of biological staining and should be removed prior to further weathering of surface [11, 12].

Strangling figs are plants that propagate through dispersion by birds. They usually grow out of crevices in walls and their roots extend downwards toward the ground [13]. They should be removed as soon as possible because growth of their roots over time may cause pressure to be exerted on the cracks, resulting in widening of cracks. Climbing figs on the other hand, are usually grown on walls deliberately for aesthetics purposes and they require regular pruning in order to maintain their aesthetical qualities (Table 3.2).

3.3. Common Defects on Façades

Common defects on façades include:

— cracking,
— joint/sealant failures,
— efflorescence,
— rising damp/water penetration,
— abrasion,

Table 3.2. Common types of biological agents found commonly on external walls [5–7, 9].

Biological Agent	Characteristics	Conditions for Growth
Algae 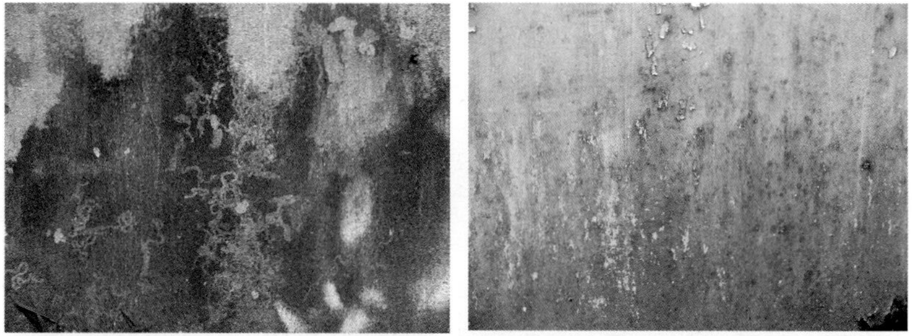	When conditions are ideal, algae will appear in 1–2 years time. It may appear initially as a film either light green, blue green or orange in color. Over time, dirt may collect over the mycelium to result in blackish and conspicuous stains	Direct and prolonged exposure to sunlight. Surface should be wet during rain, the degree of exposure and the amount of moisture available will dictate the type of algae that will colonize
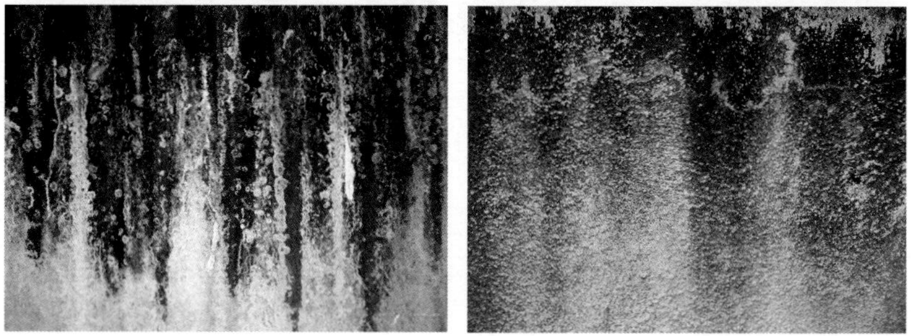	Common species of algae in Singapore are *Trentepohlia Odorata, Chlorococcum* and algae comprising of *Scytonema, Schizothrix and Anacystics*	Sloping surfaces or on the ledges of window sills, balconies and verandahs. Usually brick, concrete or rendered surfaces that are highly absorbent and textured

(*Continued*)

Table 3.2. (*Continued*)

Biological Agent	Characteristics	Conditions for Growth
Moss 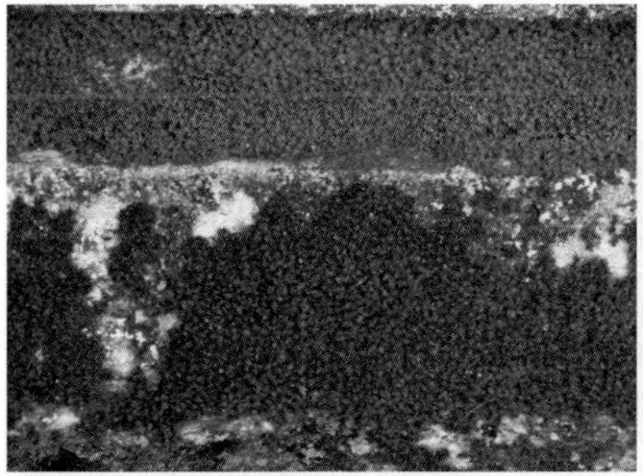 	Patches of small, mat-forming plants. It traps dirt and soil over time to result in a black carpet Has good water holding ability	Humid areas such as cracks that retain moisture and dirt. Usually appear after algae have proliferated Further improves conditions that are suitable for higher plants such as ferns and small plants.
Ferns  	Higher plant species. Roots are easily visible. They retain water well and improve conditions for further biological infestation Common fern species are *Pteris Vittaata and Nephrolepis Biserrata*	Usually appear on walls of old buildings. Cracks in walls that have accumulations of soil and abundant moisture

(*Continued*)

Table 3.2. (*Continued*)

Biological Agent	Characteristics	Conditions for Growth
Other plants  	They may be typical ornamental plants usually kept in pots	May be found together with moss and ferns
	They should be removed completely when growth is detected. That area of the wall should be restored by patching and painting to inhibit future growth	Growth is hastened by deposition of seeds from nearby potted plants placed along window sills and balconies of the same building

(*Continued*)

Table 3.2. (*Continued*)

Biological Agent	Characteristics	Conditions for Growth
Strangling/climbing figs  	Rapid growth. Roots are usually large in girth and fill up the crack that they originate from. Branch roots usually extend downwards to ground so as to draw water and nutrients from soil Common species of strangling includes *Ficus Religiosa, Ficus Microcarpa and Ficus Benjamina.* Common specie of climbing figs is *Ficus Pumila*	Growth originates from cracks in walls and from ledges (Wee, 1986). Lack of maintenance of façade will allow figs to thrive and cause serious damage to the façade Over time, figs and extending roots may totally cover the building façade

— corrosion, and
— delamination

as shown in Fig. 3.2, Table 3.3 shows the possible cause and the role of the parties involved.

Among the common causes (Figs. 3.3–3.5) responsible for these defects are:

(a) Action of wind pressures underestimated at the design and construction stage. Cladding elements can be affected by short duration high-speed gusting, local high wind pressure and suction effects.
(b) Insufficient knowledge on construction/fixing of the cladding elements. Even if suppliers of these fixing elements have already tested the elements to a certain standard, the actual use of such fixings may still need to be verified by appropriate tests depending on the situations.
(c) Insufficient knowledge on the durability of fixings; the importance of design in affecting durability is often overlooked.
(d) Failure to carry out necessary maintenance or incorrect identification of the cause of defect resulting in more damage done.
(e) Failure of joint sealants due to: (1) poor joint design; (2) inappropriate choice of sealant; (3) poor surface preparation and workmanship in installation of the sealant; (4) rapid thermal/moisture changes resulting in excessive movement; (5) ultraviolet radiation and high temperature and humidity.
(f) Inaccurate information from manufacturers resulting in inappropriate choice and use of materials.
(g) Designer lacks the knowledge of the physical properties, performance and potential deterioration of the materials.

3.4. Exposed Brick

The material characteristics of bricks in relation to staining problems are shown in Table 3.4 [14]. Figure 3.3 shows efflorescence on brick walls

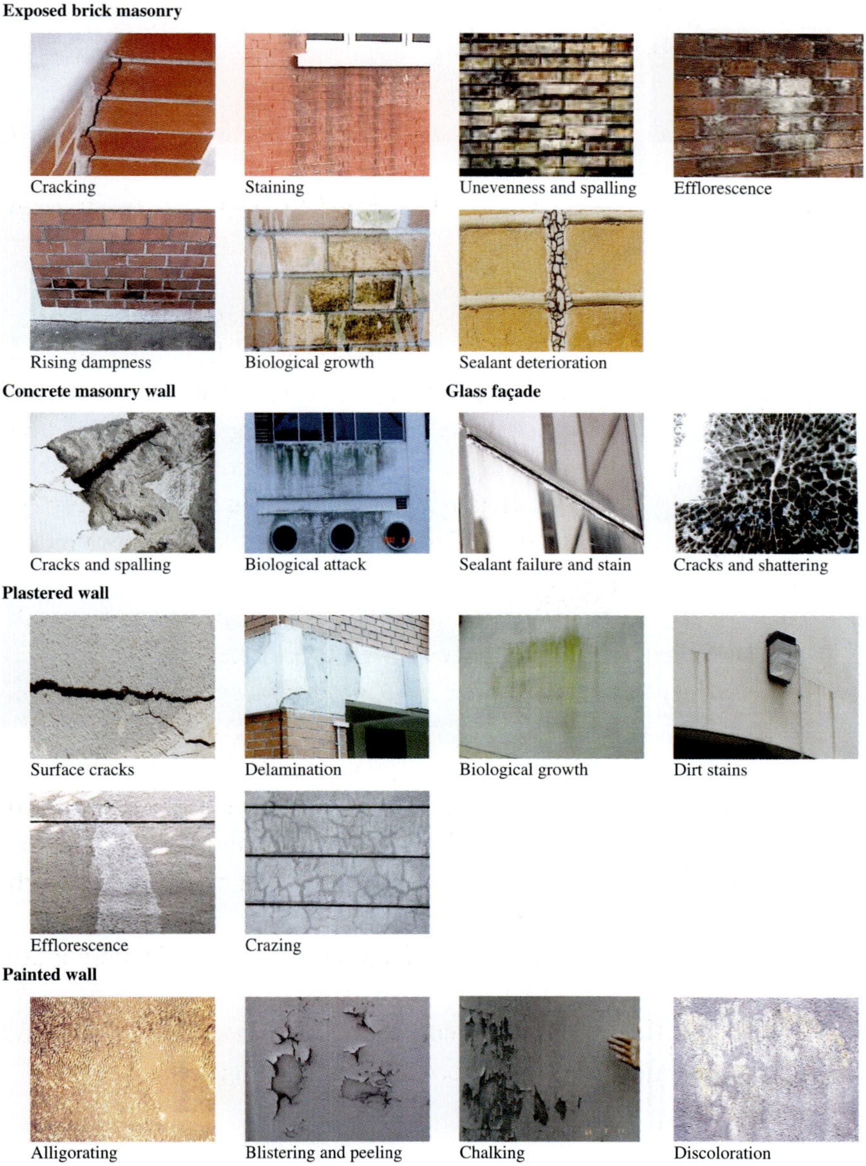

Fig. 3.2. Common defects in façade.

Biological growth

Yellowing

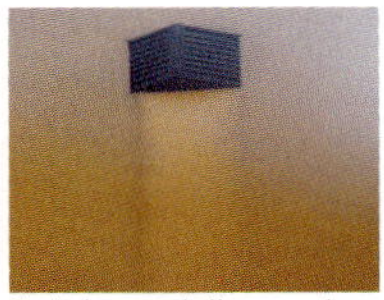
Staining and dirt streaks

Tile cladding

Cracking

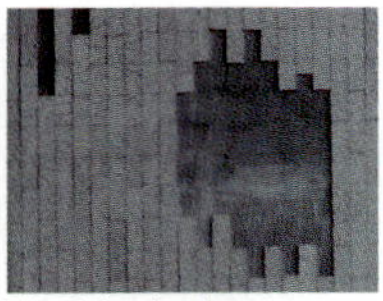
Tile delamination

Biological growth

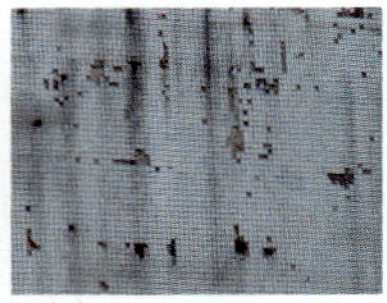
Staining

Efflorescence

Tile chipping

Natural stone cladding

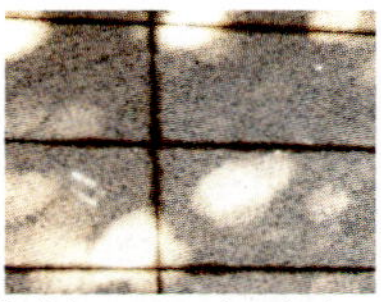
Staining: rust, chemical or dirt stain

Biological growth: fungal or algae growth

Efflorescence: salt stain, patch

Rising dampness

Cracking of surface

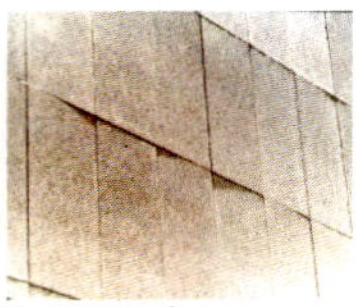
Uneven surface

Metal cladding

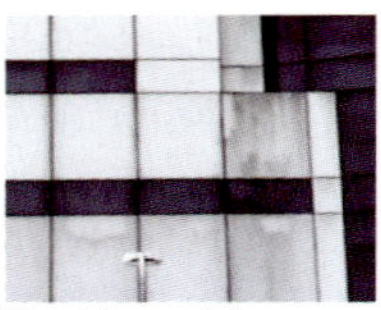
Dirt stain: mark/ discoloration

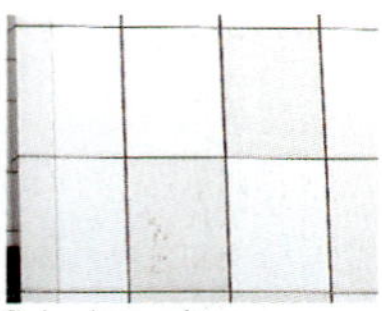
Color inconsistency or 'Chequered effect'

Deformation-buckling, warping, deflection

Coating discoloration or color change

Fig. 3.2. (*Continued*)

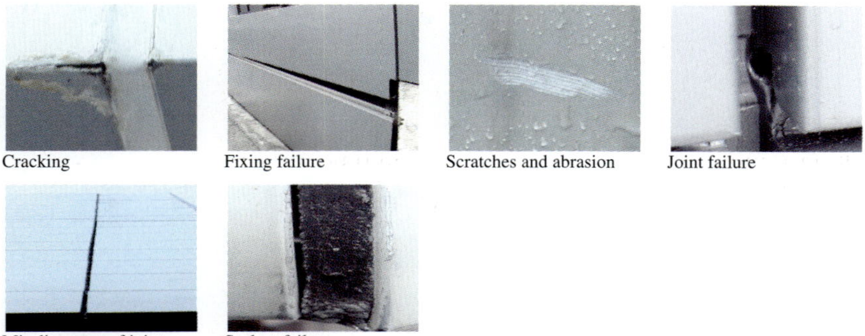

Cracking — Fixing failure — Scratches and abrasion — Joint failure

Misalignment of joint — Sealant failure

Fig. 3.2. (*Continued*)

resulting from salts dissolution. Efflorescence commonly occurs on porous wall materials such as bricks, in particular clay bricks. Clay bricks contain sulfates, which dissolve readily in the water when the bricks become wet after a rainfall. The soluble salts are transported to the surface, where upon evaporation they crystallize to become white deposits. They usually disfigure the walls temporarily and will disappear after prolonged weathering. However, continuing efflorescence may indicate that water penetration into the wall is occurring.

Brick walls may also be susceptible to lime staining. Although lime stains have similar appearances as efflorescence, they are harder to remove. Lime stains form when free lime from the mortar joints or from concrete components in close proximity is leached out by water to the surface. Upon exposure to carbon dioxide, it is converted to calcium carbonate, which is almost insoluble in water. Preventing lime staining requires the exclusion of water from wall surfaces by having good water shedding details. Figure 3.4 shows lime staining on brick walls.

Staining on brick walls may also be in the form of biological growth. Biological growth such as algae growth on brickwalls is common due to its ability to retain moisture, remaining damp for a long period of time after rainfalls (Fig. 3.5). The moisture film allows dirt to adhere easily, further promoting growth. The rough surface texture of brickwalls also retains and traps dirt particles more easily. Figure 3.6 shows dirt streaking on a painted brick wall. The surface irregularities on the wall, as well as

Table 3.3. Common defects in façade.

Defects and Coding	Possible Causes of Defects	Cause gr.			
		D*	C*	M*	E*
Exposed masonry					
Cracking (B1)	Differential structural and thermal movement between steel, concrete and brick masonry. Expansion joints should have been provided.	▲			
	Drying shrinkage of mortar. Nondurable blocks crack.		▲		
	Slight expansion of bricks on water absorption. Process is not totally reversible. As the material dries out and can cause stresses in the wall.	▲			
	Inadequate structural support, anchoring or provision for stress.	▲			
Uneven/spalled surface (B2)	Severe efflorescence caused by soluble sulfates in clay used for brick.	▲			
	Sources of sulfates may also come from acid rain or rising damp if damp proof course or DPC is absent.	▲			▲
	Upon sulfate attack mortars of low cement content deteriorates.		▲		
Efflorescence (B3)	Surface cracks on bricks or at pointing allow water to enter and exit to leach out salts.	▲	▲		
	Brick is prepared from clay that contains excess amount of soluble salts such as manganese sulfates, chlorides, silicates, calcium compounds and manganese coloring compounds.	▲			
Rising dampness (B4)	Formation of crack at the junction between the external wall and ground floor (usually structural defects, e.g. uneven settlement of foundation).	▲			
	No DPC. Or inadequate/perforated DPC.	▲			
	Ground water penetration due to bridging by earth, poor rendering, careless mortar pointing, wrong laying of floor screed, mortar dropping in cavity, etc.		▲		

(Continued)

Table 3.3. (*Continued*)

Defects and Coding	Possible Causes of Defects	Cause gr. D*	C*	M*	E*
Stain: dirt and biological (B5)	Rough surface texture allows easy accumulation of dirt particles.	▲			
	Façade design features may channelize run-off and promote staining.	▲			
	High water absorption prevents rainwater from generating a washing effect as dirt is carried and then re-deposited a short distance away.	▲			
	A thin film of rain water adhering to the surface retains dirt and spores of fungi from atmosphere.				▲
	If cracks remain unattended, seepage or dirt accumulation occurs there.			▲	
	Inadequate cleaning after each season of heavy rainfall. Often influenced by poor accessibility.	▲		▲	
Sealant deterioration (B6)	Joints are many in numbers or improper detailing to prevent exposure.	▲			
	Poor application of the sealant to joints.		▲		
	Poor material selection for the sealant.	▲			
	Poor joint geometry resulting in sealant being forced out during compression.	▲	▲		
Pre-cast concrete (PC) cladding					
Cracks (B7)	Structural issues such as differential structural movement of joint between column and wall, foundation, etc.	▲			
	No site checking of quality of delivered product. Careless storage or rough handling may also damage the panels.	▲	▲		

(*Continued*)

Table 3.3. (*Continued*)

Defects and Coding	Possible Causes of Defects	Cause gr.			
		D*	**C***	**M***	**E***
	Inadequate compression joints between floors to take deformation of floor slab subjected to creep.	▲			
	Chloride attack in marine environment.				▲
	Sulfate attack when sulfates from groundwater or acid rain penetrate and react with the concrete constituents. Provision of DPC helps.	▲			▲
Corrosion of rebars and spalling (B8)	Alkalinity of concrete is lost through carbonation. This can be prevented by adopting controlled curing at factory.	▲			
	Chloride attack from atmosphere. It reduces alkalinity of concrete.				▲
	Improper detailing (e.g. conc. cover) of rebar in design or construction.	▲	▲		
	Inadequate structural provision for deflection control.	▲			
	Unforeseen structural movement and settlement in later period.				▲
Efflorescence (B9)	Façade features may concentrate runoff and keep certain areas damp.	▲			
	Unattended cracks allow seepage and salt leaches out at that part.			▲	
Biological and dirt stain (B10)	Dirt accumulates in gap of two-stage horizontal joints between precast panels or at design features such as ledges and protrusions.	▲			
	Rust bleeding (B2).	▲	▲		▲
	Concrete absorbs and retain thin film of rain, dirt and organic particles. Algae and fungi thrives there. Regular cleaning helps.			▲	

(*Continued*)

Table 3.3. (*Continued*)

Defects and Coding	Possible Causes of Defects	D*	C*	M*	E*
Plastered wall					
Surface crack (B11)	Expansion joint not provided to combat stress at openings in wall or differential movement of structures or in adjacent apron.	▲			
	A change in substrate materials, causing differential absorption and thermal properties. A movement joint should be present.	▲			
	Insufficient thickness of plaster.		▲		
	Plastic shrinkage of mortar mix — Wrong mix proportion.		▲		
Delamination (B12)	Dirt on poorly cleaned background acts as bond breaker for plaster coating. Often the topcoat, with the undercoat of splatter dash adhering to it, come away cleanly. Powdery coating may also be found on delaminated side of detached plaster.		▲		
	The curing agent used for early stripping of formwork, if not totally removed before plastering, affects adhesion of plaster to concrete.		▲		
	Incorrect application technique and handling of plaster, e.g. hardening of plaster by delayed application.		▲		
Biological and dirt stain (B13)	The rough/porous texture of the plaster collects dirt and other organic material acting as nutrients for biological growth.		▲		
	Façade features may concentrate runoff and keep certain areas damp.	▲			
	Inadequate cleaning. Often internal angles are not reachable by gondola.	▲		▲	

(*Continued*)

Table 3.3. *(Continued)*

Defects and Coding	Possible Causes of Defects	Cause gr. D*	C*	M*	E*
Efflorescence (B14)	Soluble salts may be from plaster, clay bricks and their bedding joints or blocks of the backing wall.	▲	▲		
	Unattended surface cracks allow seepage and leaching of salts.			▲	
Crazing, web/ map cracking (B15)	High suction of substrate if proper wetting of substrate is not done.		▲		
	Higher water cement ratio of plaster mix, resulting in shrinkage when the plaster is setting.		▲		
Painted wall					
Alligatoring (B16)	Top coat applied when base coat is still wet and thus lacks grip.		▲		
	Incompatibility causing poor bonding between substrate and paint, e.g. shiny and glossy substrate or extremely hard coat over soft primer coat.	▲	▲		
	Use of low-grade paint with poor adhesion qualities.	▲	▲		
Blistering and peeling (B17)	Water seepage to the underneath of the paint film through unattended cracks. With time, blisters become bigger as more water is retained.			▲	
	Moisture migration through the wall, e.g. leaks in plumbing, use of dryer or humidifier, high level of humidity in the internal environment of kitchen and bath. To prevent this vapor barrier should be used.	▲			
	Applying paint on warm or wet (moisture content >6%) surfaces.		▲		
Chalking (B18)	For a paint with poor weather-resistance, its film disintegrates easily and becomes loose.	▲			

(Continued)

Table 3.3. *(Continued)*

Defects and Coding	Possible Causes of Defects	Cause gr. D*	C*	M*	E*
Discoloration (B19)	Paint or color sensitive to adverse environmental conditions.	▲			
	Paints of vibrant colors with organic particles are susceptible to UV rays. Discoloration is rapid for exposed areas compared to shaded ones.	▲			▲
	Acids and alkalis in the air affect the quality of paint, especially the water sensitive pigments may dissolve in rain and get washed off.	▲			▲
Yellowing (B20)	Aging of paints.				▲
	Presence of industrial pollutants (e.g. ammonia fumes) or very high ambient temperature.				▲
	Using paints with certain ingredients, e.g. epoxy and alkyd resins, which are easily affected by light, heat or environmental contaminants.	▲			
Biological and dirt stain (B21)	Wall design features concentrate runoff at some areas and cause excessive dampness — esp. seen on light surface.	▲			
	Rust bleeding — Usually from poor RCC detailing.	▲			
	Inadequate cleaning. Often internal corners are not reachable by gondola.	▲		▲	
	Anti-fungal paint is not used when there is a proximity of vegetation.	▲			
Tile cladding					
Cracking mostly at joints (B22)	Fixing tiles on poorly cured backing leads to shrinkage of backing, shear stress along interface and tensile stress in tiles.		▲		
	Inadequate movement joints to cater for cyclical expansion and contraction of tiles due to temperature and moisture, esp for large tiles.	▲			

(Continued)

Table 3.3. (*Continued*)

Defects and Coding	Possible Causes of Defects	Cause gr. D*	C*	M*	E*
	Normal service load for a long time or overloading.	▲			▲
	Differential settlement of building foundation or other members.	▲			
Delamination — Bulging or cracking may precede (B23)	Inadequate expansion joints or joint not brought up to tiles surface.	▲	▲		
	Excessive structural movements, e.g. deflection.	▲			
	Manufacturer's instruction not followed for binder.		▲		
	Inadequate curing of substrate before tiling.		▲		
	Delay between the application of adhesive and the laying of tiles or contaminated tile/substrate leads to a loss of adhesion.		▲		
	Efflorescence remains unattended.			▲	
	Porous tiles if not soaked properly before laying, absorb water from mortar backing and affect the bond strength.	▲	▲		
	Moisture ingress through unattended cracks dissolves some compounds in adhesive to make it weak.			▲	
	Inadequate keys provided on the substrate leads to poor tile bonding.		▲		
Biological and dirt stain (B24)	Dirt retained in mortar joints (more porous) get washed onto tile surface.	▲		▲	
	Stain from incompatible or inferior grout or droppings.		▲		
	Stain streaks originate from unattended voids due to delamination. Runoff tends to be diverted to these areas and the dirt retained is washed away and re-deposited on the tile surface.			▲	
	Wall detailing and design features promote staining.	▲			

(*Continued*)

Table 3.3. (*Continued*)

Defects and Coding	Possible Causes of Defects	D*	C*	M*	E*
	The grouting between the tiles is porous, it absorbs water and remains damp. In presence of dirt, biological growth takes place.			▲	
	Prolonged presence of a film of moisture on dirty tile surface and the biological agents in air.			▲	▲
Efflorescence (B25)	Water seepage through unattended cracks causes leaching of lime products from backing wall with high salt content.	▲		▲	
Tile chipping (B26)	Use of poor quality tiles.	▲			
	Rough handling or poor workmanship in installation.		▲		
	No protection of finished trades — At corners of building, especially at pedestrian level, tiled walls should be protected.		▲		
Natural stone cladding					
Cracking (B27)	Inadequate expansion joint to cater differential movement of backing and panel.	▲	▲		
	Inadequate slab thickness prevents proper/safe fixing.	▲			
	Too large area of slabs leaves insufficient number of expansion joints.	▲			
	Insufficient joint thickness or use of hard mortar to absorb movement.		▲		
	Stones are susceptible to thermal shock caused by alternate cycle of sunshine and rain.	▲			
Uneven surface (B28)	Warping due to nonuniform expansion–contraction of minerals in freshly cut stones upon water absorption.	▲			
	Insufficient expansion joints cause slabs to push against each other and few slabs may jut out.	▲	▲		
	Very thin or uneven bedding thickness.		▲		

(*Continued*)

Table 3.3. *(Continued)*

Defects and Coding	Possible Causes of Defects	Cause gr. D*	C*	M*	E*
Rising dampness (B29)	Capillary suction of ground water in porous stone.	▲			
	Inadequate plinth protection.	▲			
Efflorescence (B30)	Rainwater penetrates through unattended cracks into backing and reacts with potassium sulfate in cement used for backing.			▲	
	Water penetration through joints that are too narrow for effective application of mortar or pointed badly.	▲	▲		
	Rising dampness.	▲			
Stain: rust, chemical, dirt, biological (B31)	Caused by iron oxide present in stone. The oxide spreads out upon exposure to moisture and sunshine, esp. seen on light colored stones.	▲			
	Absorption migration and surfacing-up of chemicals in adhesives and sealants through porous stones.	▲			
	Water retention on stone by capillary action in absence of DPC.	▲			
	Rainwater contains high amount of industrial pollutant.				▲
	Cladding stone has high porosity or clay minerals.	▲			
	Parts of the façade may be sheltered from direct sunlight and remain damp for longer periods.	▲			
	Inadequate cleaning.			▲	
Metal cladding					
Corrosion (B32)	Poor UV resistance of the coating.	▲			
	Inferior quality due to poor factory application. Should be site checked.		▲		

(Continued)

Table 3.3. (*Continued*)

Defects and Coding	Possible Causes of Defects	Cause gr. D*	C*	M*	E*
	Damaged (scratches/dents) protective coating due to abrasive actions during handling.		▲		
	Unplanned cutting of panel forms weak points at the support, perforation and at the bent edges.		▲		
	Galvanic corrosion by dissimilar metals used together in design or at site.	▲	▲		
	Use of inappropriate/abrasive cleaning agent or method.			▲	
Discoloration/stain (B33)	Inappropriate design features affecting flow of run-off, e.g. ledges, protruding fixtures, misaligned joints and louvers.	▲			
	Wrong material selection and detailing as per exposure conditions.	▲			
	Inadequate cleaning.			▲	
Chequered effect or color inconsistency (B34)	Poor choice of color.	▲			
	Inferior make (varying temperature, orientation, duration, base material and paint batches during fabrication). Before installation any visual defect should be noted.		▲		
	Failure to ensure that panels are cut and formed from sheets placed in same orientation — esp. one with metallic paint containing flakes.		▲		
	Failure to match color tones during installation.		▲		
	Wrong choice of aluminum alloy as per specific requirement.	▲			

(*Continued*)

Table 3.3. (*Continued*)

Defects and Coding	Possible Causes of Defects	Cause gr. D*	C*	M*	E*
Deformation: buckling, warping, deflection (B35)	Bulging or depression in panels due to soft nature of aluminum or mishandling. It affects aesthetic value.	▲	▲		
	Insufficient expansion joint — Wrong location or large spacing.	▲			
	Inadequate thickness of panel, lack of stiffener/core especially at vulnerable locations such as corners, street level, etc.	▲			
	Rigidity of fixings within panel — All three sides should not be fixed.	▲	▲		
	Deformation of frame structure with time.				▲
Cracking (B36)	Insufficient provision of expansion joint	▲			
	Inadequate thickness of panel, especially after routing and folding forms weak point. Lack of stiffener/core and overlapping.	▲	▲		
	Inadequate support causes stress built-up at folded edge.	▲			
	Rigid fixings. Stresses due to thermal expansion and external loading cannot be relieved through them.	▲	▲		
Fixing failure (B37)	Fixing details are pre-tested by manufacturer. Field adjustment or use of insufficient/ improper tools/accessories causes such failure.		▲		
Scratch and abrasion in coating (B38)	Poor coating quality.	▲			
	Material not suitable for rough usage is employed in areas susceptible to abrasive human activities or impacts.	▲			
	Incorrect or abrasive cleaning.			▲	
	Accidental impacts on panels during storage, handling or installation.		▲		

(*Continued*)

Table 3.3. *(Continued)*

Defects and Coding	Possible Causes of Defects	Cause gr.			
		D*	C*	M*	E*
Discoloration or color change (B39)	Degradation of organic materials (e.g. plastic) and organic coating of poor quality.	▲			
	Panel material and coating susceptible to atmospheric pollution in a particular location or orientation. May lead to hardening, crazing, cracking, powdering and chalking of coating.	▲			▲
	Oxidation and chemical attack due to exposure to UV radiation and water contaminated with atmospheric pollutants.	▲			▲
	Use of incorrect cleaning agents.			▲	
Joint failure and seepage (B40)	Insufficient joint overlap — Noncompliance with specification.		▲		
	Imperfect fit between top and bottom panels — Fixings are not fitted properly.		▲		
	Drainage details are not accurate.	▲			
	Fixing failure.		▲		
Misaligned joint (B41)	Field adjustment due to nonadherence to detail.		▲		
	Required alignment tolerance is not maintained.		▲		
	No special consideration for edges and corners.	▲			
Sealant failure (B42)	Orientation of building affected by solar irradiance, temperature, ozone, oxygen, moisture, air pollutants and microbial agents cause weather crazing, spalling and hardening.	▲			▲
	Improper joint width-to-depth ratio of sealant bed and allowance for movement between substrates.	▲	▲		

(Continued)

Table 3.3. (*Continued*)

Defects and Coding	Possible Causes of Defects	Cause gr. D*	C*	M*	E*
	Adhesion failure due to dirt, grease and other foreign substances present on improperly prepared substrate.		▲		
	Poor mixing and wrong/no primer.		▲		
	Incompatibility of sealant with substrates may cause failure, staining, etc.	▲			
Glass curtain wall					
Stain (B43)	Moisture attached to glass (hydrophilic) traps dust, dirt. Needs cleaning.			▲	
	Microscopic irregularities on glass surface capture dirt and dust.	▲			
	Façade features allow dirt particles to accumulate and be re-deposited by rainwater runoff. Inadequate cleaning.	▲		▲	
	During rainfall, dust/dirt trapped in tiny gaps of glass frames/on sealants may trickle down to the glass panels below.		▲	▲	
Shattering/crack (B44)	Poor make: presence of impurities, esp. nickel sulfide in tempered/toughened glass shows thermal expansion–contraction.	▲			
	No/poor design consideration for differential movement of glass façade relative to the supporting building frame when temperature and moisture changes.	▲			
	Inadequate horizontal and vertical movement joints produce distress.	▲	▲		
	Inadequate sealant depth to accommodate movement and framing members comes in contact with each other.	▲	▲		
Sealant failure (B45)	UV exposure resulting in gaps in glazing groove.	▲			▲
	Substrates if not cleaned thoroughly lead to poor adhesion of sealant.		▲		

(*Continued*)

Table 3.3. (*Continued*)

Defects and Coding	Possible Causes of Defects	Cause gr. D*	C*	M*	E*
	Failure of seal at the frame due to movement in the too rigid system.	▲			
	Sealant not finished properly.		▲		
Window					
Seepage (B46)	No measures to prevent wind driven rain.	▲			
	Panel design is susceptible to seepage.	▲			
	Bad drainage details.	▲			
Dirt and stain (B47)	Difficult access — Need outdoor access system, hence cleaned only when the whole façade is cleaned at a comparatively longer interval.	▲			
	Poor maintenance practice.			▲	
General					
Difficult access (B48)	Building profile (plan/projection) hinders easy reach by access system.	▲			
	Access to façade/window from outside only.	▲			
Maintenance is unusually complex (B49)	Façade is made of various elements with different inspection and cleaning cycle along with varied cleaning method.	▲			
	Façade material is not easily available for replacement.	▲			
	No scope of partial removal or replacement — Very high downtime.	▲			
	Inefficient access system — Poor coverage, too many types or numbers.	▲			
	Very tall structure with extensive exposure and needs high-tech access.	▲			▲

D* — Design; C* — Construction; M* — Maintenance; E* — Environment.

Fig. 3.3. Staining due to efflorescence.

Fig. 3.4. Leaching of salts from mortar joints.

Fig. 3.5. An exposed brickwall affected by biological staining.

Table 3.4. Characteristics relating to the staining problems of brick façades.

	Bricks				
Properties	**Lightweight**	**Common and Fair Face**	**Engineering**	**Sulfate Resistant**	**Facing**
Porosity (%) (volume of pores/ total vol. of whole material)	Porosity of brick is attributed to its fine capillaries. By virtue of the capillary effect, the rate of moisture transport in the brick is ten times faster than in other building materials.				
Water absorption [(saturation wt.–dry wt.)/dry wt. of material]	12–14%	8–14%	6–7%	7–8%	6–7%
Surface texture	Mortar joints and coarse surface of brick create depressions and channels to affect runoff flow.				
Color	Uniform deep red or cherry color (due to ferric oxide). Impurities in clay can produce greenish-yellow/buff colored bricks. Brick color is permanent and will not fade during weathering process.				
Resistance to chemical attack	Clay in bricks makes it susceptible to chemical attacks due to the presence of sulfates of calcium, magnesium, sodium and potassium. Rainwater dissolve calcium salts in bricks to produce efflorescence.				
Resistance to biological growth	Inherent salts, water absorbed and retained in bricks support biological growth.				
Relevant standards	SS CP67 Part 1: 1997, BS 3921: 1985, BS 5628:3: 2001, BS 6270-3:1991, ASTM C216-02, ASTM C1400-01, ASTM C1403-00.				

Fig. 3.6. Staining due to dirt.

the window ledge concentrate runoff in discrete streams. This causes stain streaks to build up at areas where runoff flow is usually directed to run over.

3.5. Concrete

The material characteristics of concrete in relation to staining problems are shown in Table 3.5. Figure 3.7 shows large patches of stains on a concrete façade. This staining is due to the uneven washing off of dirt during rain. Dust accumulated on the façade is washed down during rainfall. If rain intensity is low, the washing effect may not be thorough. Concrete, being porous, readily absorbs moisture during rain through capillary sorption. Upon saturation of the concrete wall, a film of water laden with dirt will form on the wall surface. As water dries up, dirt stays on the façade, forming dark stains. Staining on concrete façade can also be due to the growth of algae (Fig. 3.8).

3.6. Plaster and Paint Surface Coatings

Two common surface coatings are plaster and paint. The characteristics of plaster and paint are summarized in Tables 3.6 and 3.7, respectively [18–21]. Plaster and paint coatings serve as decorative, functional and

Table 3.5. Characteristics relating to the staining problems of concrete façades.

Properties	Concrete
Porosity (%) (volume of pores/total vol. of whole material)	Water/cement ratio determines the porosity of hardened cement paste at a given stage of hydration (Fig. 3.12) [15]. Generally exceeding 15%.
Water absorption [(saturation wt.–dry wt.)/ dry wt. of material]	Dependent on air voids (1–3% of volume). Moisture causes blotches/streaking after rainfall [8, 9]. Moisture penetration may corrode steel reinforcement in concrete to form rust stains.
Surface texture	Smooth to rough depending on aggregates and face molds used for casting. Textured and profile surfaces mask stains better [15].
Color	Wide-ranging colors depending on color added to matrix. Coats of paint can be given to achieve the desired color [15–17].
Resistance to chemical attack	Dependent on quality of concrete. Rainwater dissolves calcium salts in concrete to produce efflorescence.
Resistance to biological growth	Damp concrete surfaces traps dirt and promotes biological growth.
Relevant standards	BS 6270-3:1991, BS 5628:3: 2001, AS HB 84:1996.

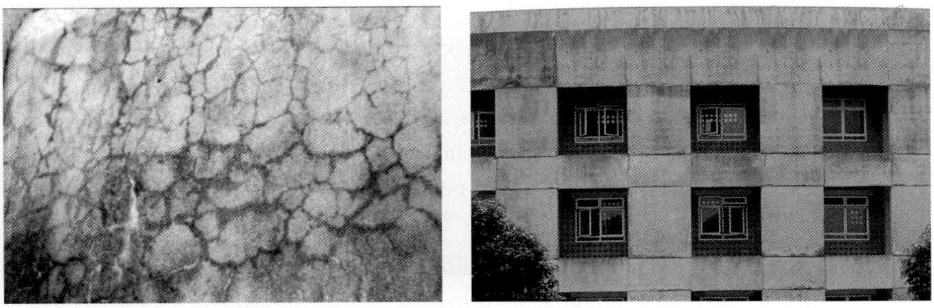

Fig. 3.7. Cracking and staining on concrete façade.

protective coverings which upon spreading over a surface, dry into a continuous solid film.

Although many coatings are able to "breathe" out any moisture trapped beneath the coating, the rate of evaporation may not be fast

Fig. 3.8. Algae growth on concrete façade.

Table 3.6. Characteristics relating to the staining problems of exposed plaster façades.

	Exposed Plaster	
Properties	**Textured Finish**	**Plain Finish**
Porosity (%) (volume of pores/total vol. of whole material)	Depends on size and proportion of aggregates.	
Water absorption [(saturation wt.–dry wt.)/dry wt. of material]	Determined by the water/plaster ratio of the mixes. If plaster is made denser, the water absorption level will reduce.	
	Flow of rainwater over the surface is distributed. Staining is thus more distributed.	Flow of rainwater over surface is concentrated. Staining is thus more concentrated.
Surface texture	Traps and retains dirt easily. Textured and rough surface may mask stains.	Stain streaks may show easily.
Color	Different colors can be obtained by adding color pigment to mix. Light color shows off stains.	
	Uniform appearance for paints can be easily achieved.	Uniform appearance for paint may be difficult to achieve.

(*Continued*)

Table 3.6. (*Continued*)

Properties	Exposed Plaster	
	Textured Finish	**Plain Finish**
Resistance to chemical attack	Prone to chemical attacks due to the aggregates used. Gypsum plaster may suffer attacks from Portland cement due to formation of calcium salts.	
Resistance to biological growth	Water may get trapped in the plaster layer and accelerate biological growth. Organic material present in the plaster mix may encourage biological growth.	Prone to biological attacks though less than textured surfaces. Constituents in plaster mix may encourage biological growth.
Relevant standards	BS 8000-10:1995, BS 5262:1991, BS 1191-2:1973, BS EN1015-19:1999, BS 00/106850 DC prEN 13914-1, BS 00/103730 DC prEN 998-1, ASTM C897-00, ASTM C932-98a.	

enough and this provides a breeding ground for algae and fungi. When the color of plaster or paint is light, staining would be readily visible.

Figures 3.9 and 3.10 show dark stains on plaster-and-paint surfaces arising from dirt deposition. The textured surfaces of plastered walls are capable of trapping and retaining dirt particles. When runoff flows over the surface, it flows along depressions and crevices of the rough surface. Any dirt that is picked up along its path is re-deposited. The flow pattern causes the resultant dirt stains to be distributed. On smooth plaster surfaces, staining may also occur but the staining pattern is likely to be less concentrated as dirt particles are washed along in more even and concentrated flows. The washing effect on a smooth wall is greater and more effective in times of heavier rainfall.

Certain organic materials that support biological growth may be present in the aggregates that make up plaster. The use of such composition may result in biological staining (Figs. 3.11 and 3.12).

Painting is commonly done on plaster walls to enhance its aesthetic qualities. Paints are relatively permeable to liquids and gases. Depending

Table 3.7. Characteristics relating to the staining problems of painted façades.

Properties	Paint Systems on Plastered, Concrete or Brick Walls								
	Acrylic Co-Polymer Emulsion	**Veova (Modified Acrylic) Emulsion**	**Epoxy**	**Acrylic Emulsion**	**Solvent-Based Acrylic**	**High-Performance Acrylic Emulsion**	**Two-Pack Poly-Urethane**	**Texture Coating Water-Based Acrylic Topcoat**	**Texture Coating Poly-Urethane Topcoat**
Porosity (%) (volume of pores/total vol. of whole material)	Depends on overall system.								
Water absorption [(saturation wt.–dry wt.)/dry wt. of material]	Depends on overall system.								
Surface texture	Flat textured	Luxurious sheen emulsion that gives a smooth and silky appearance.	Textured	Decorative, high build, textured, nongloss.	Satin finish	Smooth	Various texture may be created with different types of brushes.	Textured	Textured
Color	Matt	Sheen	Low gloss/ gloss	Matt	Semi-gloss	Low sheen	Good gloss retention	Semi-gloss	Gloss
Resistance to chemical attack	Good	Very good	Excellent	Good	Very good. Good durability to weather.	Very good	Excellent. Good weathering resistance.	Very good	Excellent
Resistance to biological growth	Good	Good	Excellent	Good	Very good	Excellent	Excellent	Very good	Excellent
Relevant standards	SS CP 22:1981, SS 5, SS 7:1998, SS 34:1998, SS 150:1998, SS 345:1990, Good Industry Practices- Painting (BCA) [20].								

Fig. 3.9. Dirt staining of Shanghai plaster.

Fig. 3.10. Dirt staining on a smooth plaster-and-paint wall.

on the type of paint system used, the water absorptivity and chemical resistance of the overall wall system can be improved.

3.7. Tile

The material characteristics of tiles with regard to staining problems are summarized in Table 3.8 [22]. Figures 3.13 and 3.14 illustrate examples

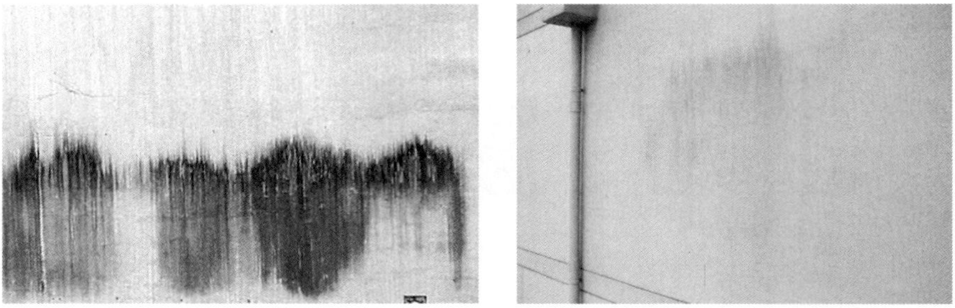

Fig. 3.11. Biological staining on plaster walls.

Fig. 3.12. Staining due to algae infestation.

of staining arising from localized dirt deposition by rainwater on tiles. The delamination of small dimension tiles in Fig. 3.13 has resulted in voids or crevices, which trap dirt and dust to cause staining during rainfalls [23, 24]. Mortar infill at joints has a higher water absorption value than tiles. This allows dirt to adhere more easily to the damp surfaces. Where tile dimension is small, the wall area that is made up by mortar joints will be large and staining from the joints will be more visible.

Figure 3.15 shows cases where severe lime leaching has taken place on tiled façades. In these cases, the mechanism of lime leaching is almost similar. Rainwater that seeps through cracks, chip-offs or mortar joints of

Table 3.8. Characteristics relating to the staining problems of tiled façades.

	Tile				
			Mosaic Tile		
Properties	**Ceramic Tile**	**Homogenous**	**Nonvitreous**	**Vitreous**	**Impervious**
Porosity (%) (volume of pores/ total vol. of whole material)	0–70% depending on glazed/unglazed.	Almost zero	Up to 30%	>3%	Almost zero
Water absorption [(saturation wt.–dry wt.)/ dry wt. of material]	3–6%	0.04–0.5%	>7%	0.5–3%	<0.5%
Surface texture	Glazed tiles are smooth surfaced unlike unglazed tiles. The smaller the dimension of the tiles, the more mortar joints there are and the overall surface will be rough and irregular.				
Color	Glazed layer gives tile its color.		Color of mosaic tiles goes all the way to tile. Mixing of tile colors offer visual aesthetics.		
Resistance to chemical attack	Inert to chemical attacks.				
Resistance to biological growth	More likely to occur on joints between tiles.				
Relevant standards	BS 5385-2: 1991, ASTM D5343-97, ASTM C482-02 (1996), ASTM C650-97, ASTM C1378-97, SS CP68: 1997, AS 4459.2-99, AS 4459.3-99, ISO 10545-3: 1995, ISO 10545-14: 1995.				

Fig. 3.13. Debonding accompanied with dirt staining of mosaic tiled façade.

Fig. 3.14. Cracking and staining of tiled façade.

the tiled wall into the cementitious backing wall or substrate will leach out the soluble salts. The penetrated water will emerge again at joints, crack lines or chip-offs and upon evaporation, crystallizes to form white deposits. The material properties of tiles do not contribute to the staining in such cases. As compared to efflorescence, lime leaching takes a longer time to form but once formed, it is harder to remove. Rectification should involve the prevention of water flow over the façade.

Fig. 3.15. Lime staining on tiles.

3.8. Cladding and Curtain Wall Façades

3.8.1. *Natural Stone*

Natural stones are formed under different natural conditions and hence no two pieces bear the same composition and appearance. A good quality natural stone can be aesthetically pleasing and durable when used as a façade material. Table 3.9 shows a checklist for assessing the quality of natural stones. Some examples of commonly used natural stones are limestone, sandstone, marble, slate, quartzite and granite, with granite being the most commonly used stone façade material. Granite is a crystalline igneous rock containing mainly quartz and feldspar. Granite is generally hard and heavy and it is one of the most durable materials used for

Table 3.9. Checklist for a good quality façade stone.

Quality	Performance
Appearance	Fine and compact texture. Veining and color variation should not be excessive as they indicate softness.
Structure	Structure of stone should be homogenous without presence of voids, cracks or patches of loose or soft aggregates. The cross section of the stone should not be dull in appearance.
Weight	Indication of porosity and density. The heavier the stone, the lower is its porosity, the denser is the stone structure and hence the higher its stability.
Toughness	Toughness is the measure of impact that the stone can take. The stone used should be relatively tough to withstand impacts and high wind loads.
Hardness	More important for floor tiles than façade tiles.
Permeability	Permeability should be low so that the degree and depth of moisture penetration into the stone is low. Permissible water absorption is given in Table 3.4.
Absorptivity	Honed and textured stones have more open pores on its surface than polished stones and hence absorptivity is higher and staining is more likely.
Weatherability	Stone should resist against wear and tear due to weather elements. When exposed to sunlight, it should not fade. Dark color stones tend to fade more easily. It should not disintegrate when subjected to thermal shocks and chemical attacks. Marble slabs show a distinct distortion when subjected to heat. When the stone is exposed to uneven heating, the side that is heated may expand and warp. On cooling, the slab may not return to its original shape.
Workability	Fabricating qualities should be consistently good so that cutting and dressing to the required shape and size is economical.

external cladding. It can be expressed in many finishes ranging from polished to rough. Granite tiles are usually installed on the tile bed in large pieces using mechanical means to reduce the incidence of debonding. The cost of granite may be initially high but the product is long-lasting and requires little cleaning. Granite is suitable for use on buildings facing the sea. They are not affected by the weathering action of sand particles blown by wind. Granite is also acid fumed and smoke-proof and is also suitable for use on buildings in areas where pollution levels are high.

Figure 3.16 shows the primary and secondary causes of failure for natural stone.

Figure 3.17a illustrates examples of staining problems on natural stone surfaces. The types of staining as observed in both figures can be defined as residue rundown. Due to the porous nature of natural stone, dirt residue that is washed down might get collected in the pores of the stone and upon drying, dark stains are formed. Table 3.10 explains the characteristics of stones in relation to staining problems. All stones absorb water. However, its water absorptivity may reduce upon sealing and polishing. The absorption of runoff may occur directly on the surface of the façade or through expansion joints and then penetrate into the granite tile. It is hence important to seal all six sides of the stone panel, but at the same time ensuring breatheability. The large patches of staining shown in Figs. 3.17 and 3.18 are due to the effect of dampness on the minerals of

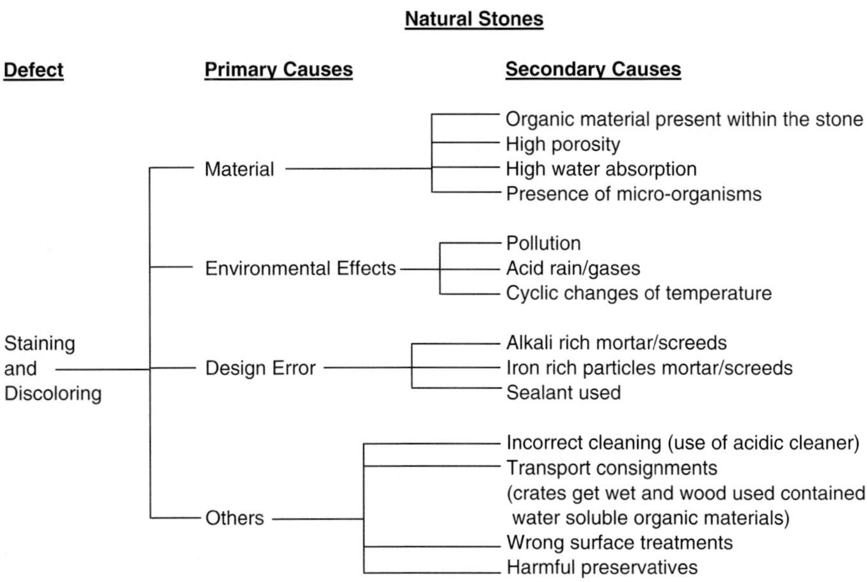

Fig. 3.16. Primary and secondary causes of failure for natural stone.

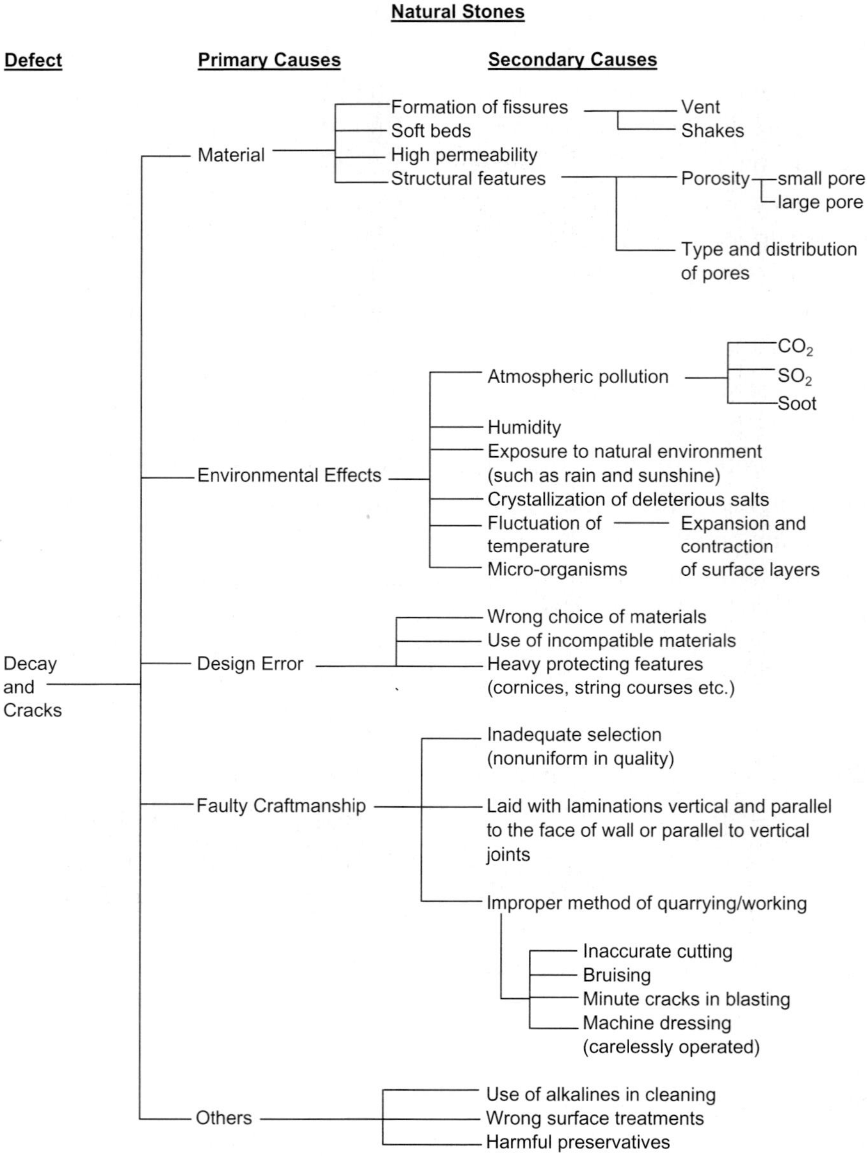

Fig. 3.16. (*Continued*)

(a)

(b)

Fig. 3.17. (**a**) Staining arising from residue rundown. (**b**) Wetting pattern after a rainfall.

porous granite. In Fig. 3.19, staining is shown to have arisen from lime leaching. Being a permeable material, natural stones allow water to seep into the backing wall. Water may also reach the substrate by penetrating through cracks and mortar joints. Water dissolves lime and other soluble salts from the concrete backing and leaches out the lime and other soluble salts from the concrete backing. The leaching of these compounds to the façade surface would form white stains on the façade upon drying.

Table 3.10. Characteristics relating to the staining problems of natural stone façades.

Properties	Natural Stone			
	Granite	**Marble**	**Sandstone**	**Limestone**
Porosity (%) (volume of pores/total vol. of whole material)	0.4–1.5%	0.6–2.3%	0.5–35.0%	0.6–31.0%
Water absorption [(saturation wt.-dry wt.)/dry wt. of material]	After sealing stones with impregnator, water adsorption values may be reduced. In its sealed and polished form, runoff generates quickly to flow on the stone façade in discrete streams. However, the stone may still be absorbent even after polishing and sealing.			
	0.2–0.5%	0.2–0.6%	0.2–9%	0.2–12%
Surface texture	Polish to give fine and evenly granular texture. Flame to give rough texture.	Takes fine polish. Requires regular re-polishing.	Weathers well when free from lime and iron. Due to high porosity, requires sealing to prevent dirt penetration into pores.	Due to high porosity, requires sealing to prevent dirt penetration into pores. Surface will weather irregularly due to mineral composition to result in roughness.

(*Continued*)

Table 3.10. *(Continued)*

Properties	Natural Stone			
	Granite	**Marble**	**Sandstone**	**Limestone**
Color	Stone color may change by applying chemical dyes.			
	Wide range (depends upon color of feldspar). Usually dark pink or deep red. Granites with red and black minerals may fade.	Wide range due to the variability of accessory minerals.	White to buff. Full range of earth colors. Iron oxide present in sandstone may develop rust spots upon oxidation.	Buff, gray limestone remains clean in portions facing rain but retain a film of soot in sheltered areas. This results in a strong color contrast. Yellows after long exposure to atmosphere [25].
Resistance to chemical attack	Good resistance against acid. May produce rust spots if contain ferrous mineral compounds.	Poor resistance against acid.	Good resistance against acid.	Poor resistance against acid.
Resistance to biological growth	Do not promote biological growth. Stones with large pore size may be attacked more easily.			
Relevant standards	BS 8221-1:2000, BS 8221-2:2000, BS 8298:1994, ASTM C1248-93 (1998), ASTM C1496-01, SS CP 67, BS ISO 11600: 2002, DIN 18540.			

Fig. 3.18. Rising dampness causes staining at ground level.

Fig 3.19. Lime leaching from mortar joints or from concrete backing wall.

Lime leaching is more serious along the horizontal joints since water enters and exits more easily than at vertical joints.

3.8.2. *Metal*

The characteristics of aluminum and steel are compiled in Table 3.11.

The main characteristics of metal cladding are:

- high strength but lightweight,
- noncombustible,

Table 3.11. Characteristics relating to the staining problems of metal façades.

	Metal		
		Aluminum	
Properties	**Steel**	**Fluorocarbon (Polyvinyl Fluoride)**	**Anodized Aluminum (Electro Deposited Oxide)**
Porosity (%) (volume of pores/ total vol. of whole material)	Low porosity due to the close and regular arrangement of atoms in structure. Due to its hydrophobic nature, rainwater forms discrete streams on its surface.		
Water absorption [(saturation wt.–dry wt.)/ dry wt. of material]	≈0. Runoff is generated almost immediately during a rainfall. Rainwater can provide a strong washing effect to cleanse the façade of general dirt.		
Surface texture	Smooth. Hydrophilic nature causes a thin layer of moisture to form on metal. This moisture layer attracts and retains atmospheric dirt.		
	Grained or smooth.	Stucco, grained or smooth.	
Color	Generally large color range. Light color shows off stains. Color loss more likely for bright colors. Dark colored panels absorb heat better and experience greater expansion.		
	Large range of colors including exotic colors.	Anodized or with paint applied coat (PVDF).	
		Unlimited range. Excellent panel-to-panel match.	Limited. Noticeable color contrast between panel to panels.

(*Continued*)

Table 3.11. (*Continued*)

Properties	Metal		
		Aluminum	
	Steel	Fluorocarbon (Polyvinyl Fluoride)	Anodized Aluminum (Electro Deposited Oxide)
Resistance to chemical attack	Can be tarnished or corrode under blistering sun. Requires frequent maintenance to retain shine.	Coating usually applied to protect base metal and increase durability. Scratches/abrasions may be inflicted during maintenance works to damage coating and expose base metal. Excellent resistance to pollution and salt sprays. Will not be stained by mortar. Durability of 20 yrs.	Stains more easily. Mortar stains surface. Whitening upon salt sprays. Durability of 15 yrs.
Resistance to biological growth	Thin film of water traps dirt and organic particles. Organic particles may also be present in coating to support biological growth. Less prone to biological attacks than masonry walls.		
Relevant standards	ASTM D523-99, ASTM D2244-02, AS 1562.1 1992, BS 6496:1984, BS 6213:2000.		

- weather resistant,
- ductile, and
- offers a wide variety of trapezoidal profiles and colored finishes.

Stainless steel, bronze and copper have also been widely used as cladding materials. Aluminum cladding has overtaken steel in terms of popularity due to its strong, light, stable and adaptive qualities. With the right finish, aluminum will retain its appearance for many years even with little maintenance. Aluminum can be finished by a number of methods e.g. hardwearing fluorocarbon coat (PVDF), anodizing — A process whereby a hard patina that prevents further deterioration of the surface is produced.

Figure 3.20 shows the primary and secondary causes of failure for metal.

During heavy rainfalls, rainwater that impacts on the façade will generate runoff to bring about an efficient washing effect. However, the washing effect during a light rainfall is weak and may result in localized staining arising from dirt absorption, erosion and deposition. The degree of staining depends on the façade design and wetability of the façade material. Aluminum has an absorption coefficient of almost zero.

Figure 3.21a shows the vertical streaks of stains below the whole length of a ledge. This is due to the airborne dirt accumulating on the horizontal window ledge and later washed down by the runoff. These streaks are more visible against the light colored panels. Figure 3.21b shows more severe dirt staining problems on fluorocarbon coated aluminum cladding. Long-term exposure to polluted air and chemical rain coupled with little or no maintenance has allowed the stains to build-up. Chemical attack from cleaning solutions or abrasions from the use of wrong cleaning agents and equipment may also cause deterioration of the surface.

Figure 3.22 shows staining associated with recessed or protruding joints, detailing faults and other factors. Figure 3.23 shows warping of metal cladding. Chemical rain brought about by highly polluted air in urban and industralized areas is one of the causes of staining. Erosion of the façade over time due to chemical rain has degraded the façade and weakened the coating's resistance to chemical attacks, thus making it easier to attract dirt. Deterioration of sealant between metal panels can also cause staining (Fig. 3.24).

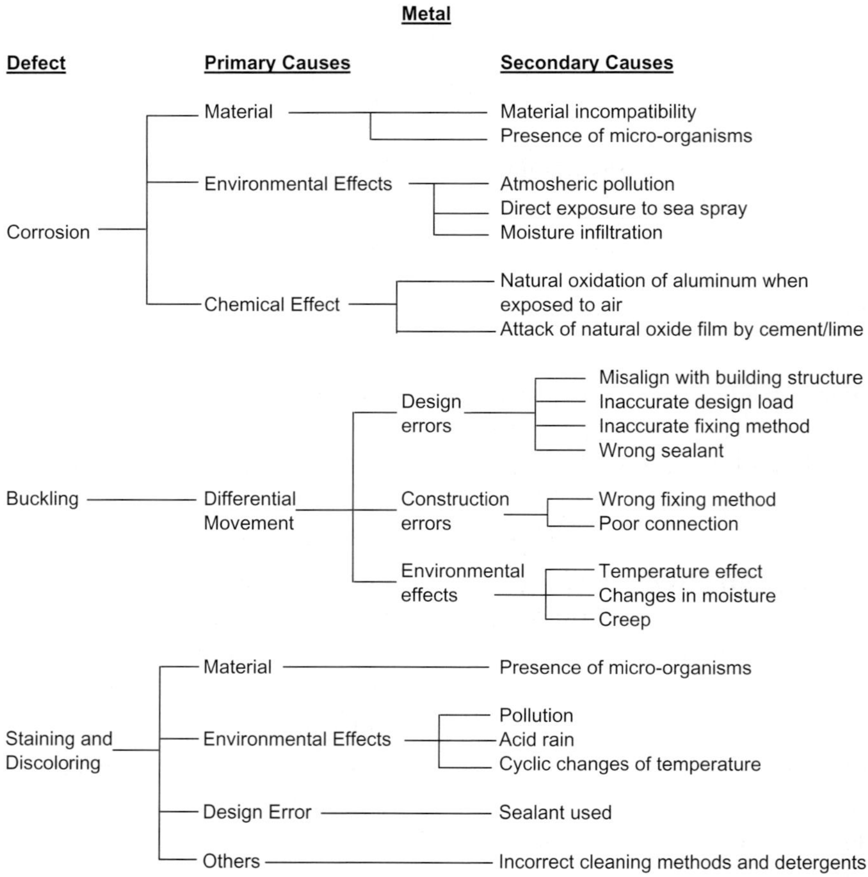

Fig. 3.20. Primary and secondary causes of failure for metal.

Biological growth such as algae and moss may also stain metallic façades. Algaes thrive on the thin film of moisture on the panels. Their growth may be supported by the presence of organic particles in the atmosphere or by the organic substances present in the fluorocarbon coated metal panels (Figs. 3.25 and 3.26).

(a)

(b)

Fig. 3.21. (**a**) Stain streaks originating from a window ledge. (**b**) Dirt stains on light colored aluminum panels show off even more.

Fig. 3.22. Staining on enamel steel.

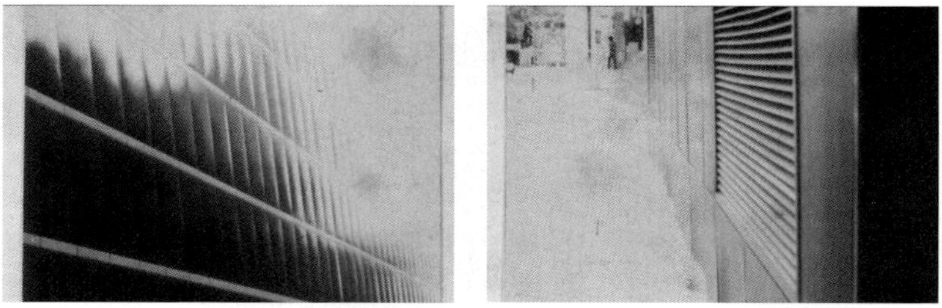

Fig. 3.23. Warping of metal cladding.

Fig. 3.24. Sealant failure resulting in black stains.

Fig. 3.25. Biological growth on façade at ground level and in proximity to vegetation.

Fig. 3.26. Biological growth on façade panels at fourth storey and in proximity to trees.

3.8.3. *Glass*

The use of large glazing panels on façades has become popular due to improvements in technologies which enable glass to span many floors but yet withstand high wind loadings and also due to building owners becoming more willing to spend lavishly for higher aesthetical qualities for their buildings. Glass is usually used at vision areas as infill materials for curtain wall systems.

The material properties of glass with respect to staining problems are summarized in Table 3.12. Figure 3.27 shows the primary and secondary causes of failure for glass. Figure 3.28a shows dark streaks below the joints. The open joints coupled with glass's hydrophilic nature trap dirt. When rainwater runoff flows over the joints, the dirt is washed out to stain the glass. Like metal, glass has an absorption coefficient of zero. Hence, during heavy rainfall, impacting rainwater and its runoff is able to effectively wash off general dirt stains from glazed surfaces. This allows glass to have a "self-cleansing effect". However, when the rainfall is light and intermittent, the runoff generated flows in discrete streams and does not provide an efficient washing effect. This results in dirt streaks forming

Table 3.12. Characteristics relating to the staining problems of glazed façades.

Properties	**Glass**
Porosity (%) (volume of pores/total vol. of whole material)	Not applicable.
Water absorption [(saturation wt.–dry wt.)/dry wt. of material]	≈0. All of the rainwater forms runoff. Hydrophobic property results in discrete streams and prevents even washing. May result in nonuniformly stained walls on wall areas directly below glazed areas. Walls below such glazing should be materials with color or texture which can mask the stain streaks.
Surface texture	Very smooth. Microscopic cracks exist on surface. Hydrophilic nature causes a thin moisture to form on glass.
Color	Can be coated. Usually gray, bronze or green. Coating should be balanced with daylight transmission and views to outside. Dark coats can mask stains.
Resistance to chemical attack	Affected by lime, caustic soda, ammonia and strong acids usually from cleaning solutions. Airborne acids, alkali and solvents from industrial sources may damage the coating of glass [27].
Resistance to biological growth	Inert to attacks. But film of moisture may promote growth.
Relevant standards	BS 9521:1995, BS 4254:1983, BS 5516:1991, BS 6262:1982, BS 5889:1989, ASTM C719-98, ASTMC603-97, ASTM C510-90, ASTM C792-98, ASTM C793-02, ASTM C1087-00, ASTM C1193-00.

almost immediately. Over time, the aesthetic quality of the glazed façade is affected. Intermediate roofs or projections without adequate drainage system or features to "throw out" the rain water, will result in accumulated dirt on the roof surface and this will be washed down onto the façade surface, resulting in obvious streaking (Fig. 3.28b). The pattern of the streaking is determined by the rain flow pattern, which in turn, is determined by the design profile, material, wind, etc.

In cases where sealant are used between glass panels, its chemical make-up and electrostatic nature may attract and retain dirt. When rain

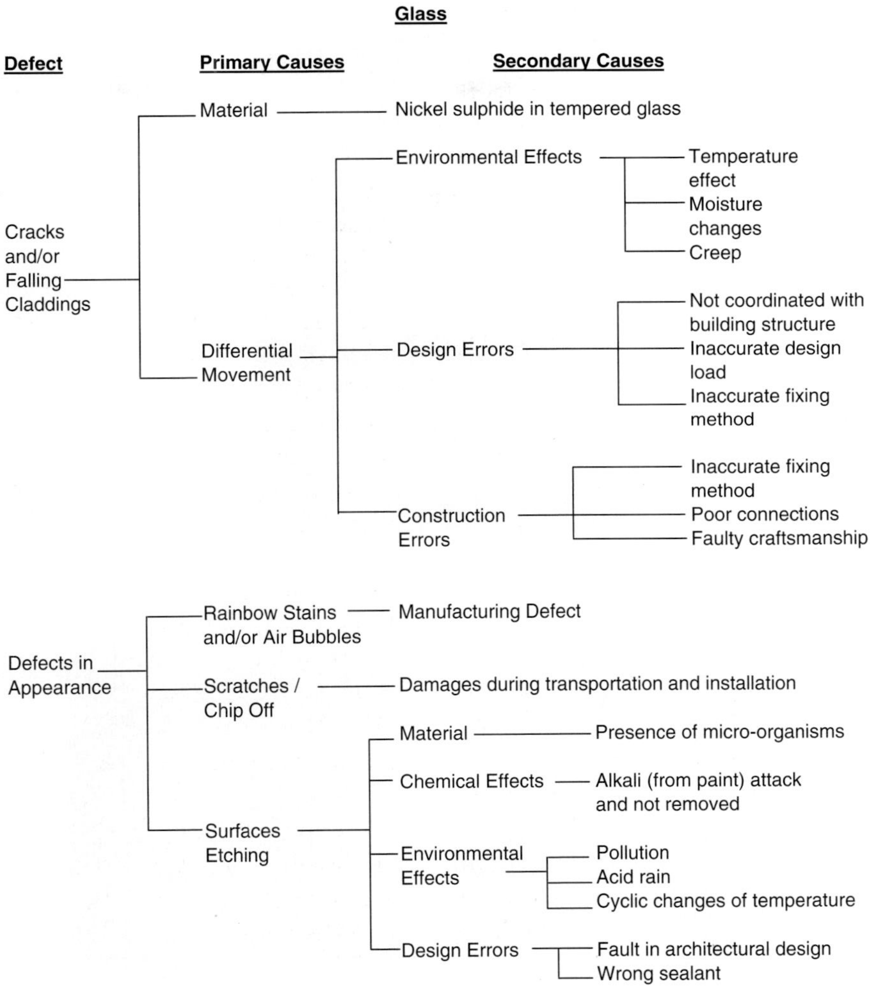

Fig. 3.27. Primary and secondary causes of failure for glass.

water runs over the sealants, loosely held dirt particles will be dislodged and carried downwards to form stain streaks that originates from the sealant joint. Disintegration of sealants worsens the staining problem (Fig. 3.29). Exposure to heat and moist conditions in the form of rain, dew and high humidity has caused many sealants to fail prematurely. The hydrolysis of the polymer chemically produces hydroxyl radicals in the

(a)

(b)

Fig. 3.28. (**a**) Staining on glass originating from open joints. (**b**) Accumulation and washing off of dirt from the sloping roof causes staining on the glass façade.

presence of sunlight. This product is detrimental to the structure of sealants and may result in leaching of plasticizers and stabilizers [26]. This may result in the upward migration of plasticizers and stabilizers in the sealant as shown in Fig. 3.30. Figure 3.31 shows staining on glazed surfaces due to the deterioration of sealant.

Fig. 3.29. Sealant between the glass panels retains dirt and causes staining to be more severe.

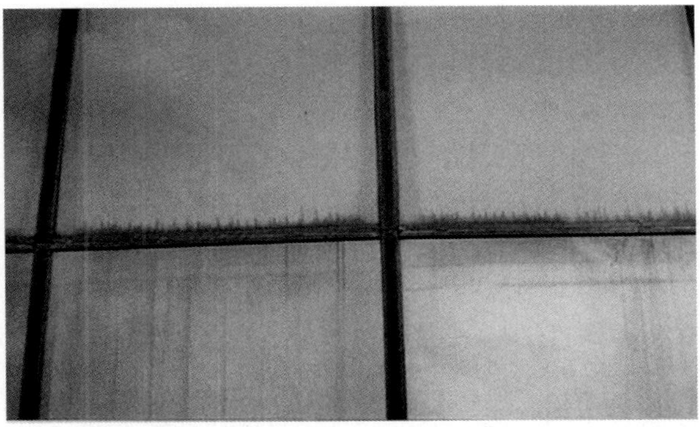

Fig. 3.30. Upward migration of plasticizers and stabilizers in the sealant causes the stains shown.

In some cases, corrosion of other components that make up the façade may cause staining to glass (Figs. 3.32 and 3.33). When cementitious adhesives are used in a glass façade, rainwater hitting the façade may carry the soluble salts in the adhesive to the surface, which will form efflorescence on drying. Façades that are exposed to direct sunlight, prolonged rainfalls and in close proximity to vegetation may be prone to algae attack (Fig. 3.34).

Fig. 3.31. Staining of glass due to corrosion of sealant.

Fig. 3.32. Staining of glass due to corrosion of transoms and mullions.

Shattering of Glass

Shattering of glass such as tempered glass shower screens (Fig. 3.35) and laminated glass wall panels is a serious threat to safety.

Fig. 3.33. Staining from weepholes results in a regular staining pattern throughout the façade.

Fig. 3.34. Algae growth on glass.

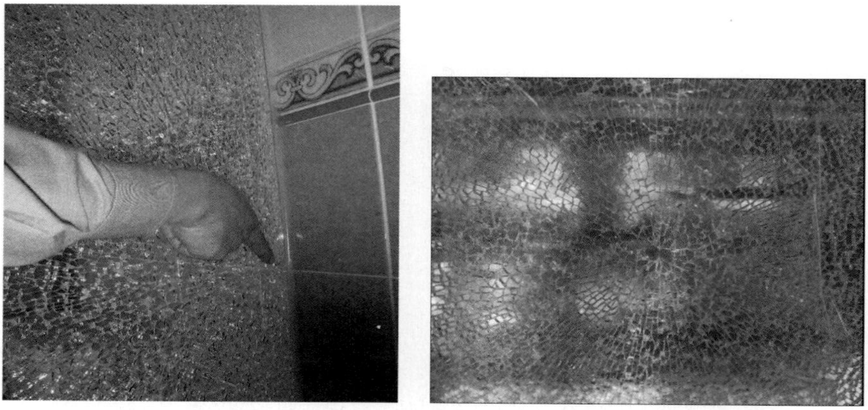

Fig. 3.35. Shattering of tempered glass shower screen.

The shattering of glass may be due to one or a combination of the following factors:

(a) Inherent glass imperfection

- Nickel sulfide (NiS) impurity
- Expansion of volume under room temperature
- About 0.1–0.5% glass formed with impurity

(b) Improper installation and handling

- Chip or crack due to knocking

(c) Support system and fixing details

- Poor hinges, handles and catches
- Contact with metal part of hinge/slider/handle

(d) Renovation works by residents

- Improper re-installation

(e) Operation and maintenance

- Collision, impact, e.g. wind, abrasion

One of the common reasons for the sudden fracture of tempered glass is due to the expansion of the nickel sulfide (NiS) stones as the result of a slow crystalline transformation from alpha-NiS to beta-NiS. During tempering, the NiS stones are completely transformed to the high-temperature state under the heating process but afterwards, the glass is cooled too quickly while going through tempering and heat strengthening to allow the reverse transformation to the low temperature beta-NiS. This reverse transformation occurs later over periods of time ranging from few minutes after thermal treatments to more than 10 years after the glazing installation. The transformation is accompanied with about 2–4% in volumetric expansion of the said impurity. Consequently, small cracks may develop from the inclusion and if these cracks penetrate the central tensile zone of the glass, the resulting release of energy will cause the glass to break spontaneously. This transformation can be accelerated by heat and hence, external glazing exposed to solar heat tends to break earlier in service than internal glazing. Nickel-rich contaminants such as stainless steel are present in the glass melt materials, or if the nickel is added in the manufacture of tinted glass. It may combine with sulfur in the glass batch or furnace fuels to form NiS impurities. Sulfur bearing compounds are known as the basic ingredients in the manufacture of glass. These impurities or stones are minute in size generally ranging from 50 to 150 microns and usually spherical or slightly spheroidal (Fig. 3.36).

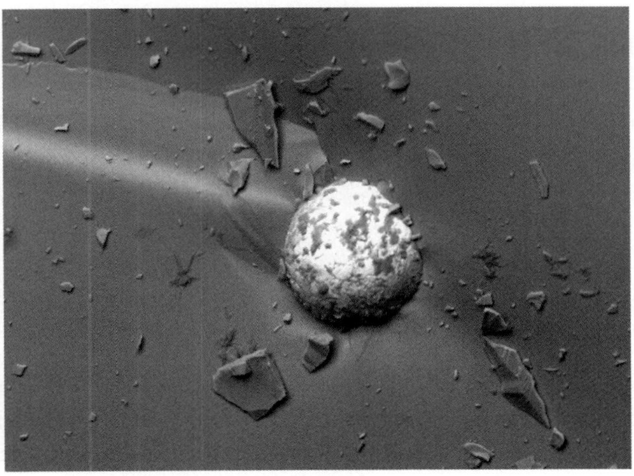

Fig. 3.36. Backscattered image of Nickel Sulfide embedded in the glass (image magnified at 450 times) [28].

3.8.4. *Sealant Failures*

Extensive study on the degradation effect of heat and moisture on sealants have been carried out [29–33].

The causes of sealant failures depend on the types of sealant used, the installation and service condition. Sealants as a material can fail due to age and weather exposure. This type of failure is often characterized by discoloration, crazing and stiffening of the sealant surfaces and is the result of the individual or combine effects of solvent evaporation, ozone attack, migration of plasticizers, ultra-violet radiation, etc. Failure can also occur if there is a large joint movement before the seal is fully cured. In other cases, sealant failures occurred because the sealants were required to take cyclic movement beyond their limits. Table 3.13. Common causes of sealant failures.

Adhesion failure: The most common type of failure due to the loss of bond between the sealant and the substrate (Fig. 3.37).

Cohesion failure: It is a failure within the sealant material. The effect of movement is likely to make the failure progressive (Fig. 3.38).

Table 3.13. Common causes of sealant failures.

Reasons of Sealant Failures	Possible Causes
(a) Material factor	Wrong choice of sealant. The usage of unsuitable sealant material for a given substrate.
(b) Quality of installation	Poor workmanship due to lack of skillful labor training.
(c) Joint design	Incorrect type of joint required with respect to dimensions, location, joint spacing related to expected movement, shape and nature of backing material.
(d) Sealant dimension	Improper or inadequate sealant dimension in the joint due to incompatibility
(e) Sealant manufacturer	Error in quality control or poor formulation.
(f) Applicator or user fault	Unable to communicate with the supplier and manufacturer or lack of sealant knowledge properties.
(g) Environmental factor	Ultra-violet, moisture and temperature.

Fig. 3.37. Adhesion failure of sealant.

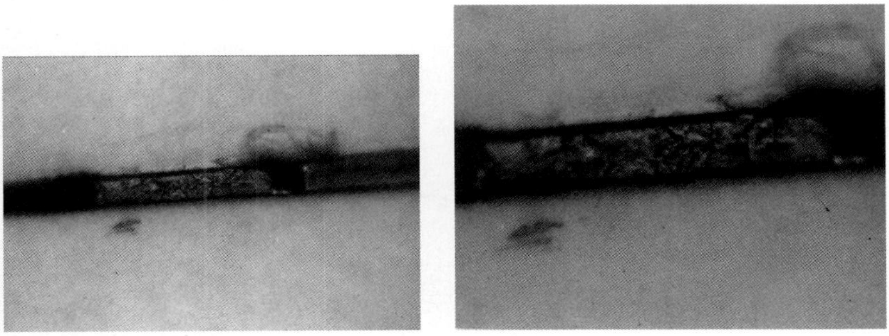

Fig. 3.38. Cohesive failure of sealant.

Hardening/weathering/crazing: Hardening of the sealant leads to formation of cracks crazing. It can be brought about by loss of solvent or plasticizers and by the action of ultra-violet radiation and ozone (Fig. 3.39).

Softening: Sealant failure causes by softening may be due to improper curing or by thermal/photochemical degradation of the polymer (Fig. 3.40).

Other common types of sealant failures commonly found on building façades are staining (decomposition), chalking, cracking, loss of glossy and loss of elasticity. There is an enormous range of material available for sealing purpose in construction, representing the widely different

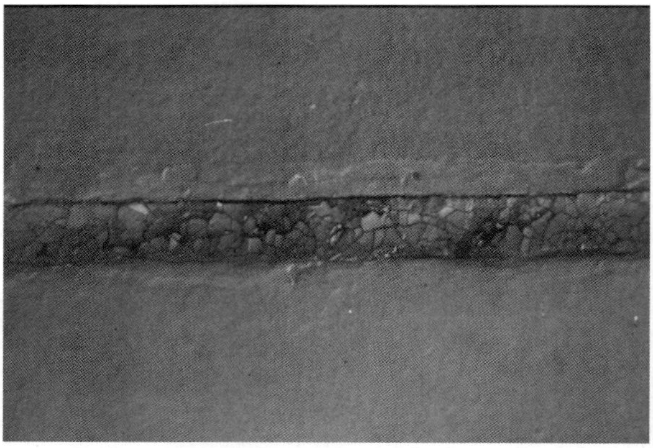

Fig. 3.39. Sealant hardening and crazing.

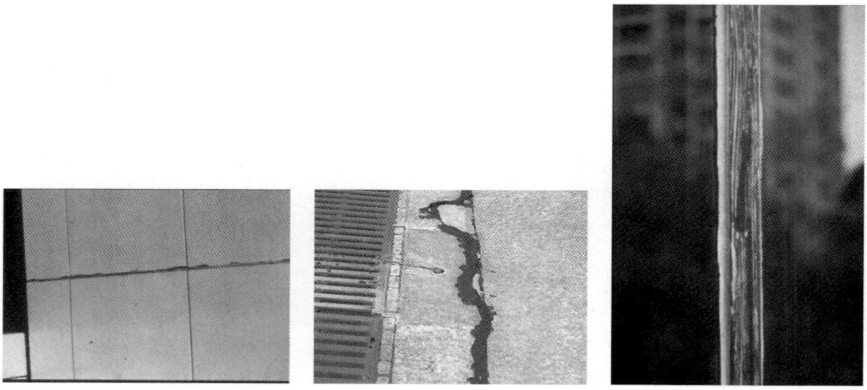

Fig. 3.40. Sealant softening and melting.

requirement of each individual situation. The suitability of a sealant for a particular situation has to be carefully assessed. To prevent early premature sealants failure and promote better performance of sealant used in the building façade, the following points give some guideline measures:

(a) Design the proper and adequate size of the joints with respect to the joint movement, location and environment.

(b) Proper selection of sealant in regard with the expected dynamic structural or thermal movement, weatherability environment and the substrate.
(c) Adequate knowledge of the sealant properties through sealant specification studies.
(d) Quality control on the installation of sealant through proper skillful training.

Sealant degradation can be defined as the process of deterioration of material resulting from the combined effects of atmospheric radiation, heat, oxygen, water, micro-organism and other factors. The resistance to weathering depends on the type of polymer, its composition and superstructure and on the ensemble of condition exposure. The components of weather elements such as ultra-violet, atmospheric oxygen and moisture are most likely to cause significant changes in the properties of sealant (modulus, elasticity, elongation, etc.).

Ultra-violet and oxidation

The shortest wavelength of light from the sun are the ultra-violet light. Approximately 5% of the sunlight reaching the earth's surface falls within the wavelength between 280 and 400 nm and is responsible for most of the photochemical changes to the materials exposed outdoors. It may cause degradation in the sealant. The presence of oxygen together with ultra-violet radiation will initiate the degradation. Carbonyl groups are especially active in most polymer systems.

Atmospheric oxygen can oxidize polymer chains through the well-known peroxide and hydroperoxide chain reactions. Free radical has to be initiated to start this chain and these are often formed by ultra-violet light. Ozone (O_3, a destructive gas), is likely to attack sites of chemical unsaturation in polymer molecules.

It can also attack saturated sites to initiate the peroxide decomposition chain. In either case the net result in the chain scission which leads to serious reduction polymer properties.

Moisture

Moisture in the form of rain, dew, high humidity, etc may be responsible for the attack on sealant materials. These include hydrolysis of the polymer chemically, the production of hydroxyl radicals in the presence of sunlight which leads to degradation, erosion caused by freezing and leaching out of additives such as plasticizers and stabilizers. A further important action of moisture is to dissolve solid pollutants and this leads to destructive effects.

Thermal shock

Outdoor installation of sealant is exposed fully to sun and rain, leading to extreme of temperatures. Air temperature can be as high as 44°C. Although such air temperatures are very seldom reached, even higher temperature is measured on surfaces which are exposed to intense solar radiation. Changes in temperature dramatically alter the properties of the material. High temperatures may change the structure of the material or cause it to melt through random chain scission.

Sealant in the presence of external environment factor such as ultra-violet, temperature and moisture as described above can degrade in different modes. In certain cases, a few modes of degradation can exist at the same time. Various modes of sealant degradation are:

(a) *Thermal degradation*: This is the oxidation of chemical compound causing a change in the structure and bonding of the macromolecules polymers.
(b) *Mechanical degradation*: This is the macroscopic effect brought about under influence of shear forces. The stress induced processes are frequently accompanied by bond ruptures in polymer main chain.
(c) *Photochemical/radiation degradation*: This is the cause by the irradiation of polymer with ultra-violet and visible light. Polymer absorbs ultra-violet radiation and induces reaction resulting in splitting of chemical bond in the main chain of linear polymer. Chromophoric group in the macromolecules is a pre-requisite for initiation of photochemical reaction in the presence of ultra-violet range wavelength less than 400 nm.

(d) *Biological degradation*: The microbial attack by micro-organisms such as algae, fungi, bacteria which excreate an enzyme causing chemical reaction to occur in the polymer.
(e) *Hydrolytic degradation*: Polymers contain functioning group which is sensitive to the effects of water will degrade.
(f) *Chemical reaction*: A degradation process induced under the influence of chemical, i.e. acid, base, solvent and reactive gas brought into contact with polymer.

In all these modes of degradation, an intermolecular crosslinking may exist in which the bond breaking of the main backbone in the macromolecule of the polymer will cause a formation of new chemical bonds between individual macromolecules and lead to an increase in molecular size to certain kind of superstructure.

Study was conducted on combination effect of heat and water on sealant [34]. Sealant degradation due to water derives from swelling and/or hydrolysis. Both water-swelling and hydrolysis can be accelerated by high temperature. Mere prolonged heat can induce serious oxidation. The effect of the combination of water and heat on the degradation of sealants can be a complicated one when compared with only water or heat [29]. The degradation effect of hot water has been demonstrated on many occasions by the exposure of epoxide-fiber composites to so-called thermal spikes, which are intended to simulate flight conditions of military aircraft. A supersonic dash by the aircraft is found to raise the temperature of the adhesives which are in contact with water which will inturn be heated from below zero to 150°C within a few minutes. Voids are discovered at temperatures above boiling point [35].

The fact that water at high temperatures will cause more induced aging than normal moisture is demonstrated using Fourier Transform Infrared (FTIR) Spectroscopy in a study which involves the exposure of dimethylbenzylamine resin to water at 80°C for up to 90 days. Short term effects were the hydrolysis and leaching out of unreacted anhydride by hydrolysis of ester groups that occurred in samples under stress; unstressed samples were unaffected [36].

It has been shown that the loss of adhesion due to the hydrolysis of chemical bonds can be accelerated, by immersing a sealant into water at elevated temperature. For silicone, it has been shown that the temperature dependency of the adhesive bond hydrolysis closely follows Arrhenius's law [37].

Other elastic sealant types are expected to behave basically in the same manner, since the adhesion of organic sealants is achieved via siloxane bonds formed by organosilane adhesion promoters. The activation energy of the hydrolysis reaction depends on the nature of the substrate and to some extent on the type of adhesion promoter used in the sealant formulation.

3.9. Façade Access Systems

A façade access system is required for inspection, maintenance and cleaning of tall buildings. Such systems can be expensive to provide and maintain, difficult to design, dangerous to use and awkward to operate especially for buildings with high irregularity in shape. The façade access system shall be designed, fabricated and installed to satisfy minimum requirements such as:

(1) It should have a relatively long design life of 10–25 years.
(2) It should withstand all loadings (wind, dead and live) as stipulated by local codes and specifications.
(3) It should provide workers with easy and safe access to all areas of the façade for all cleaning, repair and replacement works.
(4) When in operation, it should not bring about disruptions to tenants' activities or cause damage to the façade.
(5) It should be designed to provide maximum coverage to the façade including clearing protrusions on the façade such as sunshades and hence fulfill the required cleaning cycle.

The provision of façade access system should be carried out in strict accordance with recognized codes and standards.

A building façade's maintenance expenditure is influenced by the building shape and form. For any form of buildings that have been chosen, it may have a bearing on the method of maintenance. The various types of façade access systems are shown in Fig. 3.41.

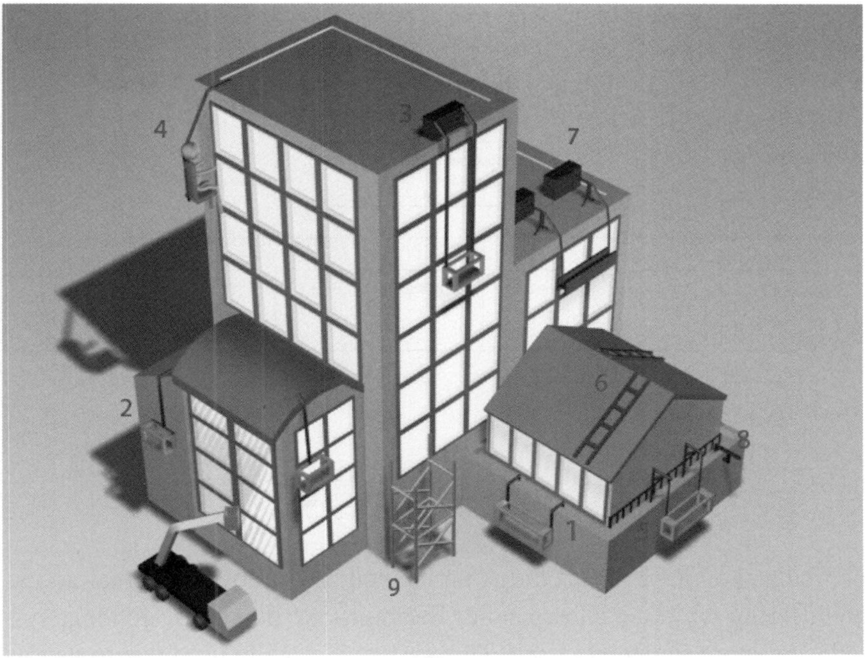

Permanent access system
1. Trolley unit/powered traveling davits
2. Mono rail system
3. Traversing trolley
4. Personal units
5. Fixed davits
6. Traveling ladders

Temporary access system
7. Counterweight suspended beam
8. Parapet clamps
9. Scaffolding
10. Booms

Fig. 3.41. The various types of access systems, both permanent and temporary [14].

3.9.1. *Permanent Systems*

Permanent access system is a planned installation which occurs at the construction stage. These access systems are usually used for maintenance of façades, window cleaning and building inspection for long-term purposes. They can be classified as shown in Table 3.14.

Permanently installed gondolas are dedicated to a specific building or structure. These systems are provided to facilitate regular inspection, maintenance and cleaning purposes. As it is specifically designed for each

Table 3.14. Classification of permanent access system.

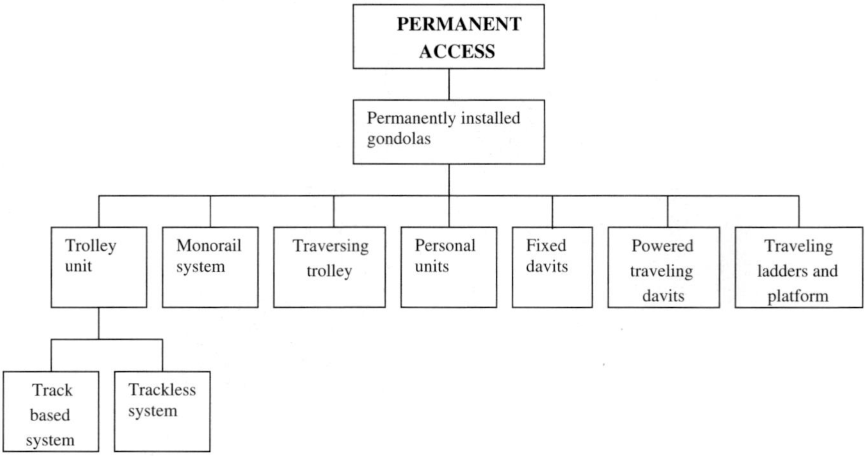

particular building, it introduces an element of safety in external access engineering which is unsurpassed. For high-rise buildings, gondola systems are more frequently used. This system is usually used due to its low installation cost. Nevertheless, this type of system does involve safety disadvantages. It is dangerous when cleaning is done under strong wind conditions.

Permanently installed gondola consists of a platform suspended from a suspension rig. The suspension rig is usually a trolley unit with a hoist, operating either on rails or on a suitable surface such as concrete track.

3.9.1.1. *Trolley Unit*

The trolley unit is also known as the Building Maintenance Unit (BMU), or a roof-car. It is a machine equipped with an integrated lifting hoist, a jib and a cradle and running on the roof of a building (Fig. 3.42).

It allows complete access to all external parts of the building at the touch of a button. Its movements are controlled safely from the push-button control panel. Access to the cradle and to its garage is from the roof and the operator has nothing to assemble or dismantle. The built-in

Fig. 3.42. A trolley unit suspended from a lifting hoist running on the roof.

safety devices control and monitor the operations of the unit to ensure safety.

The most commonly used trolley unit system used in the region are:

- Track-based system (Fig. 3.43).
- Trackless system (Fig. 3.44).

3.9.1.2. *Monorail Track*

This system is used mainly for recessed or overhanging façades, buildings with a sloped roof and for cleaning the inside of glazed atrium.

The monorail track follows closely the line of the façade, with cradles suspended from manual or powered trolleys to reach the various points of the façade or roof. It also permits access on inclined roof section (Figs. 3.45 and 3.46).

3.9.1.3. *Traversing Trolley*

Depending on the design of the building or the architect's requirement, traversing trolley may be considered for horizontal traversing on unreachable areas (Fig. 3.47).

Fig. 3.43. Trolley unit system running on tracks.

Fig. 3.44. Trackless trolley unit running on wheels.

3.9.1.4. *Personal Unit*

Personal units are access systems that allow access to the façade for only one man. They are usually used when the area to be cleaned is relatively small for minor cleaning and repair jobs such as window cleaning (Fig. 3.48).

Fig. 3.45. Monorail track along the line of the façade.

(a) (b)

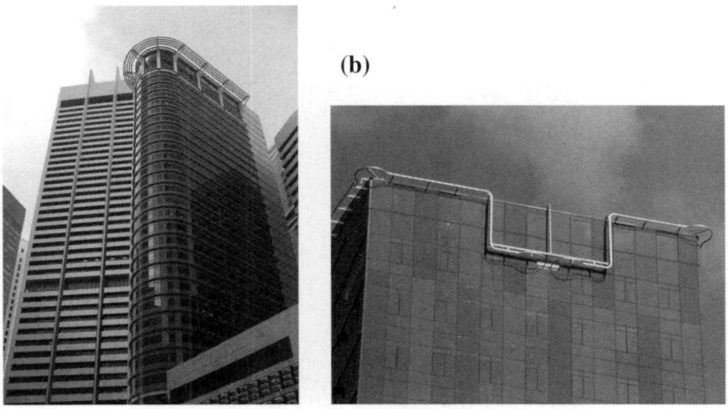

Fig. 3.46. (**a**) Special support brackets anchored to the façade and (**b**) anchored to the roof.

3.9.1.5. *Fixed Davit*

Fixed davits can be mounted on the roof slab or parapet wall. The manner of mounting will determine the ease by which the cradle can be moved from one section to another to reach the different areas of the building's

Fig. 3.47. Traversing trolley along glazed.

Fig. 3.48. A personal unit anchored on a ladder for vertical cleaning.

Fig. 3.49. Platform suspended from fixed davits.

Fig. 3.50. Powered traveling davit running on rails fixed on parapet wall.

façade. The davits can be rotated 180° to launch it off the roof onto the façade (Fig. 3.49).

3.9.1.6. *Powered Traveling Davit*

The difference between the powered traveling davit and the fixed davit is that the traveling davit is mounted on rails fixed to the parapet (Fig. 3.50). This allows the davit to move from one working position to another easily. This system is especially useful when the roof is particularly crowded with services.

3.9.1.7. *Traveling Ladders and Platforms*

These are vertically mounted or sloped, guided along rails and designed to blend in with the shape and color of the background of the building (Figs. 3.51 and 3.52). It is particularly useful for façades or building envelopes that are odd in shape and not able to take loadings.

Fig. 3.51. Traversing ladder with fixed guardrails for cleaning the exterior of a glazed roof.

Fig. 3.52. Traversing ladder for cleaning the inside of a glazed roof.

3.9.2. *Temporary Access*

Temporary access systems can be classified as follows (Table 3.15):

Table 3.15. Classification of temporary access system.

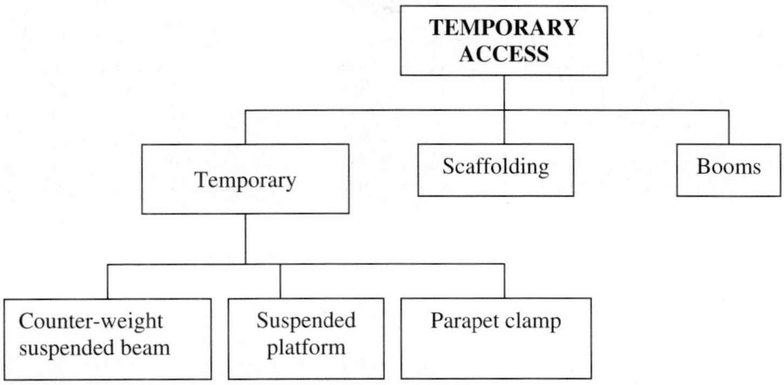

3.9.2.1. *Temporary Installed Gondolas*

Temporary installed gondolas are installed on a building or structure for carrying out specific tasks. It consists of a platform and a suspension rig which are assembled prior to use on a work site. They are dismantled and removed from site on completion of works.

It is usually used by operators for external wall repairs, painting, maintenance and inspection works.

3.9.2.2. *Counterweighted Suspension Beam*

It consists of counterweight to stabilize the gondolas. The loading allowed on the gondola will thus be determined by the counterweights used (Fig. 3.53).

3.9.2.3. *Suspended Platform*

Suspended platforms are platforms that are fixed temporarily on davits (jib). They are normally used for maintenance work in residential building or new building construction and can be moved from one place to

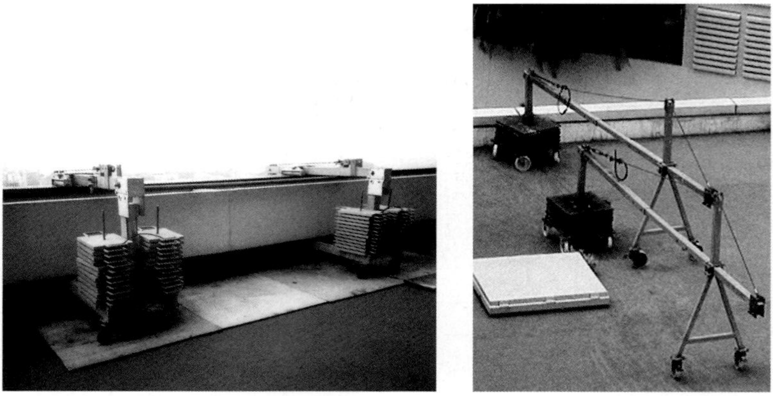

Fig. 3.53. Mobile suspension jibs with counterweight.

Fig. 3.54. Temporary suspended platform on davits: (**a**) Single deck and (**b**) double deck.

another. The platforms used can also be double-decked for increased efficiency and simultaneous work to different storey heights (Fig. 3.54).

3.9.2.4. *Parapet Clamp*

If a sufficiently solid and strong parapet (reinforced concrete or steel) is available, the parapet clamp may be used. The stability of the clamp is

Fig. 3.55. Parapet clamp as used on a parapet wall.

provided by the parapet itself. The parapet will thus need to be certified to take the loading required (Fig. 3.55).

3.9.2.5. *Scaffolding*

It can be mobile or temporarily assembled along the perimeter of the buildings. Scaffolding has the advantage of being light-weight, easy, quick to erect and economical (Figs. 3.56 and 3.57). For these reasons, they are widely used during new construction as well as for refurbishment works. It is seldom used for cleaning and maintenance works due to its limited reach and its difficulty in traversing.

3.9.2.6. *Boomlifts*

For low rise cleaning and maintenance works of up to 60 m, boomlifts are most often used. They are extremely mobile and can be hired on contract basis. A strong foundation and unobstructed access path is required when using the boomlift (Fig. 3.58).

3.9.2.7. *Façade-Cleaning Robots*

The use of robotics technology has overcome the dangerous and time consuming nature of cleaning works, especially that of façade cleaning (Fig. 3.59). Cleaning systems can be specially designed and built to

Fig. 3.56. (**a**) Mobile scaffolding. (**b**) Overhanging scaffolding.

Fig. 3.57. Bintagor scaffolding is very popular among developing countries due to its low cost and sturdy framing members [11].

Fig. 3.58. Articulated boomlifts for external façade's maintenance.

Figs. 3.59. Cleaning robots on glass façades (*Courtesy of*: Fraunhofer Institute for Factory Operation and Automation IFF).

tailor to the specific requirements, such as the surface and geometry of a façade.

The advantages of utilizing cleaning robots for façades include:

(1) Complete system for automatic façade cleaning.
(2) Low operating costs with no personnel costs.
(3) Constant availability that enhances flexible cleaning cycles, cleaning of especially dirty façade areas or cleaning on demand.
(4) Usable on various types of façades.
(5) Secure movement without guide rails on the façade.
(6) Easy to operate.
(7) Efficient, economical and ecological cleaning if recycled water is used.
(8) Short cleaning times for large surfaces.

3.9.3. *Building Forms*

A building's overall form has a direct impact on the ease of cleaning of building façades. Buildings with simple and conventional tower forms are likely to incur lower cleaning costs since accessibility to the façade would be easier. On the other hand, access system of buildings with amorphous shapes may not allow workers full and safe access to the façade. It is common to find buildings whose access systems do not offer adequate coverage to its external wall areas but yet are expensive and dangerous to maintain and operate. Such systems are usually designed at later stages of the project where they are made to fit the form of the building.

In ensuring maximum and efficient coverage of external wall areas, the conceptualizing of a building's overall form should be done with adequate consideration to the design, provision and future operation of its access systems [38, 39]. Table 3.16 illustrates common building forms and their impacts on the provision of access systems [40].

Table 3.16. Common building forms that may affect the provision of access systems.

Building Form	Impacts on Provision of Access System
Plain rectangular 	• Regular and plain form allows ease in the provision of access system. • Roof services may be located in the center of the roof, allowing tracks or rails to be laid along the perimeter for the access system. • Minimum number of corners to turn. • Maximum coverage with numerous types of access system.
Plain circular 	• Plain form allows ease in the provision of access system. • No corners present. Access system can operate without need to be raised to turn corners. Maximum efficiency. • Working platform should be contoured to the curvature of the building. • Tracks can be laid along the circumference of the roof for the access system to run on.
Sloping face 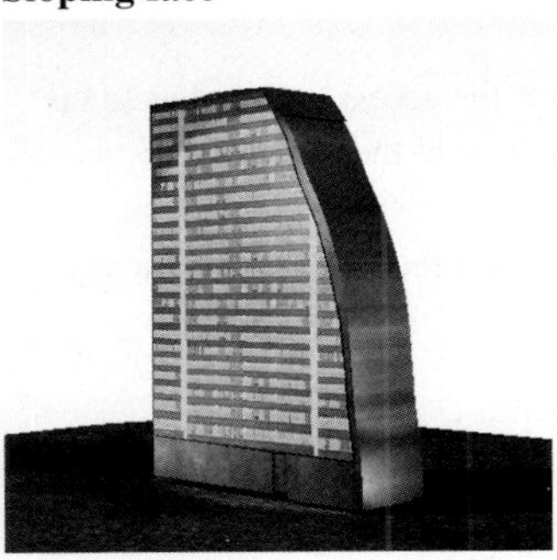	• Small roof space due to building form. • The sloping face may be accessed with a jib that extends together with the outward sloping face as the working platform lowers. • The extendable length of the jib should be able to cover the face when the face is at its largest width.

(*Continued*)

Table 3.16. *(Continued)*

Building Form	Impacts on Provision of Access System
Stepped roof 	• Roof space for services and access system is reduced. • Different access systems may need to be provided on each roof level in order to provide for full coverage of façade. • Intermediate roofs should have adequate roof space to operate the access system. • Jibs that are located on the highest roof level and are able to extend to reach all areas of the façade will be efficient and cost effective
Staggered face 	• Numerous corners require the access system to change over at each turn of corner. This may slow down maintenance times and hence reduce efficiency. • Each staggered face should be wide enough to accommodate the access system. • The access system should be equipped with luffing jibs to enable ease when turning the corner.
Sloping roof 	• Minimal roof area for storage and operation of access system. • The track or rail for the access system has to be inclined to the gradient of the roof slope. • If roof is glazed, the access system must be light-weight and within the loading that the roof can withstand.

(Continued)

Table 3.16. (*Continued*)

Building Form	Impacts on Provision of Access System
Pointed roof 	• Minimal roof area for storage and operation of access system. • Access system may be anchored to façade or housed within the roof space that is provided with openings for operating the access system. • Such building forms are usually provided with a flat area at a lower level that runs around the perimeter. • Access to the pointed roof area is possible with a traversing trolley or human rapellers.

References

[1] M. Y. L. Chew, *Construction Technology for Tall Buildings*, 3rd Edition, World Scientific, 2009.

[2] L. G. W. Verhoef *et.al.*, *Soiling and Cleaning of Building Façades: Report of Technical Committee 62 SCF, RILEM (The International Union of Testing and Research Laboratories for Materials and Structures)*, Chapman and Hall, 1988.

[3] L. Addleson and C. Rice, *Performance of Materials in Building*, Butterworth Heinemann, 1991.

[4] E. C. C. Choi, "Wind-driven-rain characteristics and criteria for water penetration test", *Proc. Int. Conf. Building Envelope System and Technology*, Centre for Continuing Education, Nanyang Technological Education, 1994.

[5] Y. C. Wee and G. Lim, "Biological growth and attack on building surfaces in Singapore", paper presented at the seminar on *Architecture, Engineering, Construction and the Law in the 80's: Plan Now and Pay Later*, 1983.

[6] K. B. Lee and Y. C. Wee, "Algae growing on walls around Singapore", *Malay. Nat. J.*, **35**, 125–132, 1982.

[7] Anon. *Specification for Algae Resistant Emulsion Paint for Decorative Purposes, Singapore Standard 345*, Singapore Institute of Standards and Industrial Research, Singapore, 1990.

[8] Y. C. Wee and K. B. Lee, "Proliferation of algae on surfaces of buildings in Singapore", *International Biodeterioration Bulletin*, **16**, 113–117, 1980.

[9] K. K. Ho, K. H. Tan and Y. C. Wee, "Growth conditions of Trentepholia odorata (Chlorophyta, Ulotrichales)", *Phycologia*, **22**, 303–308, 1983.

[10] P. Whiteley, "The occurrence and prevention of mould and algal growths on paint films", *Society of Chemical Industry Monograph*, **23**, 161–169, 1996.

[11] J. F. Ferguson, "Microbiology of paint films I", *Paint Technol*, **33**(6), 19–27, 1969.

[12] G. Lim, T. K. Tan and A. Tan, "The fungal problems in buildings in the tropics", *International Biodeterioration*, **25**, 27–37, 1989.

[13] Y. C. Wee and R. Corlett, *The City and the Forest: Plant Life in Urban Singapore*, Singapore University Press, 1986, 186pp.

[14] M. Y. L. Chew and P. P. Tan, *Staining of Façades*, World Scientific, 2003.

[15] D. Campbell-Allen and H. Roper, *Concrete Structures: Materials, Maintenance and Repair*, John Wiley and Sons, New York, 1991.

[16] A. J. Brookes, *Cladding of Buildings*, 3rd Edition, E. & F.N. Spon, 1998.

[17] A. J. Brookes, *Building Envelope*, Butterworth Architecture, London, 1990.

[18] J. B. Taylor, *Plastering*, 5th Edition, Longman Scientific and Technical, Harlow, Essex, 1990.

[19] G. D. Taylor, *Materials of Construction*, 2nd Edition, Construction Press, London, 1983.

[20] *Good Practice Guide for Paint*, Building Control Authority, Singapore, 2001.

[21] G. E. Weismantel, *Paint Handbook*, McGraw-Hill, New York, 1981.

[22] C. Palmonari, *Ceramic Floor and Wall Tiles: Performance and Controversies*, Sassuolo (Modena): Edi. Cer. S.p.A — Sassuolo, 1989.

[23] M. Y. L. Chew, "Factors affecting tile adhesion for external cladding", *Construction and Building Material*, **13**(5), 293–296, 1999.

[24] M. Y. L. Chew, "Use of intra-red thermography for assessing tile delamination on building façades", *Journal of Real Estate and Construction*, **8**, 64–72, 1998.

[25] D. Richardson, "The staining of natural stone", *Construction Repair*, March/April, 15–17, 1993.

[26] M. Y. L. Chew, "Efficient maintenance: Overcoming building defects and ensuring durability", *Conference on Building Safety*, The Asia Business Forum, Kuala Lumpur, 4 and 5 April 1994.

[27] J. S. Amstock, *Handbook of Glass in Construction*, McGraw-Hill, New York, 1997.

[28] C. W. Wong, H. C. Ta and S. C. Chew, "Spontaneous shattering of tempered glass", *Southeast Asia Building Magazine*, October 2002.

[29] J. C. Beech and A. T. Wolf, "Ageing resistance of building and construction sealant", *Durability of Building Sealants, Proc. Int. RILEM Symposium on Durability of Building Sealants*, Building Research Establishment, Garston, UK, 1994.

[30] M. Y. L. Chew, "A review of the assessment methods for the performance of sealants", *Architectural Science Review*, **42**(1), 31–42, 1999.

[31] A. T. Wolf "Ageing resistance and construction sealants — Synergistic effects, natural and artificial weathering tests", *Durability of Building Materials and Components, Proc. 7th Int. Conf. Durability of Building Materials and Components*, C. Sjostrom (ed.), Vol. 2, E&FN Spon, May 1996, pp. 139–149.

[32] J. Beech and C. Mansfield, *The Water Resistance of Sealants for Construction*, Building Sealants: Materials, Properties and Performance ASTM STP 1069, Thomas F. O'Connor (ed.), ASTM, 1990.

[33] S. Randazzo, *Sealants for Construction Joints Subjected to Continuous Water Immersion*, C. J. Parise (ed.), Science and Technology of Building Seals, Sealants, Glazing and Waterproofing, ASTM STP 1168, ASTM, 1992, pp. 57–64.

[34] M. Y. L. Chew, X. Zhou and S. H. Goh, "The effects of some chemical components of polyurethane sealants on their resistance against Hot-Water", *Building and Environment*, **38**, 1381–1384, 2003.

[35] J. Comyn, *Adhesion Science*, Royal Society of Chemistry, Cambridge, UK, 1997, pp. 109–117.

[36] M. Sarsent and J. L. Koening, "Fourier tranform IR spectroscopic analysis of the molecular structure of compatible polymer blends", M. W. Urban and C. D. Craver (eds.), *Structure-Property Relations in Polymers: Spectroscopy and Performance*, American Chemical Society, 1993.

[37] K. D. McCann, "Sealant lifetime in an immersed environment", *Dow Corning Internal Research Report*, Dow Corning Corporation, Midland, USA, 1990.

[38] E. G. Lovejoy, "Safety and security in accessibility for maintenance", E. D. Mills (ed.), *Building Maintenance and Preservation*, Butterworths, London, 1980, pp. 117–130.

[39] J. D. Harrison, "Access requirements for building maintenance", C. Briffett (ed.), *Building Maintenance Technology in Tropical Climates*, Singapore University Press, Singapore, 1995, pp. 117–138.

[40] Council on Tall Buildings and Urban Habitat, *Tall Building Systems and Concepts*, American Society of Civil Engineers, New York, 1980.

CHAPTER 4
BASEMENT

4.1. General

A basement is that part of a building structure that is below ground level and is therefore subjected to external hydrostatic pressure. Due to their exposure to surrounding soil conditions and groundwater, basements tend to suffer from a faster rate of deterioration than other building elements [1].

Protecting a basement against aggressive agents from the soil surroundings is a challenging task. Concrete and masonry being the main materials for most basement construction, are not waterproofed as a material. Waterproofing the basement may face constraints such as the method of construction. Positive side waterproofing which allows waterproofing to be installed to the outside surfaces of a basement may not be possible due to (a) space constraint; (b) method of basement construction, e.g. "cut and cover" method [2], etc.

This chapter discusses the impact of design, construction and maintenance issues on the maintainability of a basement.

4.2. Water Seepage

Water seepage of a basement could take place through one or a combination of (a) porous concrete, (b) crack, (c) joints and (d) penetrations, as shown in Table 4.1.

Concrete consists of micro pores caused during the migration of excess water during the hydration process. As these excess water dries off, they form channels or capillary pores. When subjected to water pressure, the pores allow the passage of water resulting in leakage. Porosity of concrete coupled with crack lines, allows the ingress of water and

Table 4.1. Common paths of water seepage/leakage in a basement.

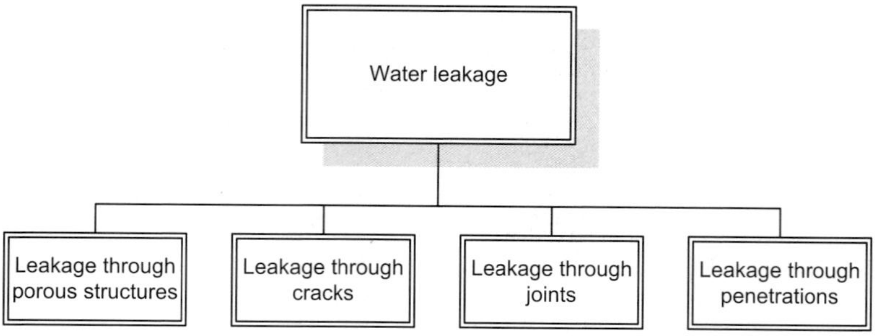

Fig. 4.1. Efflorescence on wall due seepage through crack lines.

subsequently efflorescence — A crystalline deposit of water-soluble salts (Fig. 4.1). This is exacerbated by the formation of cracks due to the weak tensile strength of concrete, against stresses resulting from shrinkage, differential settlements, etc. Figure 4.2 shows a thick efflorescence beside the black cables. Figure 4.3 shows two plastic packers being installed beside a fire protection cable trunking to receive chemical injection to seal off any paths that facilitate water entering into the inside surface of the basement wall.

In certain situations it may be impractical or inefficient to carry out chemical injection or pressure grouting due to the excessively high hydrostatic pressure. Water is allowed to seep through and channeled away. Figure 4.4 shows a scupper drain that was constructed along a basement

Fig. 4.2. Efflorescence on wall due seepage through porous concrete.

Fig. 4.3. Packers for chemical injection on a basement wall.

wall to channel water that leaks through the wall to a nearby sump pit. Water collected in the sump pit is pumped out to the main sewer line at road level.

Another easy path for water seepage in a basement is through discontinuities such as cold joints, movement joints and structural joints. Cold

Fig. 4.4. Provision of a scupper drain to channel away ingressed water.

joint is formed when there is a time lapse between two consecutive concrete pours such that the second pour started only after the first one has set. They are common in basement construction as in general, the volume of concrete for slab and wall is huge. Movement and structural joints are installed to allow movement in structural members to avoid restraint-causing cracks. However, these locations are very weak points where water can penetrate easily if not constructed carefully (Figs. 4.5–4.7).

Similar to wet area, the continuity of waterproofing may be interfered by penetration due to piping and the like if the work is not performed properly.

4.3 Risk Factors

From survey interviews and findings from past studies [3], 22 significant risk factors of basement maintainability are identified and further grouped into 12 risk parameters (Table 4.2).

(a) **(b)** **(c)**

Fig. 4.5. Water seepage through (**a**) wall joints, (**b**) ceiling/wall joints and (**c**) beam/wall joint.

Fig. 4.6. Plastic packers installed at an angle to a ceiling/wall joint to receive chemical injection.

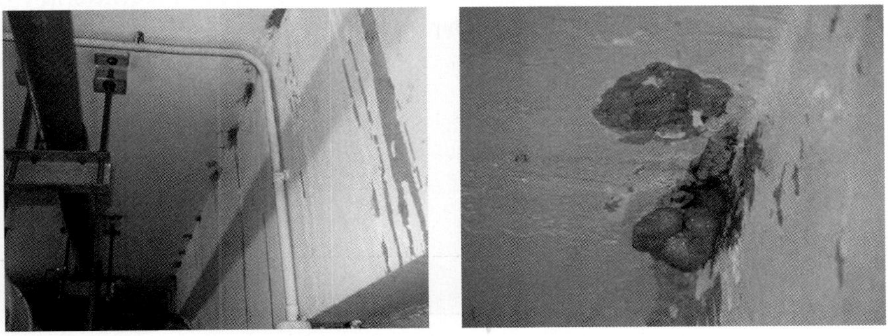

Fig. 4.7. Chemical injection at a wall-to-ceiling construction joint.

Table 4.2. Risk factors of basements.

Risk Parameters	Risk Factors	Elements
1. Waterproofing system selection	1. Nature of basement usage 2. Method of construction	— Degree of protection — Type of protection systems
2. Waterproofing detail	3. Application/design detail	— Adherence to typical design considerations and detailing
	4. Penetration detail 5. Joint detail	
3. External drainage system selection/ design	6. Design considerations	
4. External drainage system details	7. Pumping system 8. Sufficient gradient 9. Sufficient ventilation	
5. External drainage system and maintenance	10. Frequency and extent of maintenance work	
6. Building profile	11. Age 12. Depth of basement	
7. Environment	13. Hydrostatic pressure 14. Chemical composition 15. Soil types and permeability	— Groundwater table level — Noncohesive soils — Cohesive soils
8. Material selection 9. Material detail 10. Material maintainability	16. Durability 17. Sustainability 18. Cleanability 19. Inspection	— Lifespan — Testing requirements — Cleaning requirement — Inspection requirement
11. Construction quality	20. Workmanship quality	— Financial — Technical — Human resource — Quality management
12. Maintenance practice	21. Method 22. Efficiency	— Maintenance inspections, equipments and chemical used for cleaning — Frequencies of inspection and cleaning cycles

4.3.1. *Design Risk*

The failure to achieve water-tightness is the most significant problem in basements. Water-tightness of basements relies mainly on the waterproofing design detailing including the adequacy of waterproofing over penetrations, projections and joints.

For the selection of waterproofing systems, the nature of the basement usage determines the degree of protection required while the method of construction limits the type of protection systems to be chosen. The quality of application also determines the eventual degree of water-tightness attained. With proper selection of the waterproofing system, appropriate design of details and sound workmanship during application, water-tightness in basements could be improved and prolonged.

Waterproofing detailing at joints has been identified as a critical factor in achieving water-tightness as these are vulnerable points prone to water ingress. Proper detailing design will minimize the risk of such lines of weakness and, together with correct installation and workmanship, achieve water-tightness in basement structures. Inadequate detailing design will cause great risk and results in repetitive occurrence of defects at these critical points.

Although basement structures must be designed to prevent any ingress of surface runoff, this may not be completely avoided. As such, there is a need for a pumped drainage system to ensure that the basement is protected from this external source of water. The surface drainage system should adhere to the local design guidelines for minimum risk. Being a tropical country with frequent rainstorms, local basements require more regular care and maintenance to the drainage systems to prevent pipe clogging, overflows, leaks and discharge interferences. Regular inspection means problems could be detected early and rectified before developing into serious defects.

4.3.1.1. *Inappropriate Waterproofing System*

Selection of waterproofing systems of basement structures is dependent on the proposed usage of basement and the method of construction. The availability of a wide range of systems and product has made the selection rather complex. Lack of sufficient knowledge of these choices could lead to inappropriate selection of systems. Water seepage of basement due to premature failure of waterproofing system or ineffective system may be attributed to inappropriate selection.

In general, all waterproofing materials for basement can be classified into the following categories:

(a) membranes,
(b) admixtures,
(c) cementitious, and
(d) others.

Membranes consist of either preformed types which are factory manufactured and available either in sheet form or rolls, or the liquid-applied types known as waterproof coatings. Membranes work by forming a physical barrier against the passage of water and are used mainly as positive-side waterproofing systems. Admixtures are added to concrete during batching to improve the waterproofing properties. Conventional admixtures used for waterproofing of basement concrete structures include water reducing types, air-entraining types and water-repellant types. Cementitious systems mainly include crystallization types using proprietary reactive pore fillers and the flexible membrane types. Cementitious systems are easy to apply and versatile and can be used both as positive and negative-side applications.

4.3.1.2. *Poor Drainage Design Detailing*

Drainage design detailing is essential to prevent water built up behind basement walls and consequent water seepage/leakage. In the case of an active system where the external wall of a basement is to provide protection against direct exposure of high hydrostatic pressure from soil water on the structure, effective drainage design should be capable of collecting, draining and discharging water away from the basement envelope. They should consist:

(a) A perforated drainage pipe installed with perforation set downward in a bed of gravel that allows water drainage. These pipes are usually made of PVC or vitreous clay.

(b) Coarse gravel installed around and over the drainage pipings for percolation and collection of water.
(c) Sump pits or other drain fields installed to collect and discharge the drained water from pipes.

Figure 4.8 illustrates a typical system. Drained cavity system is another method used in basement construction to drain away water (Fig. 4.9). The system comprises of a cavity floor and walls with drainage channels leading to sumps, from which any water penetrating into the basement may be pumped away [4, 5]. Vapor transmission may be prevented by ventilating the cavity and providing an effective damp proof membrane beneath the drained floors and walls.

Other improper design details such as insufficient gradient of drainage channel and insufficient ventilation of cavity may increase the chances of leakages at weaker spots such as cold joints in the structure (Fig. 4.10).

4.3.1.3. *Lack of Waterproofing Detailing*

Water-tightness of basement relies mainly on the detailing of waterproofing over discontinuities such as construction joints and penetrations. The lack of these details at these discontinuities would result in eventual water seepage (Fig. 4.11). Details such as lapping at critical locations are important for waterproofing system to form an effective barrier against water seepages (Fig. 4.12).

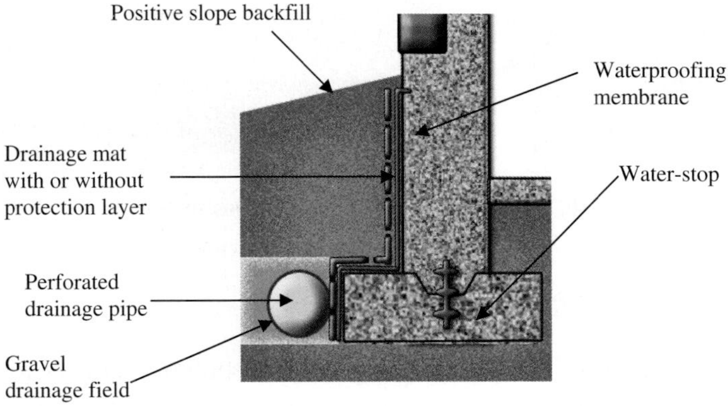

Fig. 4.8. Details of a typical drainage system.

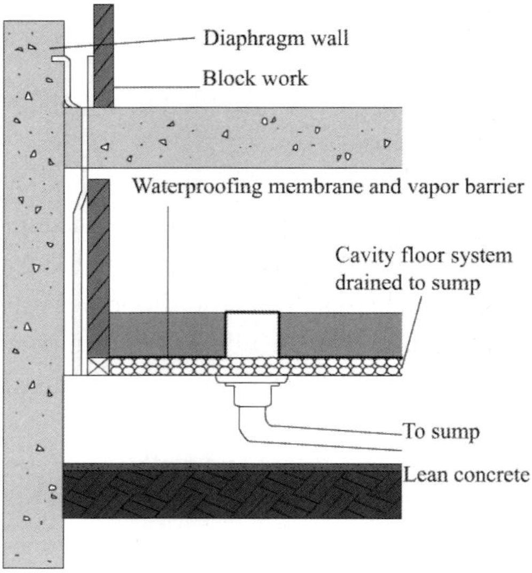

Fig. 4.9. Drained cavity system (*Source*: CP82:1999).

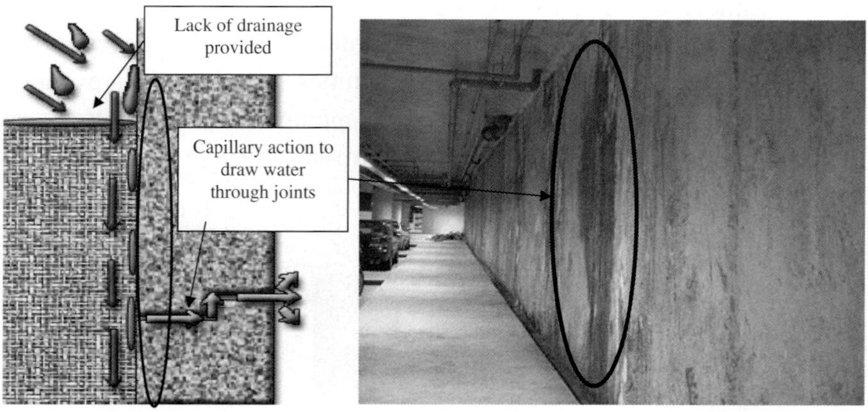

Fig. 4.10. Seepage of water through a construction joint.

4.3.2. *Building Profile Risk*

The building profile namely, the depth of basements and its age, have an effect on the degree of exposure to the external environmental risks.

Fig. 4.11. Water seepage through pipe penetration.

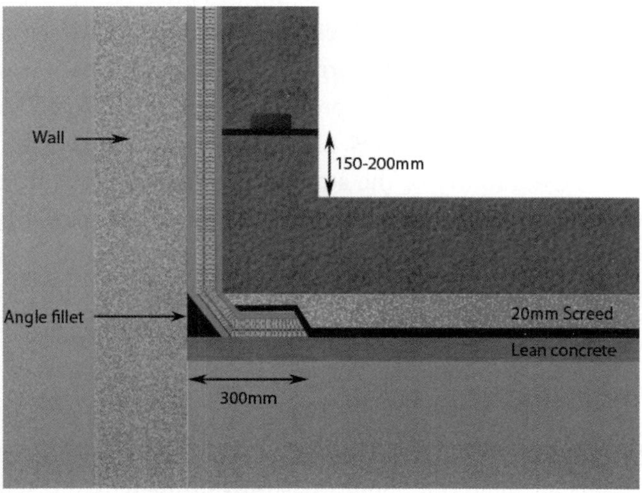

Fig. 4.12. Lapping at interfaces.

For instance, deep basements tend to be subjected to higher hydrostatic pressure and more construction difficulties, thus requiring more stringent protection to over its greater risk of water ingress.

Different shapes of a basement requires different formworking and construction methods which in turn determines the number of joints, service penetration and other weak points for water-tightness. A complex shape would generally require more joints, service penetrations with less tolerance compared to that of a regular shape.

4.3.3. *Environmental Risk*

Given the high humidity level and frequent rainy periods in tropical areas, the external environment could accelerate the rate of defect occurrence in a basement structure. Hydrostatic pressure and the soil conditions affect structural loading design, durability of structural materials, the health and safety of users and limit the construction methods used. As such, professional investigation should be sought early in the design stage to identify the environmental risks at the specific site to evaluate and consider the conditions that will affect maintainability of the basement structure.

The maximum possible amount of information on all aspects of the site condition should be collected, especially the groundwater conditions and chemical analysis should be carried out. This information is essential an consideration for the design of protection systems and the selection of materials. The presence of sulfates, chlorides, acids in the soil and groundwater affects the durability of the structural elements [6, 7]. Extreme and aggressive ground conditions pose great risk in the durability of the materials used in basement construction which will lead to occurrence of defects.

4.3.4. *Material Risk*

With good workmanship, appropriate material detailing and proper maintenance, a selected material should perform up to its expected lifespan. However, the actual lifespan and performance can be undermined by interactions with other materials in the structure. Careful considerations are needed to ensure that the material selected is suitable for the basement environment and compatible with other materials used, such that the overall lifespan would not be compromised due to premature failures of incompatible materials.

The selection of durable basement materials is considered a crucial factor pertaining to basement maintainability since the chances of replacement are rare and difficult, as in the case of the waterproofing materials. The basement material selected should have the ability to resist defects in order to perform its intended function through its design life. The inherent physical properties of materials determine the permeability of the basement to water and also the durability of the basement construction. To select durable materials and ensure that they are able to sustain their functional purposes, it is important to examine its detail specifications with recognized standards. The performance of materials can be assessed through testing and approvals for properties pertinent to maintainability.

4.3.5. *Construction Quality Risk*

Construction quality, also known as workmanship, has been identified as one of the most crucial causes in basement failures by many industry professionals. Findings from interviews carried out for this study highlighted the main issues on unskilled laborers and cost-driven contractors which led to negligence in workmanship. Even with a properly designed waterproofing system of suitable materials, the required performance of the basement structure would not be achieved if the works are not executed properly.

Building defects arise from the failings in the systems that control human activities over the entire procurement process of design, construction and maintenance [8, 9]. To prevent it and thus reduce its occurrence, effective measures must be in place to tackle each of its possible causes.

Incentives and controls, e.g. in the form of accreditation from regulators is efficient in raising capability and provide recognition to those committed to deliver quality works. An example is the Accreditation Scheme for Waterproofing Specialist Contractors [10] launched by the Singapore Concrete Institute (SCI) and the Waterproofing Trade Association in 2004. Contractors are encouraged to participate on the scheme so as to improve their quality and competitiveness to enhance the firms' images and professionalism. Building clients and developers are encouraged to specify the use of accredited contractors for their projects to ensure the level of construction quality achieved.

4.3.5.1. *Insufficient Compaction*

Insufficient compaction of concrete often contributes to increased porosity in concrete structures. Inadequate vibration creates segregation in the plastic concrete since concrete mix may experience bleeding effect where heavier components settle to displace the water component and thus voids will be formed after the water dries up. These honeycomb voids in concrete create passage for water seepage and may weaken the concrete structure substantially (Fig. 4.13).

4.3.5.2. *Improper Installation of Construction Joints*

Nonalignment of construction joints could form gaps between panels permitting water penetration (Fig. 4.14). Inadequate provision to accommodate anticipated differential movements could lead to cracking and in turn water seepage through these crack lines (Fig. 4.15).

Water stops and drainage details are common devices in controlling water seepage through construction joints in basements. Proper installation of waterproofing details at these joints to achieve optimum water-tightness (Fig. 4.16) could be achieved by the following:

(a) Internal water-stops can be secured by using wires to hold it in place.
(b) External water-stops are tied to the formwork of concrete works.
(c) Flanges of water-stops should be tied to the reinforcement steel at an interval of 300 mm along the flanges.

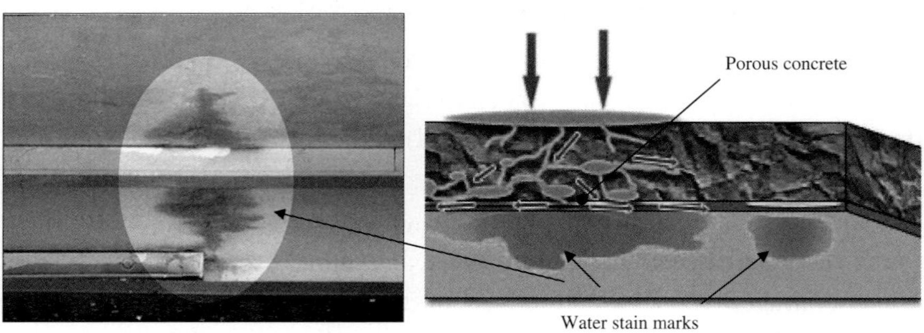

Fig. 4.13. Damp patches caused due to insufficient compaction of concrete.

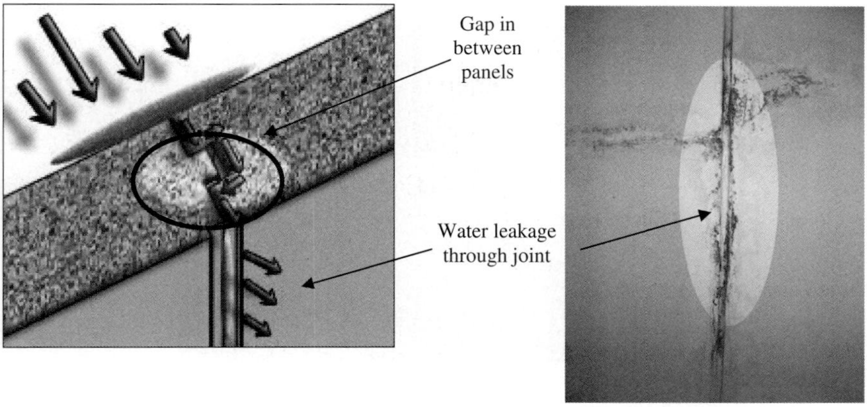

Fig. 4.14. Poorly matched prefabricated wall panels leading to easy water penetration.

Fig. 4.15. Water seepage through crack line.

(d) Bulb of the water-stop should be placed directly on to the midpoint of the joint for efficient expansion actions.
(e) Individual sections must be properly jointed by special junction pieces.

See Figs. 4.31–4.35 for more details on waterstops.

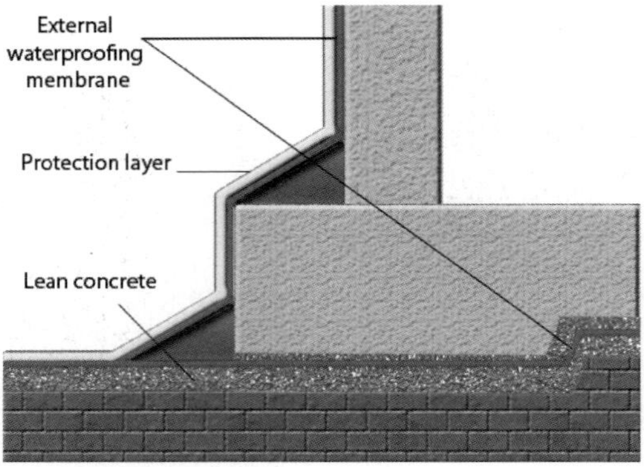

Fig. 4.16. Tanked protection system (*Source*: CP82: 1999).

4.3.5.3. *Improper Installation of Waterproofing*

Poor workmanship engaged during the installation could cause errors such as insufficient overlapping at joints for membrane type or inadequate thickness in liquid-applied membranes. Moreover, cratering, wrinkling, blistering, splitting and delaminating of the membrane are defects resulting from poor workmanship (Fig. 4.17). Figure 4.18 illustrates water seepage through a defective epoxy membrane on a ground slab as a result of rising moisture.

4.3.5.4. *Damage of Waterproofing During Construction*

There is a risk of damaging waterproofing membrane during construction due to the activities of other trades and foot traffic. In addition, waterstops, which are normally placed at the joints, could be damaged or displaced after being placed. Some could swell in size when exposed to water before casting of concrete. Premature failure of waterproofing would fail to preserve water-tightness of basement during its service life.

4.3.5.5. *Insufficient Curing of Waterproofing*

Efficient waterproofing system of perimeter walls forms a crucial barrier against all seepages/leakages. Proper curing of the waterproofing is hence

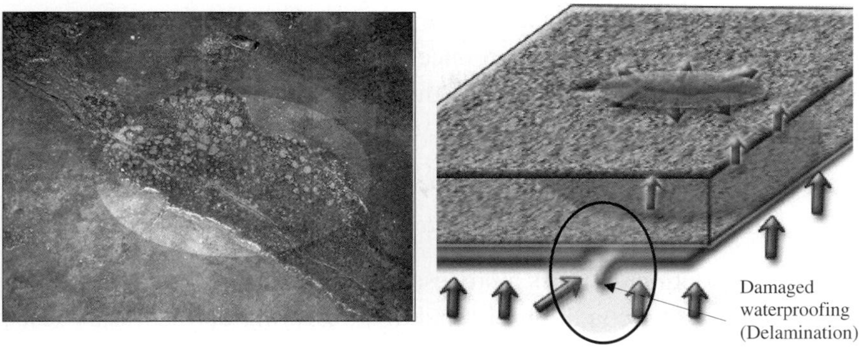

Fig. 4.17. Seepage through a fine crack.

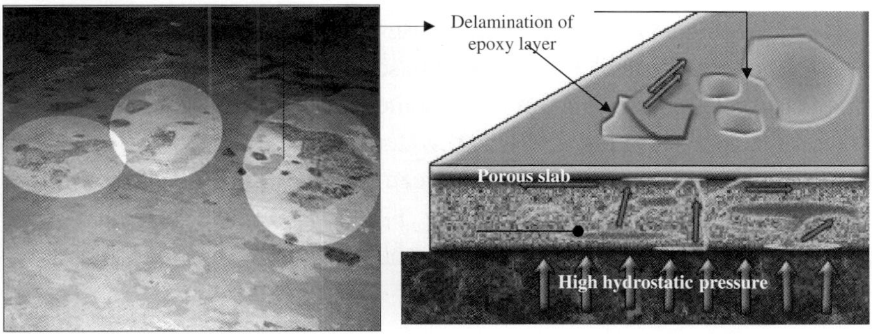

Fig. 4.18. Peeling of epoxy coating.

crucial. Insufficient curing could severely lead to excessive shrinkage cracks which in turn provide a passage for water seepages to take place.

4.3.6. *Maintenance Practice Risk*

Regular inspections and adequate maintenance can aid in the performance and condition of the basement structure throughout its whole service life. Although basement materials, like the waterproofing components, generally do not require cleaning and maintenance, once installed, visual inspection is still necessary to detect any signs of water seepage that could lead to development of other defects such as biological staining, efflorescence and reinforcement corrosion. Immediate action must be taken once such defects are detected to minimize the worsening of the defects that might result in extensive repair and even replacement works.

4.3.6.1. *Lack of Regular Inspection to Defect Detection*

Building inspection involves a process of carrying out relevant investigation of the existing condition of a building, carrying out necessary tests, evaluation of the data collected, make recommendations and prediction of the future performance of buildings [11, 12]. Visual inspections can be useful as an initial investigation of building defects through signs of material defects (Fig. 4.19).

Lack of such inspections may cause future problems which include water seepages. Typical example in a basement is water leakage at ceiling which usually originates from nonbasement features such as above grade planters and swimming pools or waterponding problem. Unattended dampness caused by these features can cause water seepages through the structure with prolonged exposure. Figure 4.20 illustrates water seepage due to extended waterponding on the slab while Fig. 4.21 illustrates

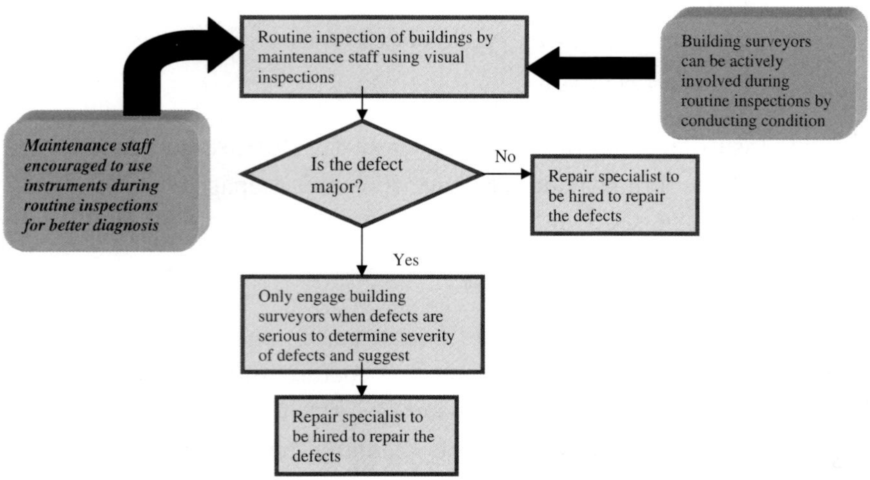

Fig. 4.19. Comprehensive inspection system.

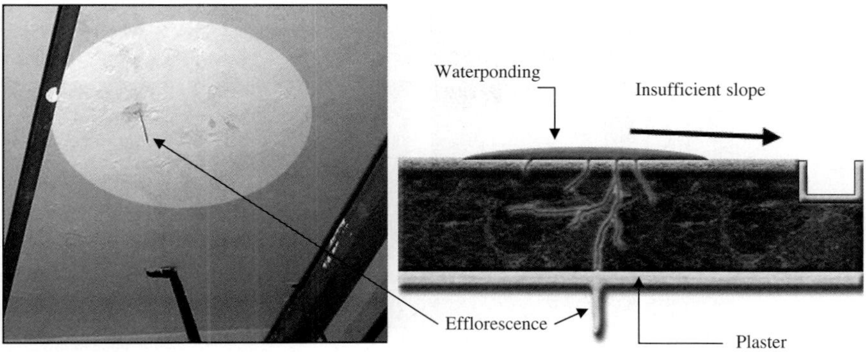

Fig. 4.20. Water seepage due to waterponding.

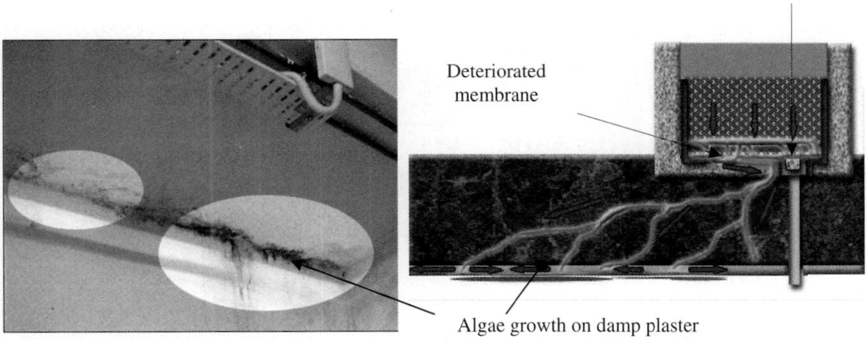

Fig. 4.21. Water seepage due to clogged drainage of the planters structure.

the case where clogged drainage in the wet-features causing excessive hydrostatic pressures against the horizontal structure.

4.3.6.2. *Delayed Repair to Initial Defect*

Water seepages of basement could often originate from inefficient repairs. For instance, leakage caused by cracks is a common problem due to delay in repairs. Leakage from cracks is not an instantaneous formation. Over time, minor cracks on the structure propagate and widen gradually and water source could start to seep through. In certain cases, retained

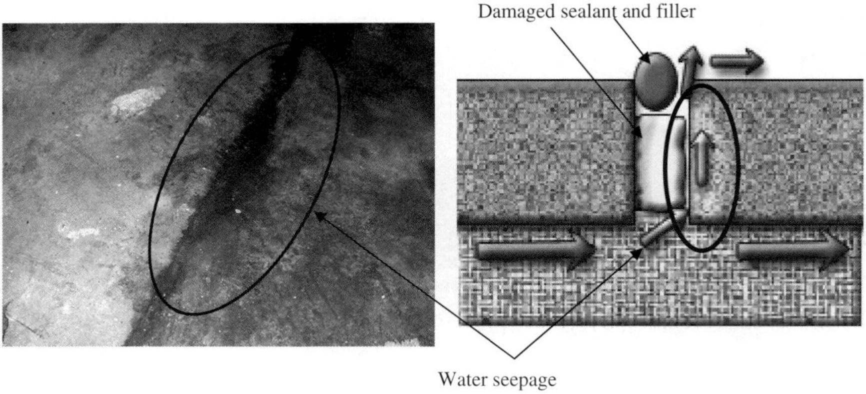

Fig. 4.22. Expansion joints detach from the sides allowing water seepage.

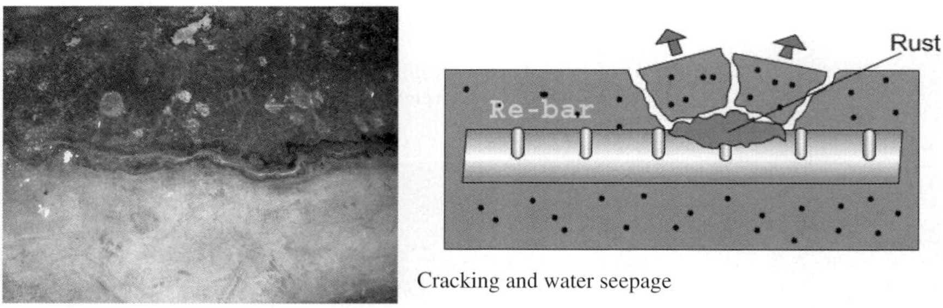

Fig. 4.23. Leaking of water through slab causing corrosion of bars.

moisture may cause corrosion of steel bars and consequently its expansion in volume resulting in cracking in the structure (Fig. 4.22).

Jointing materials such as sealant used in car parks may be easily damaged when they are subjected to vehicular traffic or natural aging or prolonged exposure to moisture [13]. When the corrective maintenance is delayed, the joint may eventually fail (Fig. 4.23).

4.3.6.3. *Deterioration of Waterproofing*

Although waterproofing systems are designed to provide protection against water, there are various limitations that may influence the performance of waterproofing system. Factors influencing the rate of deterioration include natural aging, chemicals in ground water and other external factors such as quality of application.

4.4. Common Defects in Basement

Groundwater refers to the residual water that percolates downwards to the watertable after runoff, evaporation and evapotranspiration. The hydrologic cycle of ground water is shown in Fig. 4.24. A part of the precipitation falling on the surface runs off toward the river, where some is evaporated and returned to the atmosphere. Of that part filtering into the ground, some is removed by the vegetation by evapotranspiration. Some part seeps down through the zone of aeration to the watertable. Below the watertable, the water moves slowly toward the stream, where it reappears as surface water. Water in a confined aquifer can exist at pressures as high as its source, hence the flowing well. Water trapped above the upper clay layer can become perched and reappear as a small seep along the riverbank [2].

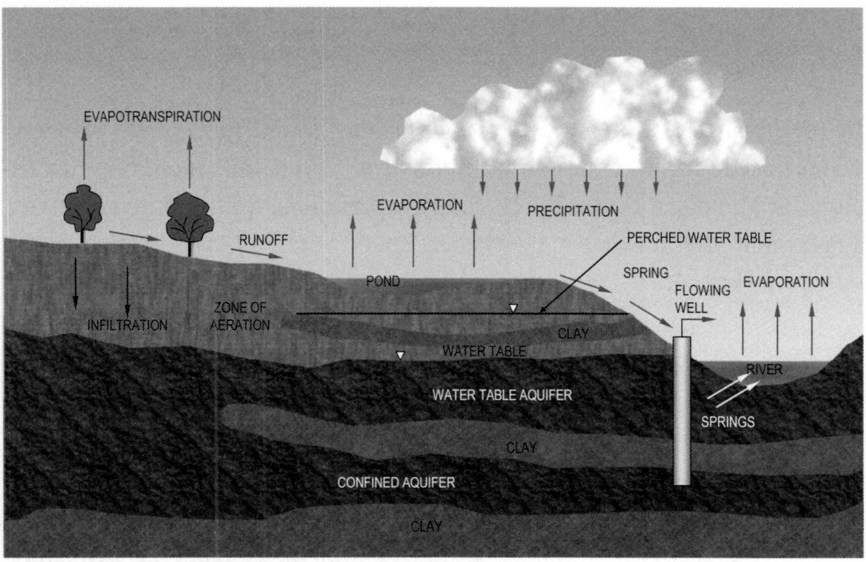

Fig. 4.24. The hydrologic cycle of groundwater.

Residual soils are product of weathering and may cover a large area. Generally formed above the groundwater table this soil is usually unsaturated. Water carrying the contaminants can move over a longer distance along soil strata before reaching the ground water table [14]. Such unpredictable environment leads to premature deterioration of underground structures. Defects mainly due to water seepage are common and tedious problem [3].

Basements are designed, constructed and maintained to keep them free from moisture and dampness. Durability is vital for a basement structure as there is limited scope for maintenance and repair [15]. The major components of basement are: structural members, waterproofing system, finishes and services. From field studies and a thorough literature review, the prevalent defects pertaining to basements are documented in Fig. 4.25 and Table 4.3 followed by analysis of causes and criticality [16].

4.5. Case Study

4.5.1. *Seepage Through Joint*

A case study on water seepage through joint is shown in Fig. 4.26.

4.5.1.1. *Problem*

In this example, waterstop at the joint was found inadequate allowing water to seep through the joint. Improper design details including insufficient gradient of drainage channel, insufficient ventilation of cavity, aggravated the leakages at weaker spots such as cold joints in the structure (Fig. 4.27).

4.5.1.2. *Good Practices*

Drainage

Drainage detailing of external wall of a basement is to provide protection against direct exposure of high hydrostatic pressure from soil water on the structure. Successful drainage design should be capable of collecting,

Structural elements

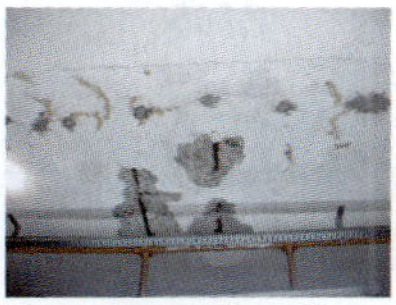
Corrosion and spalling

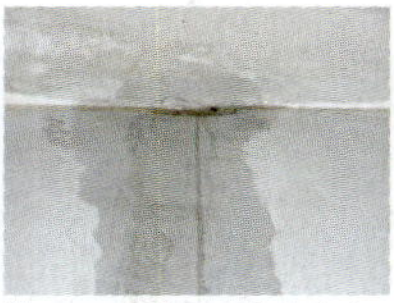
Seepage through joint

Efflorescence at joints

Seepage through cracks

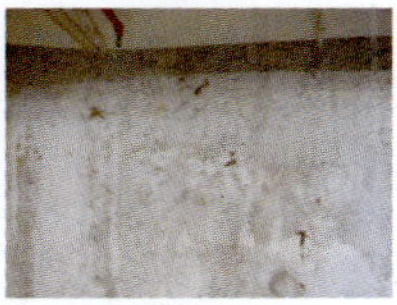
Seepage thru' porous concrete

Finishes on floor wall and ceiling

Flaking and peeling

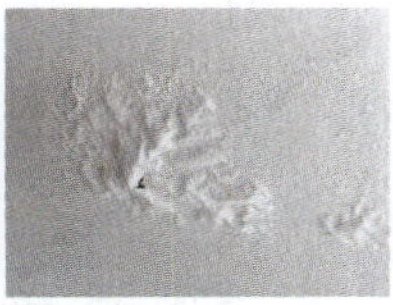
Blistering

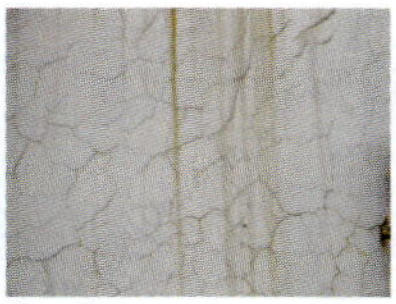
Crazing

Rust stain

Tile cracking

Damp/dirt stain

Dirt/oil stain on floor

Stains on false ceiling

Biological growth

Others

Water ponding

Water seepage through service penetration

Fig. 4.25. Common defects in basement.

Table 4.3. Common defects in basement.

Defects and Coding	Possible Causes of Defects	Cause gr. D*	Cause gr. C*	Cause gr. M*	Cause gr. E*
Structural elements					
Water leakage through joints (A1)	Poor design detailing, e.g. congestion of rebar.	▲			
	Discontinuity at joints or poorly finished joints.	▲	▲		
	Insufficient curing and vibration.		▲		
	Improper installation of water proofing or waterstops.		▲		
	Wrongly selected or bent waterstops unable to provide protection.	▲	▲		
	Aging of water proofing membrane or waterstops.				▲
Efflorescence at joints (A2)	Leaching of soluble salt or un-hydrated lime from cement. Poor quality control for cement and improper curing.		▲		
	Delayed repair of leakage and irregular cleaning.			▲	
Water leakage through concrete (A3)	Inappropriate waterproofing technique and unsuitable material.	▲			
	Deterioration of waterproofing with time or due to exposure to aggressive condition which is not considered in design.	▲			▲
	Concrete is porous due to poor mix design or bad workmanship.	▲	▲		
	Wrong location, installation of waterproofing or its damage during by other trades.		▲		
Seepage though cracks. Usually with rust staining and may lead to efflorescence (A4)	Delayed repair of initial cracks, allows moisture to reach rebars. The corrosion product has higher volume and causes spalling.			▲	▲
	Concrete is exposed to water if the waterproofing is not properly installed or damaged with time.		▲		▲

(Continued)

Table 4.3. (*Continued*)

Defects and Coding	Possible Causes of Defects	Cause gr. D*	C*	M*	E*
	Poor lay out of rebars that hinders vibration or poor compaction around rebars that leave concrete porous at certain places.	▲	▲		
Corrosion and spalling of concrete (A5)	Plastic settlement cracks if water cement ratio is excessive or poor grading of aggregate.	▲			
	Plastic shrinkage cracks if inadequate curing is done especially in warm and windy weather.		▲		
	Insufficient vibration and pouring leaves voids and honeycombs.		▲		
	Concrete is attacked by aggressive chemicals in soil. No use of corrosion resistant bars, corrosion inhibitor in such special cases.	▲			
	Delayed repair of initial cracks.			▲	
Inner cavity wall					
Damp patches (A6)	Material has high water absorption and low water resistance.	▲			
	Rising damp if cavity is not designed or constructed properly to drain out the water	▲	▲		
	Clogged and overflowing cavity. Moisture bridging by debris		▲	▲	
Finishes					
Blistering (A7)	Paint applied on wet (> 6% of moisture content) and warm surface traps moisture underneath the paint causing blistering. This problem is especially prominent in poorly ventilated basements.		▲		

(*Continued*)

Table 4.3. (*Continued*)

Defects and Coding	Possible Causes of Defects	Cause gr. D*	C*	M*	E*
Paint peeling and flaking (A8)	Unsuitable selection or poor quality of paint. A breathable system that allows excess moisture to evaporate is preferable.	▲			
	Poor surface preparation for substrate. Dirt or other contaminants reduce adhesion between the surfaces.		▲		
	Failed water proofing membrane allows water seepage that push off the paint film.	▲	▲		▲
Paint crazing (A9)	Paint is applied on wall suffering from shrinkage cracks.		▲		
	Contaminated substrate, tools, water/thinner.		▲		
Damp and dirt patch (A10)	Fatty acid of alkyd-based paint and alkali of concrete causes saponification or damp patches of soluble soap.	▲			
Rust stain (A11)	Leaching from corroded fittings or rebars	▲	▲	▲	
Tile cracking (A12)	Differential settlement of basement or shrinkage cracks of basement. No proper movement joint provided to mitigate these.	▲	▲		
	Tiling on a cracked or uneven wall.		▲		
Dirt and stains on tiles (A13)	Use of incompatible or poor quality grout.	▲	▲		
	No cleaning of tile surface after laying or dirt from other works.		▲		
	Irregular cleaning.			▲	

(*Continued*)

Table 4.3. *(Continued)*

Defects and Coding	Possible Causes of Defects	Cause gr. D*	C*	M*	E*
Tile debonding (A14)	Back of tile contaminated with dirt or dust. No key provided.		▲		
	Inadequate curing of substrate before tiling or porous tiles if not soaked properly, absorb water from substrate.		▲		
	Tiles not properly tapped in position.		▲		
Dark, damp patches on ceiling (A15)	Water seepage from upper floor (usually below planter/pool).	▲		▲	
	Condensation and water seepage from HVAC ducts damage false ceiling boards.			▲	
Dirt and oil stains on floor (A16)	Dirt and oil stains from vehicles remains unattended due to inadequate cleaning.			▲	
	Oily patches due to saponification from paints.	▲			
	Carbonation from CO_2 form vehicles is confined in one place (usually for reserved parking).				▲
	Trapped moisture if finishing coat is applied on damp substrate or water seepage from ground.	▲	▲		
Efflorescence on floor (A17)	Excess seepage through deep/wide crack.	▲	▲	▲	
	Concrete is porous or weak at poorly prepared construction joints.	▲	▲		
	Leaching of soluble salts or unhydrated lime from cement. Poor quality control for cement and improper curing.		▲		
	Delayed repair of leakage and irregular cleaning.			▲	

(Continued)

Table 4.3. (*Continued*)

Defects and Coding	Possible Causes of Defects	Cause gr.			
		D*	C*	M*	E*
Biological (algae, fungi) staining on surface (A18)	Poor ventilation and natural lighting.	▲			
	Lack of inspection and cleaning. Delayed repair of leakage.			▲	
Others					
Water leakage at service penetration (A19)	Poor detailing and installation of water proofing around the pipe, e.g. no angle fillets at sharp turn, improper lapping, etc.	▲	▲		
	Irregular inspection of leaky pipes.			▲	
	Unplanned service penetration such as hacking can damage the structural element as well as the waterproofing.	▲			
Water ponding (A20)	Entry of surface water run-off. No catchments provided.	▲			
	Poor drainage design (insufficient slope/outlet).	▲			
	Poor cleaning for water brought in by cars.			▲	
	Uneven/dented surface due to poor levelling of floors.		▲		
	Clogged internal drainage.			▲	
	Uneven ground settlement due to massive change in ground water table or poor compacting of compressible soil.	▲	▲		▲

D* — Design; C* — Construction; M* — Maintenance; E* — Environment.

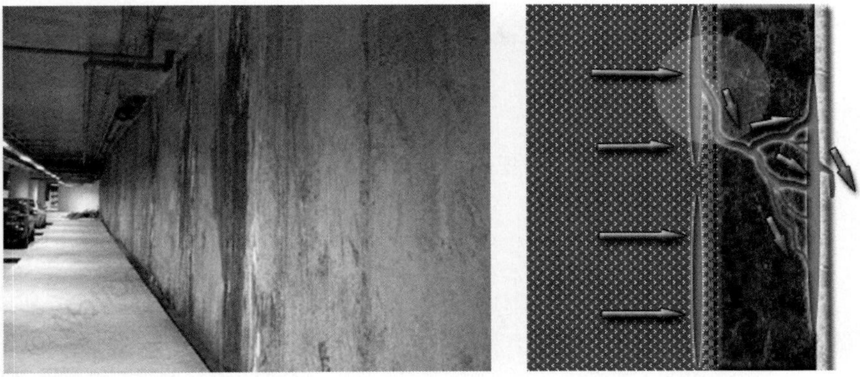

(a) (b)

Fig. 4.26. Water seepage through joint.

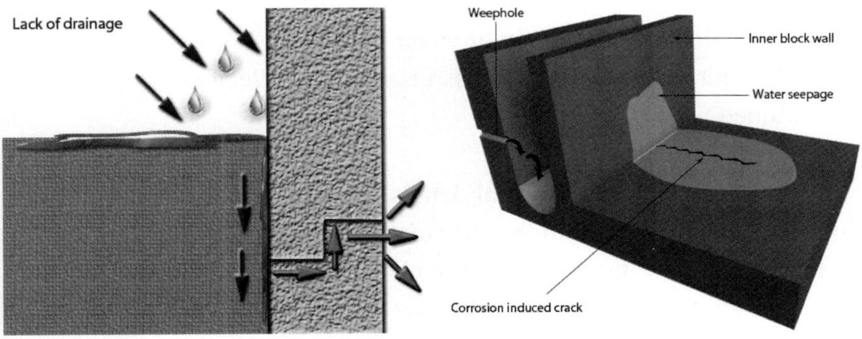

Fig. 4.27. Drainage for basement.

draining and discharging water away from the basement envelope [17–19]. They should consist of the following fundamental items:

(a) A perforated drainage pipe installed with perforation set downward in a bed of gravel that allows water drainage. These pipes are usually made of PVC or vitreous clay.

Fig. 4.28. Typical drainage system.

(b) Coarse gravel installed around and over the drainage pipings for percolation and collection of water.
(c) Sump pits or other drain fields installed to collect and discharge the drained water from pipes.

Figure 4.28 illustrates typical details. In addition, back filling should slope away about 3 m from the structure to avoid ponding (Fig. 4.29).

Drained cavity system is another method used in basement construction to drain away the water. This system is usually adopted where an external tanking system cannot be provided [2]. The system comprises of a cavity floor and walls with drainage channels leading to sumps, from which any water penetrating into the basement may be pumped away (Fig. 4.30). Vapor transmission may be prevented by ventilating the cavity and providing an effective damp proof membrane beneath the drained floors and walls.

Joint

Construction joints in basement structure should be protected by waterstops (Figs. 4.31 and 4.32). These waterstops should be compatible with the waterproofing system. They can be placed internally or externally

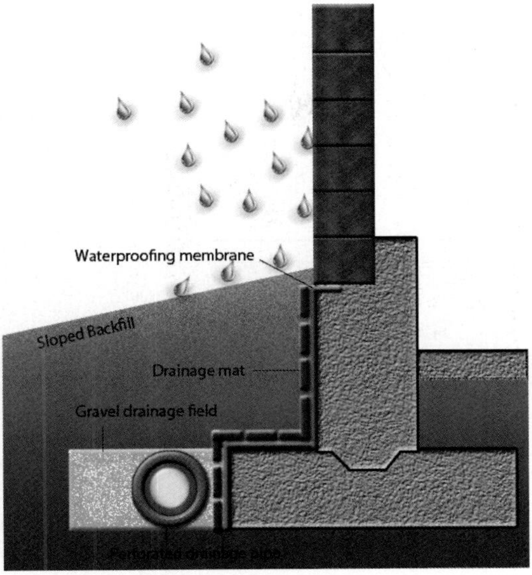

Fig. 4.29. Back fill slope.

(Fig. 4.33). However, choice of waterstop design (Fig. 4.34) should be appropriate with the characteristic of the joint to which they are applied (Table 4.4) [17, 18]. Generally, external waterstops should be at least 240 mm in width to be effective.

Waterstops must be fixed and firmly held in place until concreting is completed. Otherwise, the waterstop may fold during concreting and prevent the effective performance of waterstops (Fig. 4.35).

Internal waterstops can be secured by using wires to hold it in place [19].

- For internal waterstops, tie flanges of waterstops to the reinforcement steel at an interval as recommended by the supplier. Placed waterstops in position prior to the lying of steel in order to prevent accidental abrasion and puncturing the waterstop.
- Place bulb of the waterstop directly in the midpoint of the joint to facilitate efficient expansion actions. Avoid puncturing during steel fixing by nails and debris.

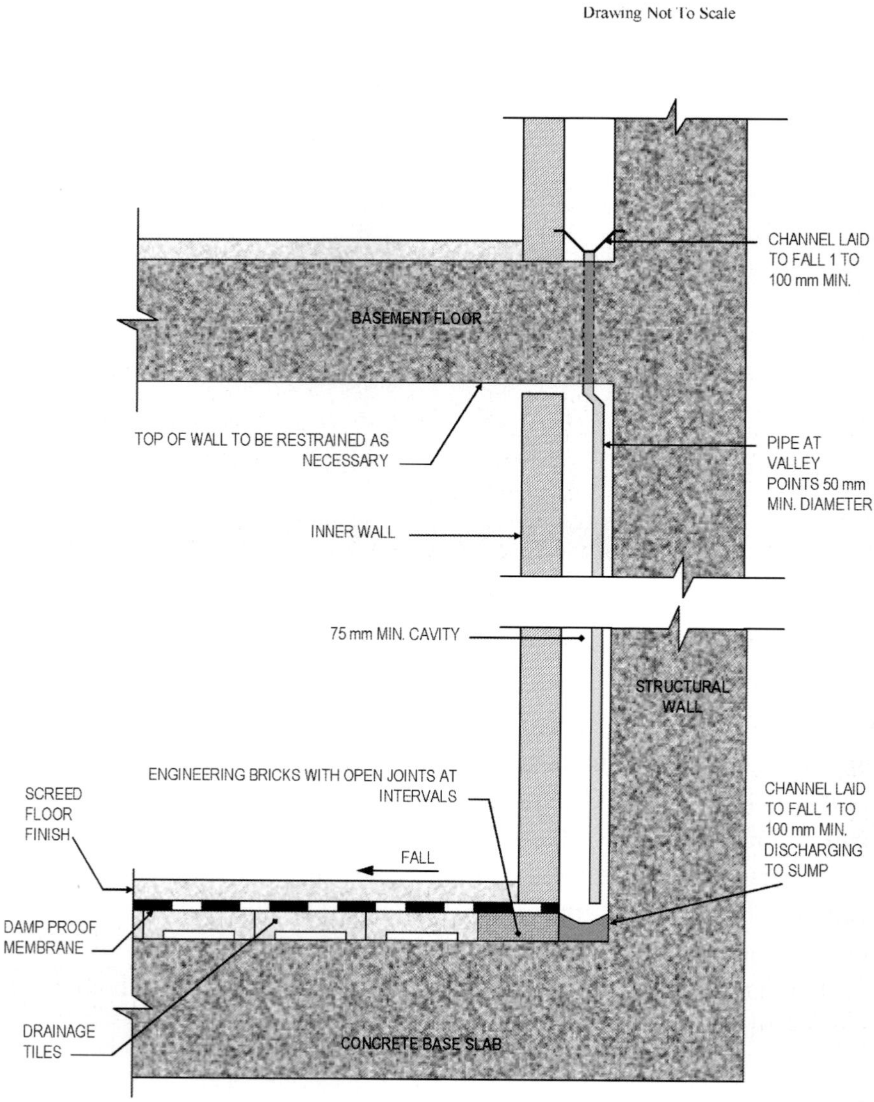

Fig. 4.30. Cavity drainage.

- When hydrophilic waterstops is used, avoid exposure to liquid prior to concreting to prevent expansion.
- Backfilling should be done with a granular material adequately compacted.

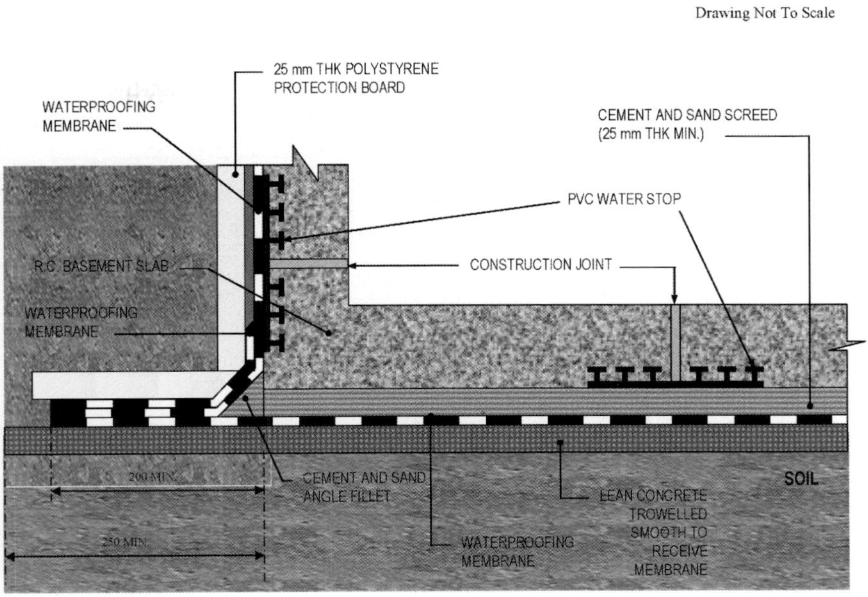

Fig. 4.31. Typical details for basement waterproofing using membrane system.

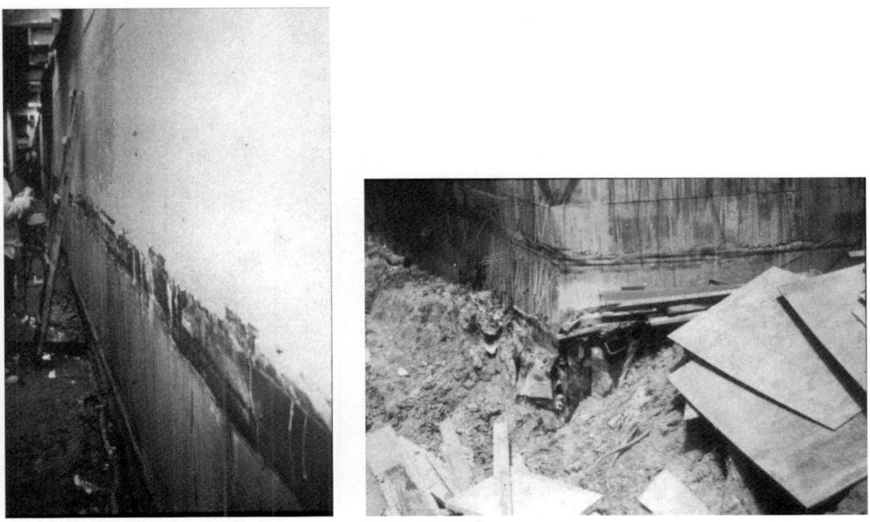

Fig. 4.32. The use of waterstop along a construction joint.

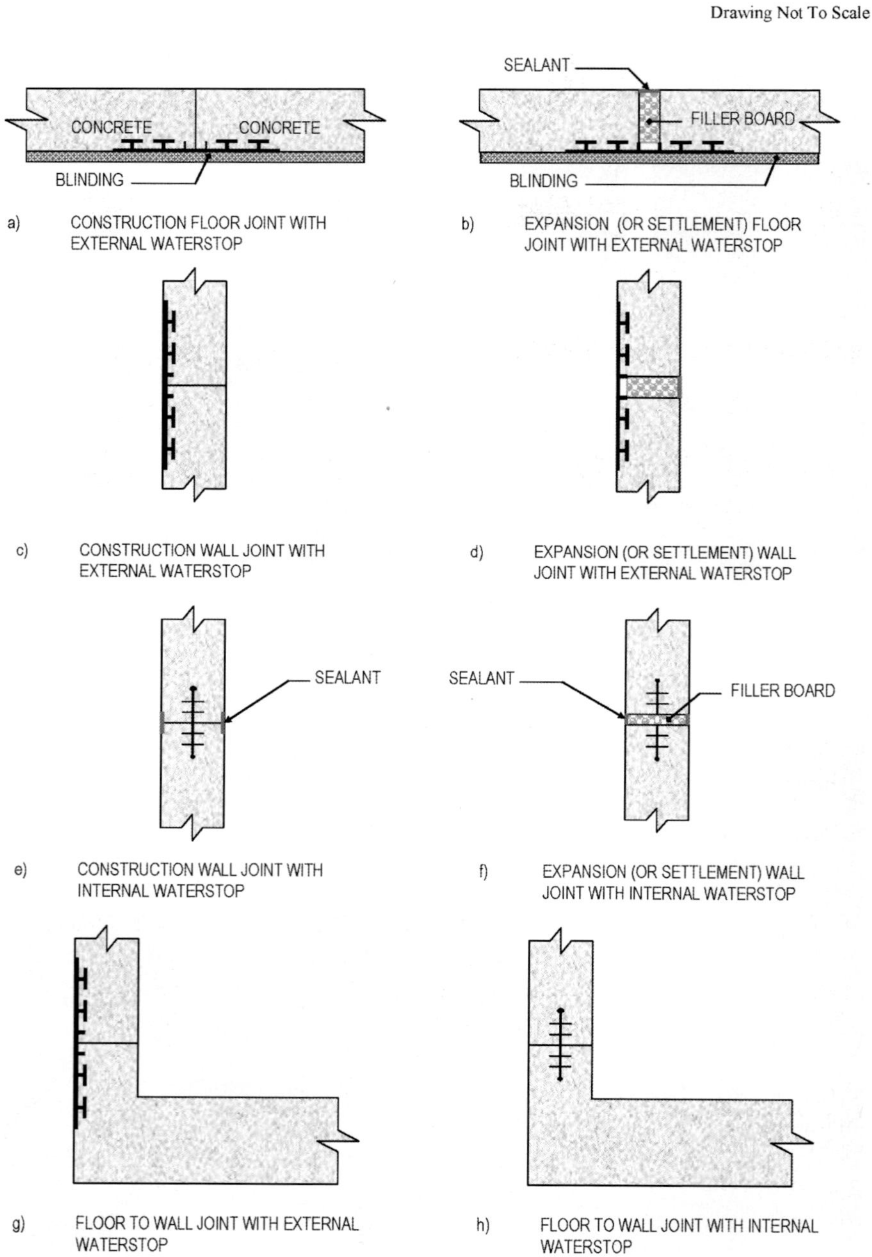

Fig. 4.33. Typical applications of waterstops.

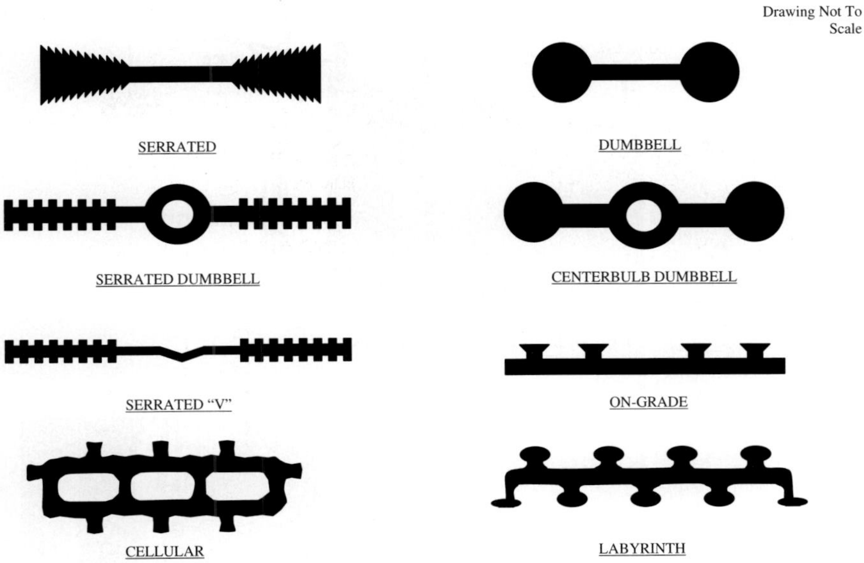

Fig. 4.34. Basic types of waterstops.

Table 4.4. Characteristics of joints.

Waterstop	Characteristics
Ribbed flat and dumbbell	Construction or contraction joints where little or no movement is expected. Ribbed shapes provide a better seal than dumbbell shapes.
Base seal	Ideal for slab-on-grade joints or walls which will be back-filled. Easy to form.
Labyrinth	Primarily used in vertical joints where little or no differential movement is expected. Does not require split forming and adds a key to the joint. Difficult to be used in horizontal joints.
Split waterstop	Most versatile type with the centerbulb that accommodates lateral, transverse and shear movement. The larger the bulb, the larger its tolerance for movements.
Tear web	Accommodates large movements. U-bulb ruptures and deforms without putting the material in tension during joint movement.

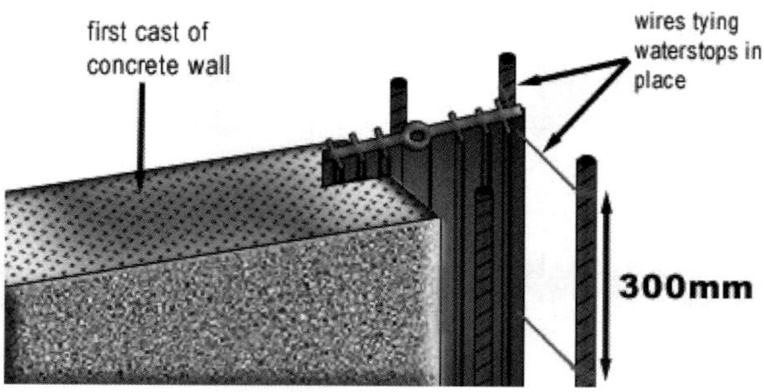

Fig. 4.35. Holding of waterstop flanges.

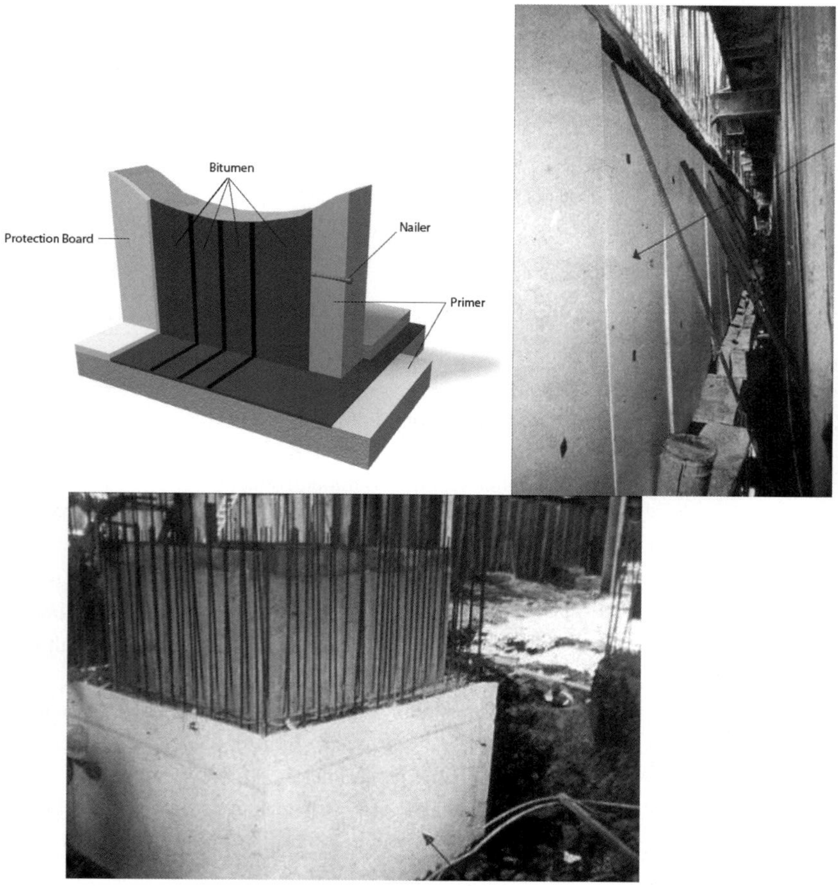

Fig. 4.36. Use of protection board.

- Protection of waterproofing membrane during backfill is very important so as to prevent any damages during construction. One common method is with the use of polystyrene foam as protective board (Fig. 4.36).

More examples can be found in the Maintainability Website (http://www.hpbc.bdg.nus.edu.sg/).

References

[1] P. C. Sandeford, "Specific problems of waterproofing, roof, basement, interior wet areas", *Waterproofing Structures in Singapore, Current Knowledge and Practices*, Singapore, 1995.

[2] M. Y. L. Chew, *Construction Technology for Tall Buildings*, 3rd Edition, World Scientific, 2009.

[3] M. Y. L. Chew and N. D. D. S. Egodage, "Benchmarks to minimise water leakages in basements", *Structural Survey*, **21**(4), 2003.

[4] Singapore Standard, *Waterproofing of Reinforced Concrete Buildings, CP 82*, Singapore Productivity Board, Singapore, 1999.

[5] I. H. Wong, "Experience with waterproofness of basements constructed of concrete diaphragm wall in Singapore", *Tunneling and Underground Space Technology*, **12**(4), 491–495, 1997.

[6] Building Research Establishment, "Sulphate and acid resistance of concrete in the ground", *Digest*, **363**, 8, 1991.

[7] C. W. Fetter, "Soil permeability", *Applied Hydrogeology*, 3rd Edition, Upper Saddle River, Prentice Hall, Inc, NJ, 1994.

[8] A. V. D. Beukel, "Cost of defects, how to diminish them", M. Moroni and P. Sartori (eds.), *Proc. Int. Symposium on "Dealing with Defects in Building"*, Varena, 1994, pp. 545–551.

[9] S. D. Kaplan, "How can we reduce the incidence of defects in buildings", M. Moroni and P. Sartori (eds.), *Proc. Int. Symposium on "Dealing with Defects in Building"*, Varena, 1994, pp. 811–819.

[10] Building Construction Authority and Singapore Concrete Institute, *Accreditation Scheme for Waterproofing Specialist Contractors*, unpublished, Singapore, 2004.

[11] F. Olubodun and T. Mole, "Evaluation of defect influencing factors in public housing in the UK", *Structural Survey*, **17**(3), 170–178, 1999.

[12] L. How Son and G. C. S. Yuen, *Building Maintenance Technology*, Macmillan, Basingstoke, Hants, 1993.

[13] G. G. Litvan, "Waterproofing of parking garage structures with sealers and membranes: the Canadian experience", *Construction and Building Materials*, **10**(1), 95–100, 1996.

[14] R. B. Rezaur, H. Rahardjo, E. C. Leong and T. T. Lee, "Hydrologic behavior of residual soil slopes in Singapore", *Journal of Hydrologic Engineering*, **8**(3), 133–144, 2002.

[15] J. Henshell, *Manual of Below-Grade Waterproofing Systems*, C. W. Griffin (ed.), John Wiley and Sons, Inc., New York, 1999.

[16] M. Y. L. Chew, *Maintainability of Buildings Under Tropical Conditions*, R-296-000-046-112/490, National University of Singapore, March 2004.

[17] SS CP 82:1999 — Waterproofing of reinforced concrete buildings.

[18] BS8102: *Code of Practice for Protection of Structures Against Water from the Ground*, BSI London, 1990.

[19] Moisture and Thermal Considerations in Basement Walls, http://irc.nrc-cnrc.gc.ca/cbd/cbd161e.html. Retrieved June 2009.

CHAPTER 5

ROOF

5.1. General

The roof forms the top part of the façade system to protect a building's interior. Its functions are thus similar to that of the façade. In addition, it has to be designed to take account of the following loads:

- dead load,
- imposed load,
- wind load,
- loads incidental to construction,
- access and roof traffic,
- deflection and differential movement, and
- lateral restraint and shear diaphragm.

There are essentially three main types of roof: (a) flat, (b) pitched and (c) shell. In general, a flat roof is chosen if the roof is to be used for storage, maintenance, etc. A pitched roof is chosen for simplicity in design and construction, for the relatively less maintenance and the good thermal insulation it provides. A shell roof is chosen mainly for aesthetic and architectural reasons. The various structural forms of roof are illustrated in Fig. 5.1 [1, 2].

5.2. Common Defects in Roof

Many of the major components of roof including the waterproofing and structural deck are beyond regular maintenance and repair. Major problems identified are: (1) water ponding and leakage; (2) sealant defect;

(i) Beams and slabs

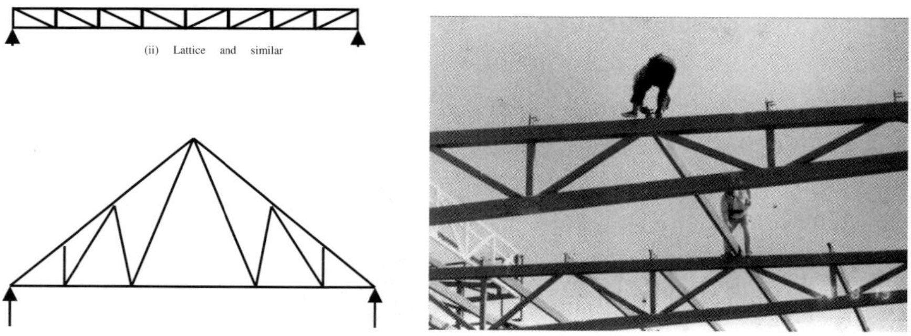

(ii) Truss and girder

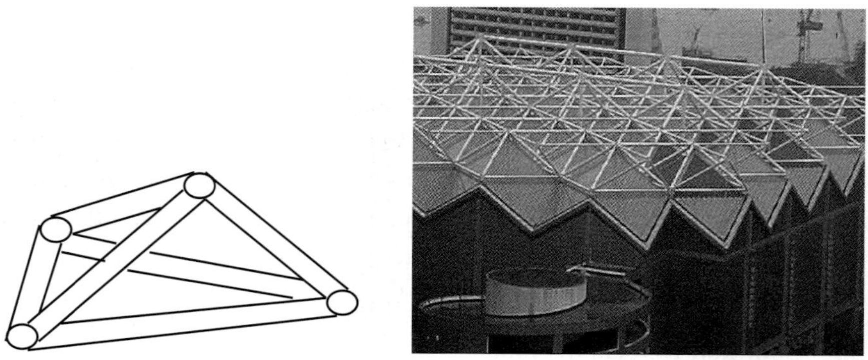

(iii) Spaced frames

Fig. 5.1. Various structural forms of roofs.

(iv) Shell roofs

(v) Folded plates

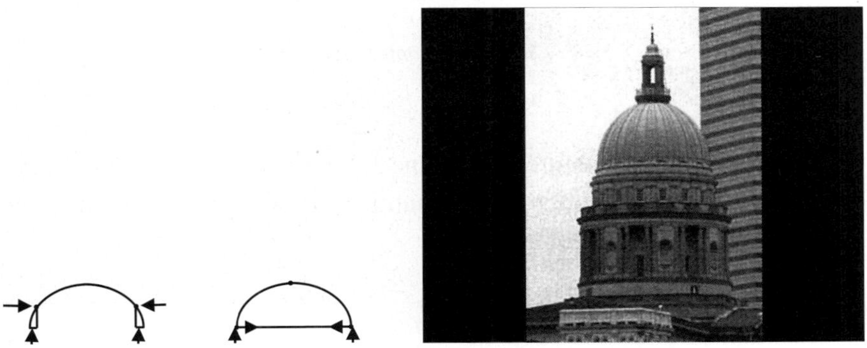

(vi) Arches

Fig. 5.1. (*Continued*)

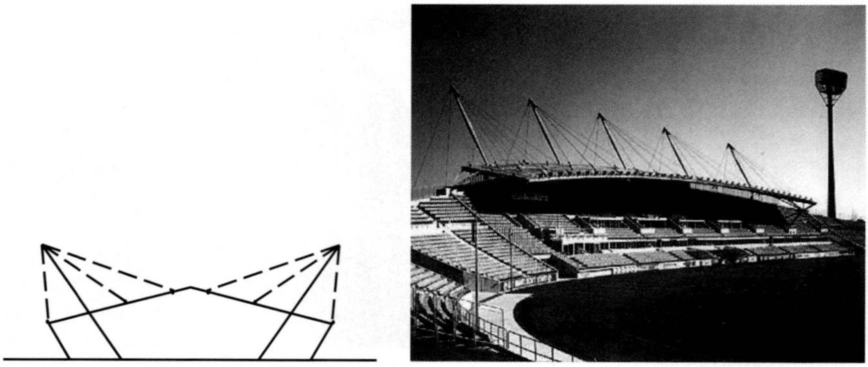

(vii) Cable-stayed (suspended) roofs

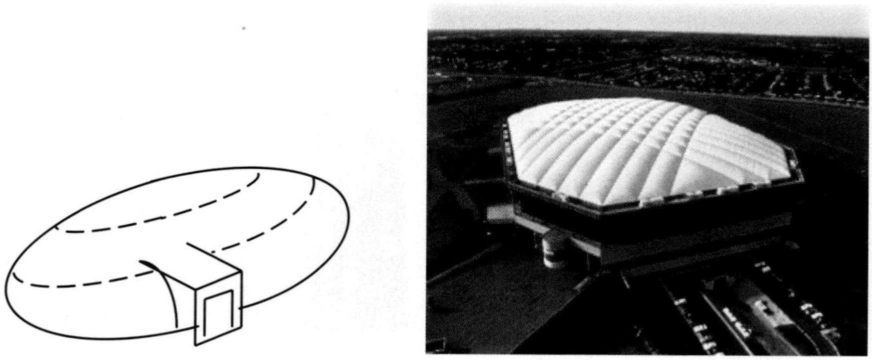

(viii) Inflated membrane structures

Fig. 5.1. (*Continued*)

(3) blistering; (4) premature membrane failure; and (5) cracking. Other defects are seepage through pipe penetration, microbial growth and tile popping (Fig. 5.2 and Table 5.1).

5.3. Roof Decking

RC flat roof is the most commonly used roofing system for tall buildings. It was classified into four basic types, namely, inverted, conventional,

Fig. 5.2. Common defects in flat roof.

Table 5.1. Common defects in flat roof.

		Cause gr.			
Defects and Coding	**Possible Causes of Defects**	**D***	**C***	**M***	**E***
Structural elements					
Cracks at roof drainage: On side/base of scupper drain (D1)	Poor workmanship in casting the deck along with precast drain.		▲		
	Plastic shrinkage due to wrong mix design and improper curing.	▲	▲		
	Adjacent structures act as thermal restraint during curing.		▲		
	Inadequate support to edges along scupper drain.	▲	▲		
Cracks at const. or expansion joint (D2)	Nonprovision/damaged metal capping exposes the sealant and concrete kerb to direct sunlight and UV radiation.	▲		▲	
	Wrong dimension (depth:width ratio) of joint.	▲	▲		
	Spacing of joints is too high.	▲	▲		
	Construction joint lacks bonding if not properly treated.		▲		
	Gap in joints is filled up during careless grouting.		▲		
	Water seepage through hardened and brittle sealant.			▲	▲
Leaking roof lead to paint peel, fungi, efflorescence at ceiling (D3)	Cracks in deck at expansion joints (D2) affect the integrity of waterproofing membrane directly attached to it.	▲	▲	▲	▲
	Omission of detailing in waterproofing.	▲	▲		
	Damaged waterproofing due to exposure — No protective layer.	▲			
	Careless maintenance of rooftop equipment, e.g. dropping of tools and erection of ladder. Co-ordination among disciplines is needed.	▲			
	Pipe penetrations are not pre-planned.	▲			

(Continued)

Table 5.1. *(Continued)*

Defects and Coding	Possible Causes of Defects	Cause gr. D*	C*	M*	E*
Cracks at base of parapet wall (D4)	Poor detailing such as coping, flashing, upturn and angle fillet.	▲			
	Omission of detailing in waterproofing.	▲	▲		
	Steep slope of surface near perimeter causes ponding at base.	▲	▲		
	Differential movement of roof slab and parapet. Poor structural design.	▲			
	Slab lifting due to inadequate thickness/excessive spacing of expansion joints with/without proper insulation.	▲	▲		
	Water logging at base of parapet wall is overlooked for a long time.			▲	
Algal growth on roof elements as ugly dark green patches (D5)	Dust/dirt deposition on surface, combined with sunlight and water helps algal growth.			▲	
	Ponding (D21) causes algal growth on horizontal surface.	▲	▲		
	Absence of coping and drip course allows rainwater to run against the parapet wall and algae grow.	▲			
	Negligence to remove early signs of algae.			▲	
	Unattended severe seepage at parapet base causes cracking and corrosion of deck and finally reaches the exterior. It causes stains.			▲	
Waterproofing: Sheet or LAM					
Blistering: Swelling of en-trapped moisture and air under sun: At between or below piles (D6)	No vapor barrier is used on deck for membrane or LAM of low water vapor transmitivity (WVT) or breathability.	▲	▲		
	LAM applied on substrate which is not properly cured and contains excess moisture.	▲	▲		
	Adhering air/rainwater on deck is not broomed out.		▲		
	No venting of waterproofing esp. hot-applied bitumen to remove excess moisture on or in the felts during construction.	▲			

(Continued)

Table 5.1. *(Continued)*

Defects and Coding	Possible Causes of Defects	D*	C*	M*	E*
	Water absorption on felts and drying shrinkage of wet felts.		▲		
	Poorly bonded roof assembly undergoes relative movement due to mechanical reasons (deflection, stress concentration).	▲			
Ridging: Long parallel un-dulation on roof — May affect runoff (D12)	Unskillful laying of membrane, e.g. inadequate stretching, too thin or nonuniform application of the hot asphalt layer.		▲		
	In the absence of vapor barrier, moisture from the building interior rises through the roof deck and condenses within the base felt. In the long run, the felts swell and buckle, forming ridges directly above the insulation joints.		▲		
	Can be avoided by using LAM.	▲			
Wrinkling: Ridge in irregular pattern (D13)	Membrane unable to handle excessive thermal movement. Can be avoided by using LAM.	▲			
	Poor adhesion of felt to deck.		▲		
Tearing of membrane: May result in leakage (D14)	Poor workmanship to achieve adequate DFT of LAM.		▲		
	Differential movement of membrane and surface due to use of partial bonding system or insufficient mopping.	▲	▲		
	Poor detailing or finishing of edge termination points are vulnerable to water flow. The edge is disturbed by such flow and once loosened, it allows water seepage.	▲	▲		
	Heavy traffic on unprotected membrane.		▲		
	Careless maintenance of rooftop equipment, e.g. dropping of tools, erection of ladder. Co-ordination among disciplines is needed.	▲			
	Shrinkage crack if membrane is installed at very high temperature.		▲		

(Continued)

Table 5.1. (*Continued*)

Defects and Coding	Possible Causes of Defects	D*	C*	M*	E*
		Cause gr.			
Surface erosion and chalking (D15)	Excessive foot traffic (esp. if nontrafficable roof) and weathering.	▲			▲
	Accelerated degradation from wrong application of coats.		▲		
Protective or surface elements					
Cracks in conc. panel (D16)	Excessive stresses built-up due to improper thermal joints. Separation layer and wire meshing are helpful.	▲	▲		
Tile popping: Severe separation and buckling of tiles from its bed (D17)	Use of closed joints or a thin layer of cementitious adhesive. The tile layer becomes very rigid against thermal movement.		▲		
	Tiles are not durable for large thermal cycles in the long run and show sign of delamination from substrate.	▲			
	Inadequately cured screed shows excessive shrinkage.		▲		
	Tiles lack full contact with the substrate due to localized absence of screed, inadequate tapping of tiles, contaminated surfaces with dust/dirt or no key at back.		▲		
Sealant					
Adhesion or cohesion failure of sealant: Loss of bond to the substrate, separations within the sealant (D18)	Wrong design of sealant joint (geometry, dimension, backer rod, etc.) where sealant is applied as a filler of movement joint.	▲	▲		
	Sealants become hard and brittle with time due to combined effects of solar UV rays, temperature, oxygen, ozone, moisture, air pollutants and microbial factors.			▲	▲
	Use of product which is not compatible or has crossed shelf life.	▲	▲		
	Improper batching and mixing of sealant components.		▲		

(*Continued*)

Table 5.1. *(Continued)*

Defects and Coding	Possible Causes of Defects	Cause gr. D*	C*	M*	E*
	Contaminants (dirt, grease and other foreign substances) on a poorly prepared substrate hinder adhesion with the sealant.		▲		
Biological growth on sealant (D19)	Water and dirt held in cracks of sealants facilitate germination of seeds and pores carried by birds or wind. If unattended, deep roots can damage the roof.			▲	▲
Bulging of sealant (D20)	Wrong joint geometry makes sealant sink if both panels contract with temperature drop. Sealant bulges out if both panels expand.	▲	▲		
	With time, sealants harden, lose flexibility and stay bulged at various spaces. Should be rectified.			▲	▲
Drainage system					
Water ponding at local depressions on roof surface (D21)	Excessive deflection either due to inadequately supported very long span or inability to support dead and live load.	▲			
	Uneven substrate, e.g. no screed on the top of precast slabs.		▲		
	Fall is nonuniform due to poor detailing or workmanship or post-construction work on finished roof.	▲	▲		
	Outlet not at lowest elevation or misaligned slope.	▲	▲		
	Installation of new equipment/elements without checking the safe load bearing capacity of deck.	▲			
	Bulging at poorly finished waterproofing at joints. Should be detected during tests.		▲		

(Continued)

Table 5.1. (*Continued*)

Defects and Coding	Possible Causes of Defects	Cause gr. D*	C*	M*	E*
Water ponding/leakage at outlets (D22)	Under-sized rainwater outlets cause build-up of excessive head of water, inadequate drainage and ponding.	▲			
	Insufficient number of outlets to cater for the flow rate. Effective run-off should be considered in calculation.	▲			
	Inadequate fall design of roof keeps water standing, esp. if the volume is not large, water flows slowly to the outlets.	▲			
	Leakage at poorly detailed/finished joint between outlet and RWDP.	▲	▲		
	Clogged outlet if no leaf cover provided, poor access for cleaning tool or irregular cleaning schedule.	▲		▲	
Lower level flooded with rainwater (D23)	The water is discharged directly at lower level without proper termination of roof drainage.	▲			
Services on roof					
Water ponding at the base of mechanical equipment (D24)	Base supports are mounted directly on roof without any concrete insulation pad to absorb the vibration from equipment. So deck deflection takes place.	▲			
	If machinery is bolted directly on finished roof, there is a small gap in between which retains water.	▲			
	Leakage from the installation such as tank and condenser.	▲			

(*Continued*)

Table 5.1. *(Continued)*

Defects and Coding	Possible Causes of Defects	Cause gr. D*	C*	M*	E*
Leakage at pipe penetration (D25)	Penetrations are not pre-planned and hacked at later stage.	▲			
	Too many pipes passing very closely are not grouped and it is difficult to seal efficiently each pipe separately.	▲			
	Poor design or workmanship for flashing at the penetration, e.g. the membrane upturn sleeve missing, <150 mm or merely finished off with a layer of cement and sand.	▲	▲		
	Poor jointing and alignments of rainwater pipes.	▲	▲		
Cracking at service support (D26)	The support is not strong enough for the imposed load.	▲			
	Concrete insulation pad is not provided to cater vibration.	▲			
	Temporary gondola system is used without adequate structural consideration for roof.	▲			
Stains under equipment (D27)	Rust dripping from corroded fittings made of poor quality or not coated with special antirust paint.	▲			

D* — Design. C* — Construction. M* — Maintenance. E* — Environment.

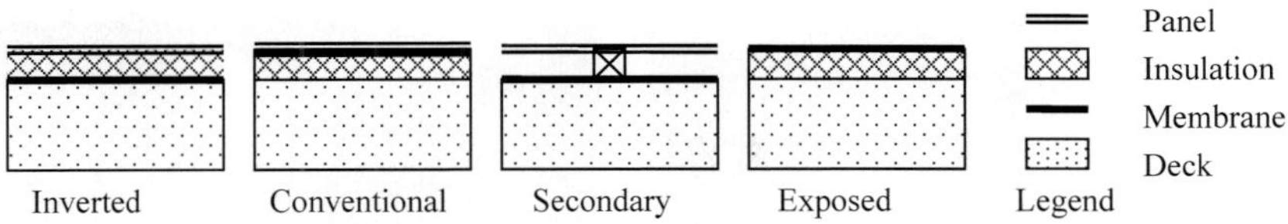

Fig. 5.3. Types of roofing system.

secondary and exposed as shown in Fig. 5.3. The regular components are: Deck, insulation, barrier, membrane and attachment system.

The construction of a flat roof is similar to that of a floor. It is made up of layers of materials to satisfy the thermal and waterproofing requirements. It is usually covered with a screed of lightweight concrete to provide the necessary fall. The screed can also serve as a protective layer to the waterproofing membrane and thermal insulation.

5.3.1. *Drainage*

The biggest problem with flat roofs in tall buildings is associated with water leakages due to drainage problems and deterioration of roofing materials such as the waterproofing membrane. Figure 5.4 shows an example of choking at drainage outlet by the accumulation of debris, leaves, etc. Figure 5.5 shows the "bubble" phenomenon on a waterproofing membrane, caused by pressure of hot air evaporating from entrapped moisture introduced during construction or migrated from elsewhere. Entrapped moisture before and after construction in concrete roofs can take several months to dry out properly.

In addition, joints in exposed sheeting materials should be arranged to facilitate the flow of water. Special attention should be paid to the design of flashings at gutters and outlets.

5.3.2. *Concrete Cracking*

Concrete is strong in compressive but weak in tensile strength. Cracking occurs when the tensile stresses exceed the tensile strength of the material. Tensile stresses on a concrete roof deck could be built up due to shrinkage, thermal and/or moisture contraction/expansion, restraint to

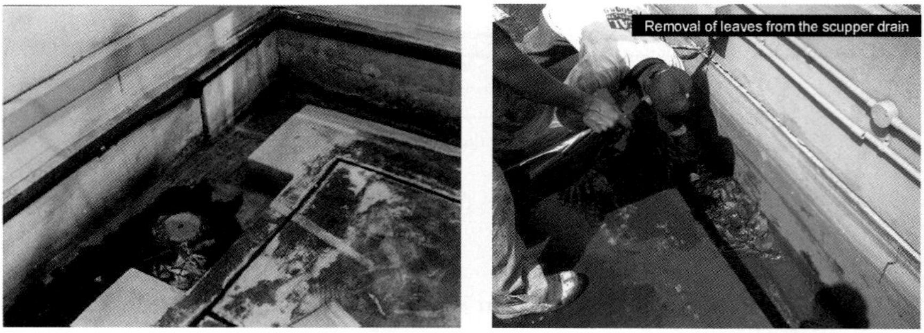

Fig. 5.4. Choking of drainage outlet.

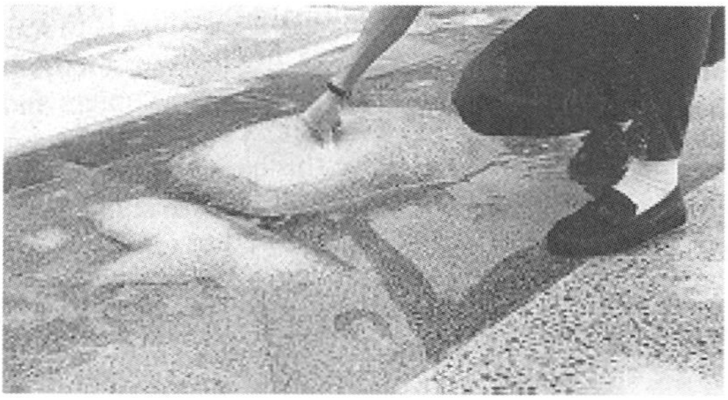

Fig. 5.5. "Bubble" phenomenon of roof membrane.

shortening/loading, differential settlement/movement, corrosion, alkali–silica reaction, etc. [3, 4].

Cracking on a roof deck is a concern as they could lead to water leakage and also provide easy access for the infiltration of aggressive substances to the level of reinforcing steel resulting in corrosion. Cracks often found on concrete roof decks are at stress concentrated areas such as openings, interfaces and joints. Figures 5.6 and 5.7 show examples of cracks developed near a drain and expansion joint, respectively.

5.3.3. *Ponding*

The accumulation of rainwater on a roof is often termed as ponding. It increases the risk of leakage and microbial growth such as algae on the roof deck. Figure 5.8 shows examples of ponding on roof decks.

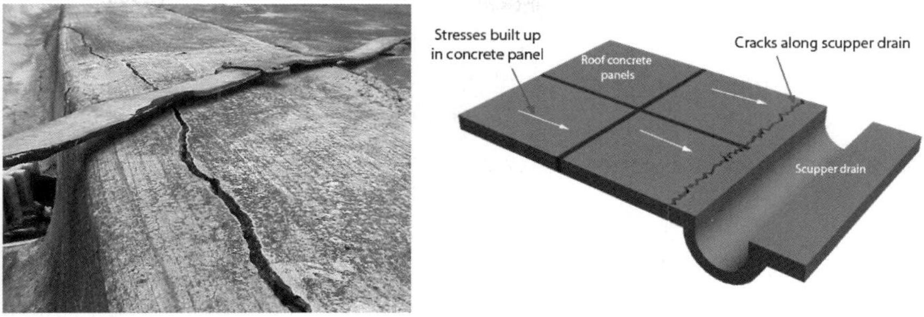

Fig. 5.6. Concrete cracking near a drain.

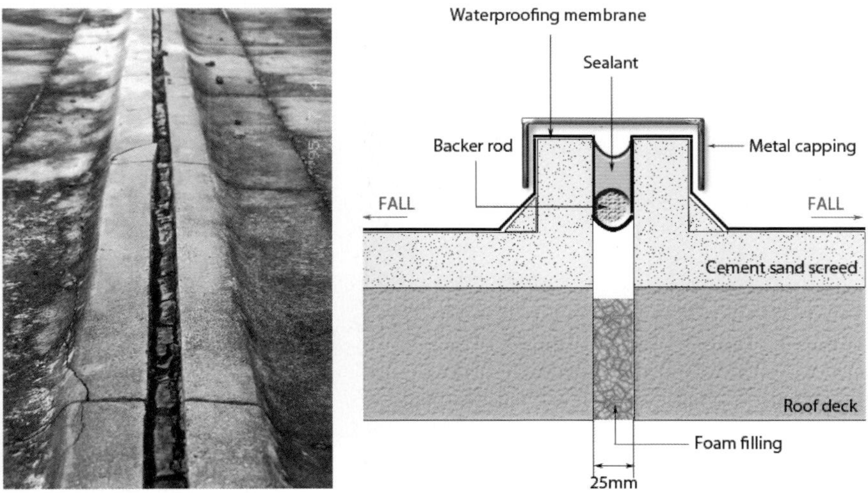

Fig. 5.7. Concrete cracking at an expansion joint.

Fig. 5.8. Ponding on a roof deck.

5.3.4. *Biological Growth*

Unintended plant growth on a roof deck could be initiated when the seeds or pores of plants are carried by wind or by birds/insects and deposited on spots conducive for growth. Figure 5.9 shows an example of an unintended plant growth on a sealant joint and a crack. Rain provides the necessary water needed for germination of seeds. After germination, the roots of seedlings create gaps in the roof structure thereby facilitating penetration of rainwater. If the roots penetrate deep enough, waterproofing membrane and insulation will be damaged.

Intended plant growth as in the case of a roof garden, needs due consideration to tackle problems such as root penetration (Fig. 5.10).

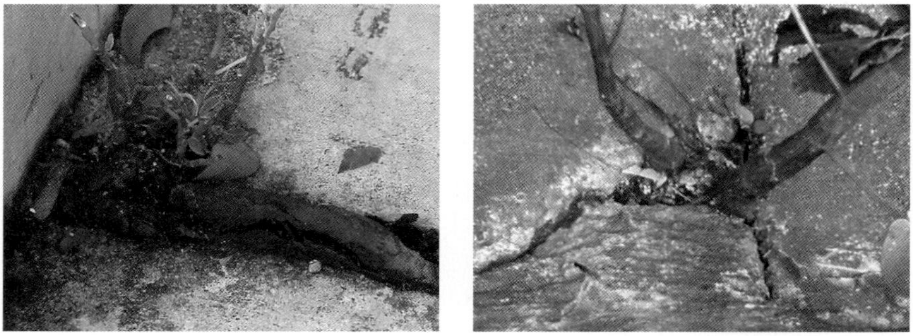

Fig. 5.9. Plant growth on a sealant joint.

Fig. 5.10. Plant root penetration into waterproofing membrane.

Fig. 5.11. Sealant deterioration, hardening and bulging.

5.3.5. *Sealant Failure*

Sealants used on a roof deck are subjected to aggressive agents including heat, water and UV radiation. The various combinations of these three factors in specific location at certain times of a day or a night during changing seasons would build different environments for sealants applied on a roof deck, making improvement of weathering resistance a very complicated task [5, 6].

Figure 5.11 shows an example of deterioration, hardening and bulging of a sealant. The inadequate sealant selected for this case hardened and deteriorated under the effect of exposure to heat and ultra-violet radiation from the sun or ozone. Without adequate joint width-to-depth ratio and sufficient cohesive and adhesive strength, the organic sealant may bulge out or sink when the concrete panels expand or contract due to thermal and moisture movements.

5.3.6. *Secondary Roof*

Figure 5.12 shows the construction details of a traditional Housing Development Board (HDB) roof system. The waterproofing layer comprises of a bituminous waterproofing treatment over the roof. It is made up of six coats of material, one layer of fiberglass and one coat of bituminous aluminum paint. The system also includes a secondary roof which consists of lightweight precast concrete panels (1,500 × 600 × 50) placed on hollow concrete stools (150 × 150 × 225) laid 225 mm above the main roof structure to serve as a heat insulation barrier (Fig. 5.13). The amount of heat and the speed with which it penetrates into the roof slab is greatly

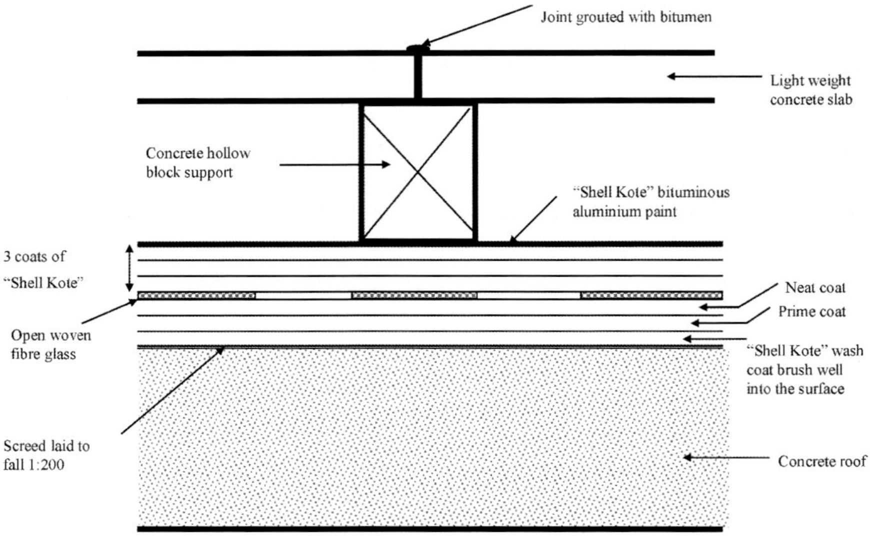

Fig. 5.12. Sectional detail of waterproof treatment over a typical HDB flat roof.

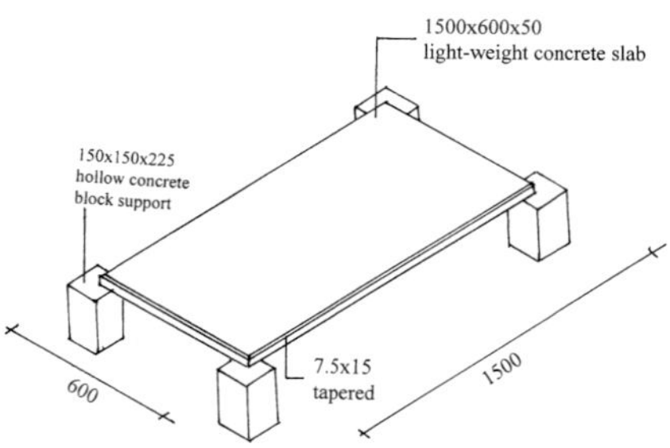

Fig. 5.13. Lightweight precast slabs supported by hollow concrete blocks for thermal insulation.

reduced by the panel as well as the air gap. The heat penetration through joints in the precast secondary roof slabs are sealed with an elastic form of hot bitumen which is leveled and cleaned and bent to the shape of a v-groove 12 mm thick. The system works as long as the concrete stools and slabs remain intact. Deterioration (breaking off) of the concrete stools have been reported in many cases and if the debris (mainly sand) is not cleaned off, serious drainage blockage may occur.

5.4. Roof Waterproofing

This section discusses the common problems associated with roof waterproofing. Figures 5.14–5.16 show respectively the priming, application of rolled waterproofing membrane and ponding (to test for water-tightness) of a typical flat roof deck.

Fig. 5.14. Priming of the deck to receive the waterproofing membrane.

Fig. 5.15. Application of the rolled waterproofing membrane.

Fig. 5.16. Ponding or water test on the deck.

5.4.1. *Blistering*

Blistering occurs when pockets of trapped air and moisture which expand under the sun and displaces the membrane to form a small blister. Air and moisture passing through dense substrate materials at slow rate of flow allow blisters to form. Under high surrounding temperatures, the vapor pressure rises too quickly for the trapped air and moisture to escape through the dense deck. Pressures will then develop, causing a displacement and stretching of the waterproofing membrane when air pocket increases in size (Figs. 5.17 and 5.18).

Where blistering is caused by water vapor and is confined, vapor vents may be installed to release the water vapor pressure underneath the roof coverings (Fig. 5.19). However, these will have a limited effect where the first layer is bonded to the roof deck.

5.4.2. *Alligatoring of Membrane*

Alligatoring is the cracking of the surfacing bitumen on a built-up roof, producing a pattern of cracks similar to an alligator's hide; the cracks may or may not extend through the surfacing bitumen (Fig. 5.20). It is

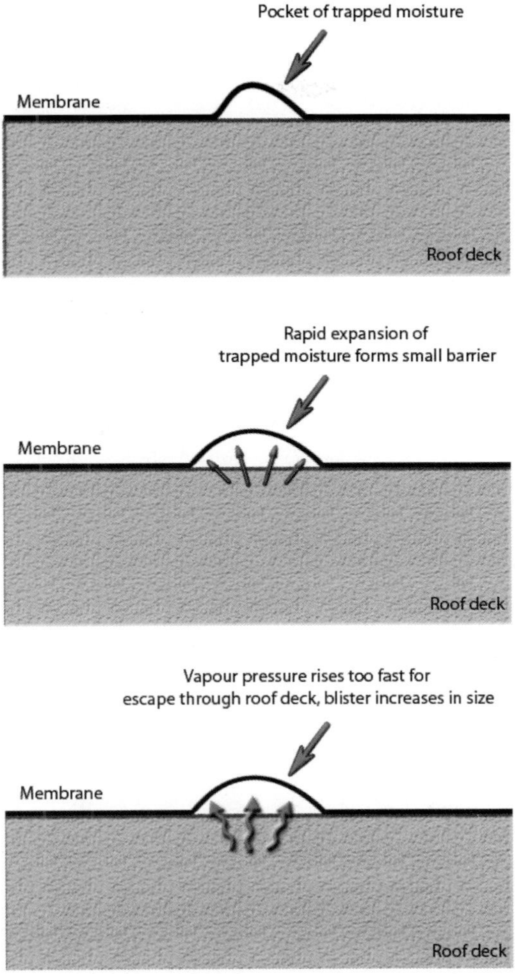

Fig. 5.17. Formation of blisters.

sometimes also referred to as crocodilling, a drastic type of crazing producing a pattern resembling the skin of a crocodile.

This defect is mainly caused by continued oxidation, aging and embrittling of the membrane which are resulted due to long exposures of membrane to the sun's high-energy ultraviolet radiation and cyclical moisture and thermal movements.

Fig. 5.18. Blistering.

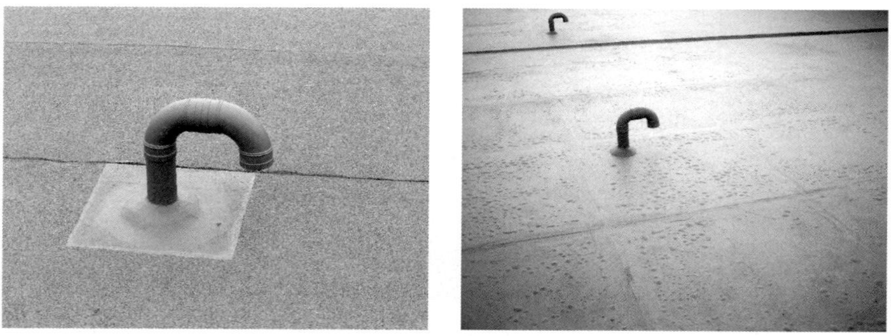

Fig. 5.19. Vapor vents to release water vapor pressure.

5.4.3. *Delamination of Membrane*

Delamination is the separation of the membrane or laminated layers of insulation from the roof deck. Water if penetrated can retain underneath the membrane (Fig. 5.21).

Common causes for delamination include the use of inadequate materials as the finishing membrane, improper surface of preparation or incompatibility

Fig. 5.20. An overall view of the alligatoring of membrane.

Fig. 5.21. Delamination of waterproofing membrane.

of components. When liquid-applied membrane is used, it is recommended to apply three coats with overall thickness of 1.2 mm minimum.

The weakness is always along the joints and hence adequate lapping is crucial. The edge termination points of waterproofing membrane upturn is also vulnerable (Fig. 5.22).

5.4.4. *Ridges of Membrane*

Ridges may appear in parallel lines of long undulation felts along the roof surface (Fig. 5.23). They may affect rainwater runoff and result in water ponding.

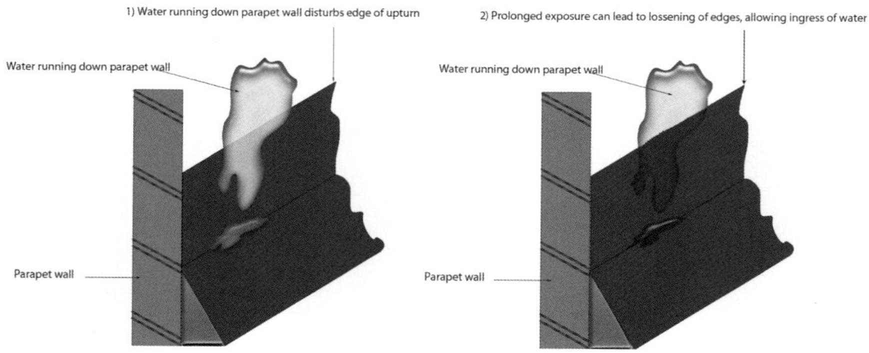

Fig. 5.22. Ingress of water through edges.

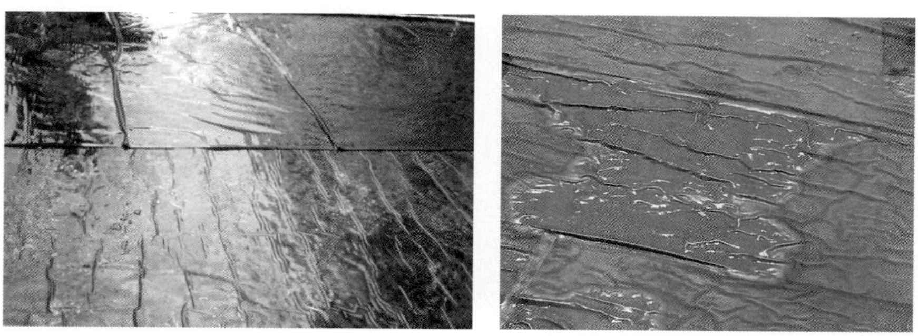

Fig. 5.23. Formation of ridges.

Possible causes of the formation of ridges include:

- inadequate stretching of the membrane during application,
- uneven application of the base coat and components underneath and
- water vapor from the roof deck condenses underneath the base felt, the felt swells and buckles, forming ridges directly above the insulation joints (Fig. 5.24).

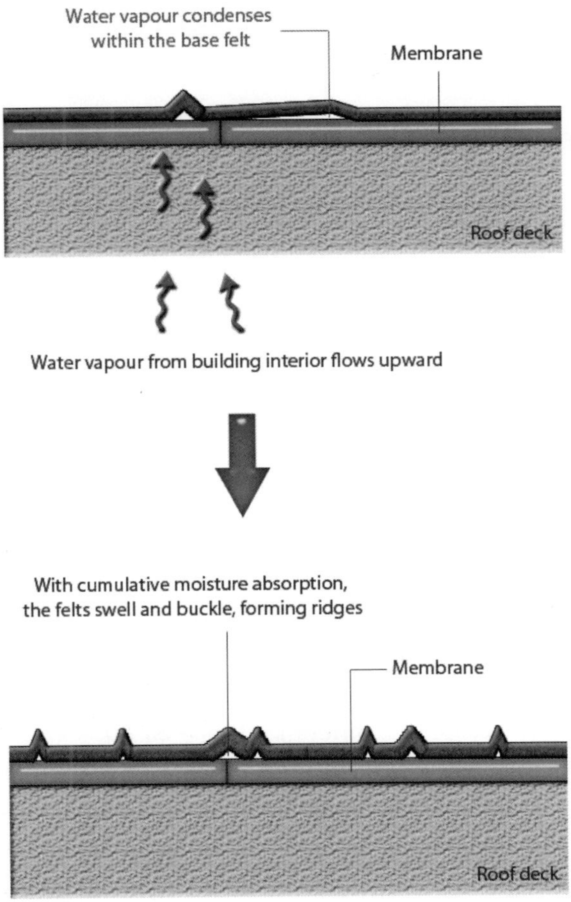

Fig. 5.24. Ridge formation due to water vapor from the roof deck.

Fig. 5.25. Patching of waterproofing membrane.

This failure is generally due to insufficient stretching of the membrane. Patching (Fig. 5.25) may be considered if the problem is not extensive. The long-term remedial treatment is to strip off the membrane and to relay the deck to prevent flexing.

5.4.5. *Water Ponding*

Ponding is the retention of water caused by the lack of an adequate gradient to fall or unevenness such as a deflection of the roof's underlying structural components (Fig. 5.26). Ponding water on a roof may deteriorate the waterproofing system resulting in water leakage, as well as algae and fungi growth.

The generally accepted minimum fall is 1 in 80, some codes recommend 1 in 40 to make allowance for any inaccuracies on the site and possible deflection of the roof structure. Ponding occurs only when there is unevenness of the slab surfaces. Regular inspection should be carried out especially after a downpour. This allows for visual inspection of possible ponding area.

When ponding happens, immediate repair is needed. Apply a layer of screed so as to level the surface and take note of the fall, directing the flow of the water to the drainage area.

Fig. 5.26. Water ponding on a roof.

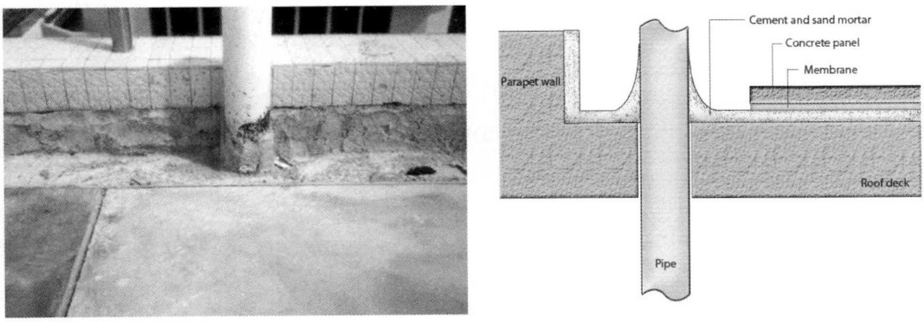

Fig. 5.27. Pipe penetration with improper design.

5.4.6. *Service Penetration Through Membrane*

Service penetration breaks the continuity of a waterproofing are the vulnerable to water seepage. Figure 5.27 shows an example of pipe penetration with improper design resulting in water seepage.

The penetration may be provided with an angle fillet around the pipe for the water proofing membrane to bend upwards smoothly. The waterproofing membrane shall finish at 300 mm above the finished roof level. At the skirting end of the cement and sand mortar, sealant shall be provided all around the pipe. The sealant helps to prevent water from seeping

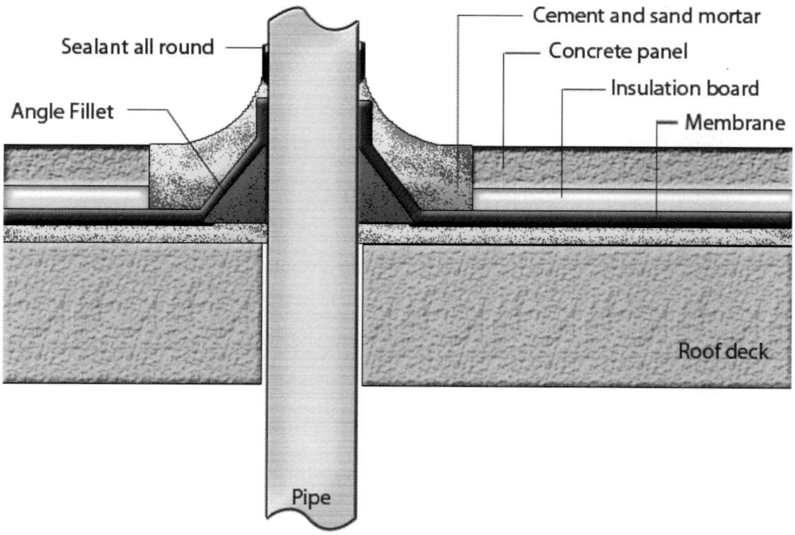

Fig. 5.28. Waterproofing at a pipe penetration. Figure showing a good design at pipe protrusion at roof.

in during any differential movement among the elements. The minimum distance between the sealant and the waterproofing membrane should be 25 mm (Fig. 5.28).

5.4.7. *Adhesion Failure of Membrane*

Figure 5.29 shows an example of adhesion failure at the end lap of preformed membrane sheets. The openings allow rainwater to seep in.

For preformed membrane, all areas laid should be checked for air trapped between it and the concrete surface or between successive layers of the membrane. This is to ensure there is no gap in between the two layers which would allow water to accumulate.

End laps of the preformed membrane should be at least 75 mm and the side laps should be 50 mm unless otherwise stated by the manufacturer. End laps between each roll should be staggered (Fig. 5.30).

The membrane should be applied to cover the whole roof and have an upturn of at least 300 mm and tucked into parapet wall (Fig. 5.31).

Fig. 5.29. Adhesion failure of preformed waterproofing membrane.

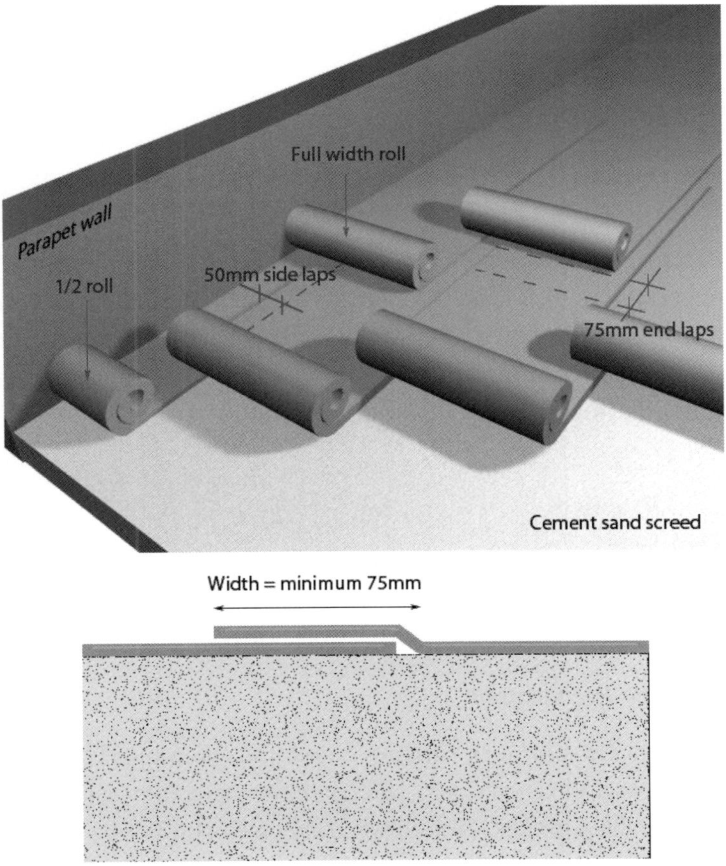

Fig. 5.30. Importance of membrane end and side lap.

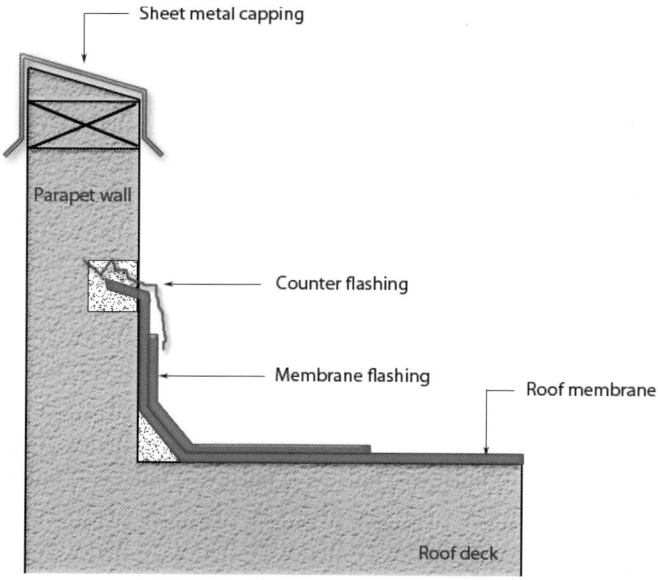

Fig. 5.31. Upturn of membrane and tuck into parapet wall.

The defect as shown in Fig. 5.29 may be rectified by scraping off the loose material granules (Fig. 5.32) on the surface of the bottom layer and reapply with more bitumen to the end lap. This is done using the torch-on method as shown in Fig. 5.33. By doing so, the opening is sealed up, preventing the flow of rainwater from getting in.

5.5. Good Practices

Some good practices for the construction of roof deck are discussed here.

5.5.1. *Surface Preparation*

Preparation of the roof deck surfaces cannot be taken lightly as without a sound, even and clean surfaces, the membrane cannot achieve a good bonding with the surfaces. In addition, rainwater will not be able to drain off properly due to the uneven surfaces although there is sufficient design of falls and drainage system.

Fig. 5.32. The loose material granules on the surface of the preformed membrane must be scraped off before lapping.

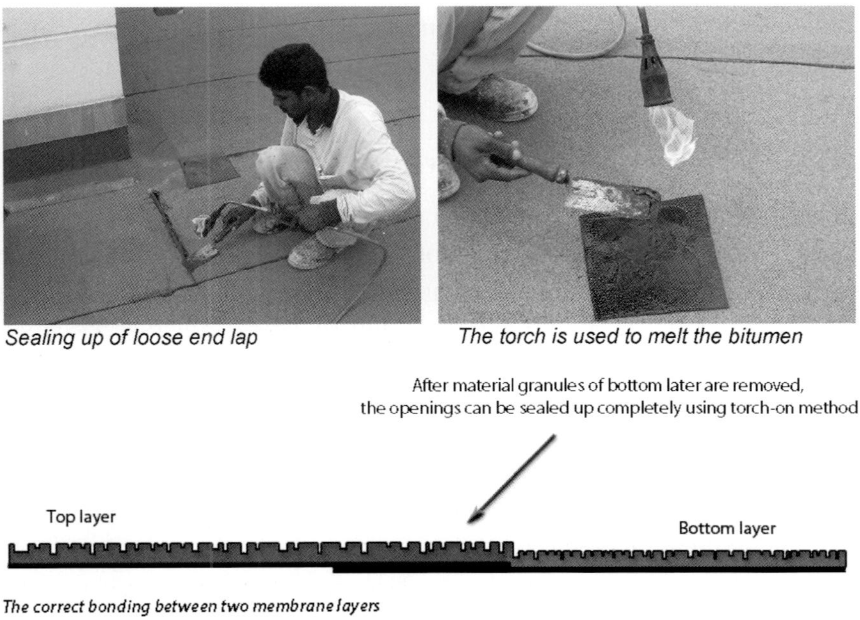

Fig. 5.33. Rectification of adhesion failure.

Fig. 5.34. Removal of dir, debris and oil.

The surface preparation starts with the removal of dirt, debris and oil. Any protrusion or substances that may damage the membrane are removed. If the roof surfaces are uneven, a layer of 25–50 mm cement sand screed can be installed to make it smooth (Fig. 5.34).

5.5.2. *Application of Preformed Membrane*

During the application of preformed membrane using torch on method, it is important that full and continuous bonding is achieved. There are namely two methods of membrane application. The following steps serve as a guide.

(a) The torch is placed near the underside of the membrane. The bitumen is heated until the molten coating runs down to form a continuous flow of hot coating in front of the roll (Fig. 5.35).
(b) The membrane is then rolled forward into the flow of bitumen whilst the torching operation continues.
(c) The flow of bitumen is maintained in front of the roll. This allows the membrane to achieve continuous bonding and the flow of bitumen ensures air is not trapped under the membrane.

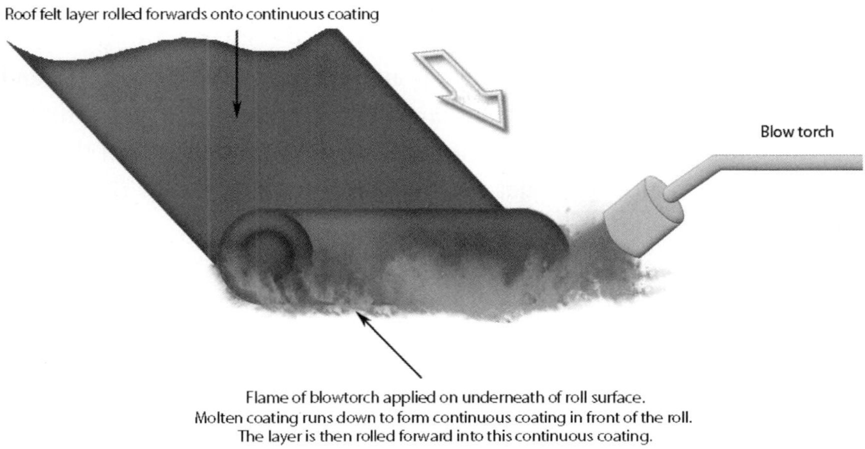

Fig. 5.35. Illustration of membrane application process.

Fig. 5.36. Sweeping loose material granules on to the bead.

When correctly applied, a small bead of bitumen should appear at the sides of the laps which indicate that the membrane is well-bonded. The bead can be concealed by sweeping loose material granules onto the bead while it is still hot (Fig. 5.36). The membrane is rolled with a minimum side lap and end lap of 50 and 75 mm, respectively.

5.5.3. *Lapping*

Preformed waterproofing membrane is applied with minimum 75 mm end laps and 50 mm side laps. The laps are staggered between layers, with the lapping joints of the second layer away from the lapping joints of the first layer (Fig. 5.37), making water seepage if any more difficult.

5.5.4. *Application of Liquid-Applied Membrane*

Figure 5.38 shows coats of fluid bitumen being applied on a roof surface with the use of brushes or rollers.

(1) The coats are applied with a minimum thickness of 1.2 mm. Two to three coats are normally applied. Two types of bonds (Fig. 5.39) for the different layers may be used.

- *Full bond*: This method involves fully bonding the felt to the roof deck and subsequent layers of felt to one another by the pour and roll technique. This method ensures against any wind uplift and forms the bond of succeeding layers creating a water-tight joint. It is used when strong wind uplift is expected.

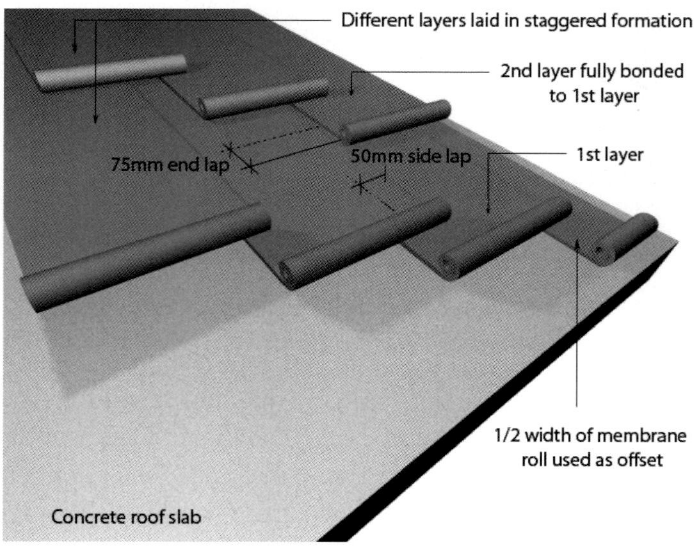

Fig. 5.37. Different layers laid in staggered formation.

Fig. 5.38. Workers applying the first layer.

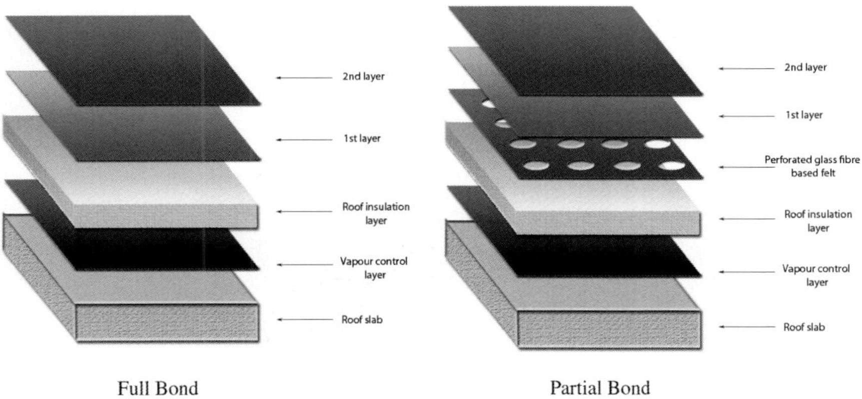

Fig. 5.39. Types of bond for liquid-applied membrane.

- *Partial bond*: An additional perforated glass fiber-based felt is used beneath the first layer. The perforated first layer is laid loose over the roof deck. The first layer of bitumen felt is then placed on top of the perforated glass fiber felt. They are bonded together by the use of hot liquid bitumen.

The hot bitumen between the bitumen felt and glass fiber-based felt is forced through perforations in the glass fiber-based felt and adheres the bitumen felt layer directly to the structural deck. A partial bond between the bitumen felt and roof deck is thus formed. Allowance of thermal movement within membrane prevents problems like delamination and cracks.

(2) After application of the first coat, the second coat has to be applied in perpendicular direction to the first coat.

(3) The second coat must be applied while the first coat is still tacky but sufficiently set so that it is not drawn off while applying the second coat.

(4) A membrane up-turn of 300 mm must be done whenever there are areas that need to be up-turned, e.g. parapet wall and pipes protrusion (Figs. 5.31 and 5.40).

(5) The finished membrane is left to cure for a minimum of 24 h prior to water ponding test.

(6) A sample board may be used as a record of the finished thickness of the applied-liquid membrane. This allows a cross sectional view of the membrane thickness (Fig. 5.41).

Fig. 5.40. 300 mm height upturn at parapet wall.

Fig. 5.41. Worker applying third coat onto the sample board.

5.6. Membrane Roof

An example that surpasses the traditional roof materials is the invention of fabric membrane. Typically, fabric membrane consists of non-natural woven materials (like fiberglass, coal fibers and polyester), resin coating (such as Poly Vinyl Chloride (PVC), Teflon and Silicon) and is "anisotropic"; that is, the orientation of the treads is weaved in two directions [7]. The membrane offers a certain level strength, permeability, resistibility (solar, dust, etc.) and is available in different colors and opacities for aesthetic purposes. Its lightweight allows for faster erection.

Over the years, a handful of structures have been integrated with fabric membranes worldwide. From temporary to permanent, small to large scale, these fabric membranes are commonly adopted in cladding and roofing systems of event structures (i.e. hanger, building atria, pavilion, amphitheatre, stadium and airport). At a broader aspect, fabric membranes have been employed for the entire building envelope including the Millennium Dome, United Kingdom and Oita World Cup Stadium, Japan.

Fig. 5.42. The use of PVC coated with PVDF (polyvinylidene fluoride) as roof membrane in Choa Chu Kang (CCK) Swimming Complex, Singapore.

The three most popular synthetic membranes used for roof structures are PVC coated polyester, PTFE (polytetrafluorethylene) coated glass fabric and ETFE (ethylenetetrafluorethylene) foil with their lifespan estimated to be that of 15–50 years [8]. Figures 5.42 and 5.43 show the use of PVC and ETFE as roof structures.

It has been reported that, with time, the PVC/PVDF membranes appear to be brittle and are far from its original condition. This might be due to the effect of UV as well as long exposure of the material to inclement weather. This can be illustrated in Fig. 5.44 where severe tearing on the membrane have resulted the peeling off of PVDF coating from the fabric. In addition, it can be seen that the steel structures are stained with dirt and algae. Figure 5.45 illustrated the apparent discoloration of the PVC/PVDF membrane, hence giving viewers an unpleasant sight.

In comparison, ETFE foil appears to perform better especially with its self-cleaning effect. With ETFE foil's low impermeability, minimum grease and dirt will be attracted to the foil's surface. Moreover the smooth surface allows the membrane to have a self-cleaning effect by itself during heavy rainfall, washing away the dirt accumulated. However, PTFE fabric membrane promotes algae growth through long exposure of wet and dry weather. One of the reasons may be because of its less even surface as compared to ETFE foil. These problems of algae staining and dirt accumulated can be rectified through regular cleaning and maintenance.

The primary structure of the "Angles" consists of 18 steel frames supporting a total area of 10,630 m^2 ETFE foil membranes. These membranes are separated into five pieces, forming a "cross" shaped at two different heights. The largest piece of membrane which is circular in shaped made up the intersection of the "cross" with four smaller pieces by the sides. Each piece of ETFE membrane consists of two parts namely (i) the pneumatic cushion membrane that is transparent at the top and bottom sandwiching a pattern printed layer of fritting, and (ii) the single layer of transparent wedge-shaped membrane along the perimeter. Each piece of pneumatic cushion membranes is tightly supported by either three or four frames with the help of 12 tension rods attached to the top and bottom of each frame. As for the curvy membrane, they are held together with pre-stressing strips in-between them with edge cables at the verge. These edge cables are then pulled through during installation.

Fig. 5.43. The use of ETFE pneumatic foil membrane for the main canopy roof in Clarke Quay, Singapore.

5.7. Green Roof

A green roof or roof garden is a roof that is partially or completely covered with vegetation and/or plants. A different waterproofing system over the roof deck is required. Problems in green roofs are unique and are most often associated to waterproofing, root penetration and drainage [9–11].

Although waterproofing materials are of higher quality today, due to better materials and manufacturing processes, defective waterproofing due to poor workmanship is still present. In addition, the suitability of a material cannot be easily verified. Building owners and professionals in general feel that roof garden may damage the waterproofing of the roof deck [12].

Fig. 5.44. Peeling, algae and dirt staining on typical structural membrane and steel frame.

Fig. 5.45. Discoloration of PVC/PVDF coated membrane.

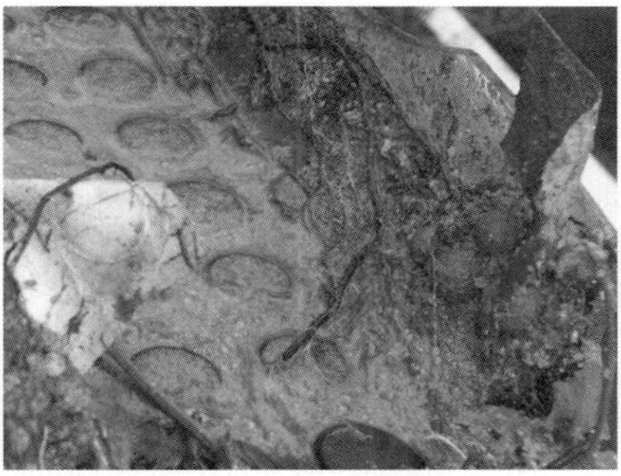

Root grown into the liquid applied waterproofing at the edge.

Roots anchored on the membrane.

Fig. 5.46. Root penetration into waterproofing membrane.

The damage ability of roots cannot be under-estimated. Tree roots penetrated concrete dam, illustrating the extent of damage which roots can cause [13]. Although tree roots are much larger than roots of plants used on green roofs, the possibility of root damage to green roof waterproofing cannot be ignored (Fig. 5.46).

References

[1] A. J. Elder and M. Vandenberg, *AJ Handbook of Building Enclosure*, The Architectural Press, London, 1974.

[2] M. Y. L. Chew, *Construction Technology for Tall Buildings*, World Scientific, 2009.

[3] A. J. Dutt, S. K. Roy and M. Y. L. Chew "Effects of wind flow on freshly poured concrete", *Journal of Wind Engineering and Industrial Aerodynamics*, **44**(1–3), 2629–2630, October 1992.

[4] S. K. Roy and M. Y. L. Chew, "Some common factors causing concrete cracking", *RILIM — IMEKO International Conference on Diagnosis of Concrete Structures*, Bratislava, Czechoslovakia, 2–6 September 1991.

[5] M. Y. L. Chew, "Curing characteristics and elastic recovery of sealants", *Building and Environment*, **36**(8), 925–931, October 2001.

[6] M. Y. L. Chew, X. Zhou and S. H. Goh, "The effects of some chemical components of polyurethane sealants on their resistance against hot-water", *Building and Environment*, **38**, 1381–1384, 2003.

[7] R. Bradshaw, D. Campbell, M. Gargari, A. Mirmiran and P. Tripeny, "Special structure: Past, present and future", *Journal of Structural Engineering*, **128**, 691–709, 2002.

[8] J. Pudenz, "Material and workmanship", K.-M. Koch (ed.), *Membrane Structures: Innovation Building with Film and Fabric*, Prestel Publishing Ltd, New York, 2004, pp. 46–65.

[9] Chemische Fabrik Grünau GmbH, *WOLFIN IB Handbook for Specifiers* (Clark H. Translated), unpublished book.

[10] K. S. Barth, *Green Roofs: Stormwater Management from the Top Down*, 2004, http://www.greenroofs.com/pdfs/archives-katrin.pdf. Retrieved June 2007.

[11] Metropolitan Area Planning Council. Green Roofs, 2005, http://www.mapc.org/regional_planning/lid/PDFs/greenroof.pdf. Retrieved June 2007.

[12] N. H. Wong, S. E. Lee, D. Cheong, G. T. Lim, B. Yuen, K.W. Tham, C. Sekhar, Y. Chen and H. Yan, *Study of Rooftop Garden in Singapore*, National University of Singapore, Singapore, 2002.

[13] B. D. Marks and B. A. Tschantz, *A Technical Manual on the Effects of Tree and Woody Vegetation Root Penetrations on the Safety of Earthen Dams*, Marks Enterprises of NC, PLLC, North Carolina, 2002.

CHAPTER 6
M&E

6.1. General

Mechanical and electrical (M&E) services in this chapter cover subjects related to:

(a) Sanitary-plumbing system — Sec. 6.2
(b) HVAC system — Sec. 6.3
(c) Mechanized transportation system — Sec. 6.4
(d) Electrical system — Sec. 6.5
(e) Fire protection system — Sec. 6.6

6.2. Sanitary-Plumbing System

6.2.1. *General*

Sanitary-plumbing system forms one of the most important building services in residential buildings. The breakdown of water service, sewerage and other sanitary-plumbing features is disruptive not only due to the inconvenience caused by the downtime, but more importantly, the possible safety and health concerns. Reported accidents include consumption of contaminated water from faculty water service system by users and poisonous gases inhalation during cleaning of sewers by cleaners.

Without proper design considerations and planning executed during the design stage, sanitary-plumbing systems will not be maintainable and would not operate efficiently. The results of ignoring preventive maintenance aspects are shorter service life, premature replacement of components and increased operational costs due to problems such as leakage of piping etc.

6.2.2. *Maintainability of Sanitary-Plumbing System*

For sanitary-plumbing systems, design changes are very costly and tedious to implement during the operation stage. Considering maintainability during the design stage ensures:

- longer lifespan of systems and components,
- lower operating costs by performing at high efficiency,
- lower unscheduled downtime, by preventing failures, and
- lower scheduled downtime, by decreasing the time required to perform a particular maintenance task.

Ignoring equipment availability, maintenance requirements and not taking the design of access for maintenance equipments into consideration are the main factors contributing to faulty designs in M&E services [1].

Inspections and adjustments tasks should be performed easily and comfortably if specific provisions for carrying out visual inspections and maintenance actions are incorporated into the original design. It is easier to lubricate assemblies and components quickly and effectively if such systems are readily available [2].

If standardized parts and assemblies are recommended and implemented at the design stage, repair time can be reduced to the minimum at operation stage which comprises the highest proportion of life cycle cost (LCC). Systems with simple and effective design will assist in reducing redundancy and in turn reducing failure rates as the systems are being customized to fit the surrounding environment and the cause of system failure is much easier and faster to understand, ascertain and detect.

The current design for sanitary-plumbing system does not indicate provisions for "green systems." There is no specific provision like LEED (USA) or BREEAM (UK) that take effective sustainable sanitary-plumbing system design into consideration. Such designs have proven to be effective in saving resources (electricity and water).

6.2.3. *Common Defects*

Apart from wasting precious energy and water, sanitary-plumbing system can become a source of fatal contamination unless it is designed, constructed and maintained properly. Visible cleanliness of water alone cannot be considered satisfactory. Particularly in densely populated cities, waterborne diseases may spread quickly unless a high standard of public health is maintained [3]. Today's design professionals emphasize notably on occupant's health [4].

Many codes of practice, standards and guides of design and operation handbooks [3–7], along with detailed consideration for certain major issues such as piping [8], valves [9], pumps [10], LLC [11], etc. are available. Major problems all round the world such as leakage resulting in water scarcity, sewage contamination of potable water causing sickness [12] and recurrent cases of Legionella [13] are reported frequently. The root cause of many of these problems is derived from corrosion and wearing of metallic parts which are common in pipelines, particularly in the consumer's plumbing fittings. Corrosion favors bacterial growth, developing high concentration of lead and copper components within the system and hence leading to dirty and contaminated water [14]. Researchers have drawn attention to various problems related to sanitary-plumbing system, such as lead contamination [15], biofilm formation [16], etc.

In a survey on residential buildings, 56 significant problems were identified [17]. Defects identified in this study are summarized in Fig. 6.1 and Table 6.1.

6.3. HVAC System

6.3.1. *General*

Central chilled water HVAC systems form one of the most important building services in large commercial buildings. Maintenance costs represent a significant expenditure over the life of a system, often argued to be more significant than initial upfront capital cost in total economic lifespan of an HVAC system. However, there are limited available data on HVAC maintenance costs for benchmarking these costs. Even the efforts

Supply piping

Damaged outdoor pipe

Corrosion of pipe

Scaling in pipe

Broken pipe brace

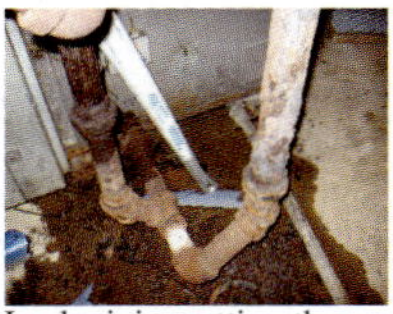
Leaky join wetting the floor

Corroded and broken valve

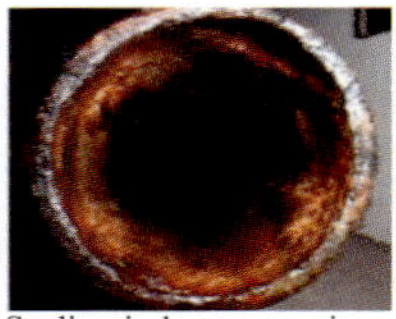
Scaling in hot water pipe

Corrosion and pitting in hot water pipe

Storage tank

Corroded tank body

Leakage

Overflow

Damaged strainer

Fixtures

Clogged floor trap

Water ponding

Slow discharge

Damaged flush

Pumps

Corrosion and scaling

No vibration isolator

Worn rubber coupling

Sewage disposal

Broken manhole cover

Corroded iron pipe

Fig. 6.1. Common defects in sanitary-plumbing system.

Table 6.1. Common defects in sanitary-plumbing system.

Defects and Coding	Possible Causes of Defects	Cause gr. D*	C*	M*	E*
Supply pipes in general					
Physically damaged pipe: esp. UG pipe (E1)	Pipe type and joint not suitable for underground condition and suffer from uneven settlement.	▲			
	Bursting of pipe under water pressure.	▲	▲		
	Inadequate longitudinal support to cater traffic vibration and load.	▲			
	Careless backfilling: Poor compaction, inadequate depth (<500 mm), use of heavy/pointed backfill.	▲	▲		
	Damaged by work on structure/other services that are too close.	▲			
Degraded pipe: Rusting, pitting, etc. (E2)	Material corrodes in contact with surrounding/supports.	▲			
	Protective coating damaged during careless installation or integrity not tested.		▲		
	Wearing of external protective coat of metal pipe exposes it to electrolytic and atmospheric corrosion.			▲	
	Aging, hardening of PVC pipe under prolonged exposure to sun.	▲			▲
	Thermal stress from close proximity of hot water pipes if very limited freedom for thermal movement is provided.	▲			
Contamination: Users fall sick (E3)	Interference of potable and nonpotable water due to poor planning/workmanship, e.g. sewers placed above supply lines.	▲	▲		
	Wrong position of draining tap allows drained water in certain parts of the building, interferes with the supply to other parts of the building.	▲			
	Unsuitable material for pipes/joints (e.g. lead found in very old building).	▲			

(Continued)

Table 6.1. (*Continued*)

Defects and Coding	Possible Causes of Defects	Cause gr. D*	C*	M*	E*
	Legionella growth esp. in hot water pipes if temp. does not reach 60°C. Water analysis can put a check on contamination.	▲		▲	
	Water stagnation or dirt accumulation at pipe joints due to wrong detailing or poor finishing of joints.	▲	▲		
	Dirt enters by seepage through wall or with dirty testing tools.		▲	▲	
	Stagnation inside large tanks if outlets are not diagonally opposite to the inlets. Lack of ventilation aggravates the problem.	▲			
	Mosquito breeding, entry of other creature, leaves, dust in tank if the net at overflow or vent pipe is omitted/damaged or cover is left open.	▲		▲	
	Underground pipe is corroded, leaky or ruptured or penetration at underground reservoir is leaky.		▲	▲	
	Seepage through pipe joints which are not water tight (wrong jointing method/ poor workmanship).	▲	▲		
	Irregular cleaning of tanks or flushing of distribution pipes. This problem is more frequent if cleaning is difficult (E7).	▲	▲	▲	
	No flush out points at the end of distribution pipes.	▲			
	Back flow and back siphonage happen due to wrong detailing (e.g. for tanks inlet pipes are at lower level than the overflow pipe).	▲			
	Sewers are nearby and leaking (damaged, corroded and poorly joined).	▲			

(*Continued*)

Table 6.1. *(Continued)*

Defects and Coding	Possible Causes of Defects	Cause gr. D*	C*	M*	E*
Leakage at joints (E4)	Pipe joints are not properly finished or wrong jointing method used.	▲	▲		
	Joining pipe of two different materials is inherently weak.	▲			
	Valve stems are not fitted tightly with gasket/washer (missing).		▲	▲	
Inadequate/poor supply (E5)	Faulty design to calculate demand, diameter of pipe, etc. Also no allowance for frictional head loss, bends and fittings.	▲			
	Direction of flow is against hydraulic gradient.	▲			
	Excessive friction loss in pipes with very rough inner surface (poor make) or at joints which are not finished smooth or have projections.	▲	▲		
	Oversized pipe or under capacity pump unable to deliver right pressure.	▲			
	Loss of pressure by faulty pumps. Regular testing can detect the fault.			▲	
	Hunched pipes, wrong hydraulic gradient or wrong size of valve orifice cause airlock.	▲	▲		
	Dirt accumulation at joints — Bad fixing detail and irregular flushing.	▲		▲	
	Excessive leakage at pipe joints (E4).	▲	▲	▲	
Noisy/turbulent flow (E6)	Design details unable to prevent water hammer, splashing or noise transmission.	▲			
	Inadequate support/anchor blocks at every bend, branch and dead-end results in hydraulic thrust.	▲			
	Hunched pipes, wrong hydraulic gradient or wrong size of valve orifice cause airlock.	▲	▲		
Difficult maintenance (E7)	Pipes are difficult to access if laid at a depth more than 2 m below ground, covers of pipe work are not easily removable or embedded in structural element or insufficient servicing space.	▲			

(Continued)

Table 6.1. (*Continued*)

Defects and Coding	Possible Causes of Defects	Cause gr. D*	C*	M*	E*
	Haphazard layout — Poor design or noncompliance with drawing.	▲	▲		
	Interrupted supplies during repair if stop valves at every level or at convenient places/on branch pipes are not planned. Hence repair at one part disturbs supply to other parts of building.	▲			
Underground and overhead atorage tank					
Corroded tank, parts (E8)	Poor application/peeling and delamination of coat (steel tank) or water seepage (concrete tank)	▲	▲		
	Parts (e.g. pipes/strainer) not corrosion resistant or changed regularly.	▲		▲	
Leakage, flooding (E9)	Cracks due to corrosion or poor structural design.	▲			
	Severe rusting of metal tanks.	▲		▲	
	Poor detailing/workmanship at pipe penetration cause leakage.	▲	▲		
	Inadequate provision to cater overflow — Poor drainage of receiving basin of underground storage.	▲			
	Damaged float valve or liquid-level indicator unable to control overflow.			▲	
Difficult cleaning (E10)	Inadequate space around the tank or access is not proper.	▲			
	Wash out pipes/sumps not provided or wrongly constructed.	▲	▲		
	Tanks are not compartmentalized — Hence cleaning results in shutting of the whole supply and causes inconvenience.	▲			

(*Continued*)

Table 6.1. *(Continued)*

Defects and Coding	Possible Causes of Defects	Cause gr. D*	Cause gr. C*	Cause gr. M*	Cause gr. E*
Hot water supply					
Water not hot enough/ takes long time to deliver (E11)	Wrong design results in excessive dead leg (the pipe length of cold water allowed to run off before hot water reaches tap spout).	▲			
	The primary and/or secondary circuit is not insulated to prevent heat loss or insulation is damaged with time.	▲	▲		▲
	Short circuiting in the piping of the primary and secondary circuit. Combustible insulation of the primary and secondary circuit of the hot water cylinder needs replacement at regular interval.	▲		▲	
Corrosion and scaling (E12)	Piping material not corrosion resistant or galvanic corrosion. Timely action (e.g. paint) is helpful.	▲		▲	
	Protective coating lost/damaged during installation or done wrongly.		▲		
	Hard water reaches a temperature >60ºC and forms scales.	▲			
General and sewage pumps					
Excessive vibration, noise over heating (E13)	Loose screws or misaligned bearings.		▲	▲	
	Impeller goes out of balance if not designed for large range of head variation. Requires alignment.	▲	▲	▲	
	Wear and tear of parts — Especially if not cleaned or lubricated. esp. if maintenance is hindered by poor accessibility.	▲	▲		
	No vibration isolator provided.	▲	▲		
	Automatic pressure release valve unable to provide circulation of sufficient water and pump becomes hot. Needs servicing.			▲	

(Continued)

Table 6.1. *(Continued)*

Defects and Coding	**Possible Causes of Defects**	**Cause gr.** D*	C*	M*	E*
Chocked (E14)	Hunched pipes, wrong hydraulic gradient or wrong size of valve orifice cause airlock.	▲	▲		
Clogged sewage pump (E15)	Long fibrous material or solids of dia. ≥ 64 mm are common in sewage and causes clogging if strainer/baffle is not installed.	▲	▲		
	Regular cleaning required.			▲	
Corrosion (E16)	Material is not corrosion-resistant or not suitable for pumped medium (esp. for sewage pumps).	▲			
	Corrosion in parts is neglected. No cleaning of rust, painting or lubrication of affected part.			▲	
No more auto inter-change of pumps (E17)	The auto control is done through many elements working together — The float switch, lead/lag selector switch, hour run meters for each pump or high-level alarm. Any of them goes faulty and not rectified.			▲	
Sanitary appliances					
Clogged/soiled fixture (E18)	Inadequate or harsh cleaning of fixture and trap. If the material in not vitreous or surface is scratched, cleaning is difficult.	▲	▲	▲	
	Backflow or back siphonage. If the preventive devices (stop valves, vacuum breaker, check valve, etc) are damaged with time or from mishandling. Parts should be replaced after certain interval.			▲	
	Inadequate trap-seal or venting to prevent backflow.	▲			

(Continued)

Table 6.1. *(Continued)*

Defects and Coding	Possible Causes of Defects	Cause gr. D*	C*	M*	E*
Ponding below/around fixture (E19)	Inadequate grading of flooring towards the WC pan.	▲	▲		
	Connection of fixture and discharge pipe is not watertight. Flexible type connectors (collar, flange) or seals can handle fixture movement and remain watertight if installed properly.	▲	▲		
	Connector between flush pipe of cistern and pan is damaged with time.			▲	
	Gap between fixture and walls is not sealed.		▲		
	Omission of the waste pipe/trap under the bath or overflow preventer.	▲			
	Poor maintenance of waste pipe/trap, especially if not accessible.	▲		▲	
Ponding on floor/slow drainage (E20)	Outlets are inadequate in number or too far away from fixtures.	▲			
	Floor grading is inadequate or in wrong direction.	▲	▲		
	Clogged floor trap from inadequate cleaning or entry of solid waste through misfit grating. Construction waste should not be dumped.	▲	▲	▲	
Wall-hung fixture dislodged (E21)	Supports and bracket are not strong enough or not securely fixed to the wall to sustain the load.	▲	▲		
	Severe corrosion of screws or bolts. Material is not corrosion resistant and initial rusting is neglected.	▲		▲	
Auto devices no more auto (E22)	Wear and tear of operating parts with time and careless use (repeated/hard pressing of sensor).			▲	▲

(Continued)

Table 6.1. (*Continued*)

Defects and Coding	Possible Causes of Defects	Cause gr.			
		D*	C*	M*	E*
Sanitary piping					
Leakages (E23)	Joining pipes of different material without adaptor.	▲			
	Absence of elastomeric seal at joints or sleeve at slab/wall penetrations.	▲			
	Poor workmanship at joints. Not even tested to rectify.		▲		
	Corrosion — Unsuitable material or damaged coating.	▲	▲	▲	
Slow/no discharge (E24)	Inadequate size or gradient of branch or vent pipes.	▲	▲		
	Dirt entering through damaged/removed floor trap.			▲	
	Solid construction waste clogs the pipes.		▲		
	Irregular/inadequate cleaning especially if accessibility is poor (E26).	▲		▲	
Discharge of foul air into the building (E25)	Nonprovision/wrong design of ventilating pipe.	▲			
	Trap seal loss of fixtures for waving out (pressure fluctuation) for short/straight vent.	▲			
	Induced siphonage due to absence of trap ventilating pipes.	▲			
	Self siphonage of undersized discharge stacks.	▲			
No access for maintenance (E26)	Mismatched position of access cover/cleaning eyes and pipe joints.	▲	▲		
	Pipes covered by building elements.	▲			
	Pipes laid with inadequate clearance from structure.		▲		
Sewerage ejectors/Solid diverter tank					
Fowl smell (E27)	Poor ventilation — Vent pipe and means of forced ventilation such as exhaust fans not provided.	▲			

(*Continued*)

Table 6.1. *(Continued)*

Defects and Coding	Possible Causes of Defects	Cause gr. D*	C*	M*	E*
	Waterless appliances are not maintained well — Cartridge of urinal or moisture level in waterless closets.			▲	
	Irregular cleaning — Poor housekeeping practices or cleaning is not regular if there is unsafe working condition, poor accessibility, etc.	▲		▲	
Incomplete discharge (E28)	Wear and tear of the isolating and air release valves of air release pipe or two way valve of diverter tank.				▲
Ponding/leakage (E29)	Water seepage through pit cover — Not properly secured/stolen.	▲	▲	▲	
	Check valve not working to prevent surge of backflow.	▲	▲	▲	
	Severe corrosion of tank body.	▲			
Sewer and manhole/Inspection chamber					
Choking and overflow (E30)	Solid matter is getting deposited — Improper gradient.	▲	▲		
	Solid construction debris enters the pipe.		▲		
	Irregular cleaning — Poor accessibility or maintenance practice.	▲		▲	
	Size of the downstream sewer causes low flow velocity upstream. This esp. happens when new sewer is added to existing network.	▲			
	Clogged screens. Not cleaned after rainstorms.			▲	
Damaged cover (E31)	Forceful handling of cover which is in mismatched position or not flush with ground level.		▲	▲	
	Unable to withstand the vehicular load and breaks.	▲			
	Excessive corrosion — Poor make or overlooked.	▲		▲	

D* — Design; C* — Construction; M* — Maintenance; E* — Environment.

to collect and analyze maintenance data by organizations like ASHRAE, has met with limited success [18]. Nevertheless, a large maintenance budget have to be put aside annually for maintenance of the HVAC system for every building that uses it.

Operating costs typically represent the largest expenditure over the useful life of a mechanical system, as the HVAC system consumes a large amount of energy monthly. HVAC systems may consume more than 50% of the total energy in commercial buildings [19]. Research works on optimizing the HVAC systems in terms of system designs were reported [20–22].

Without proper design considerations, HVAC systems will not be maintainable and would not operate efficiently. The results of ignoring preventive maintenance are shorter service life, premature replacement of the components and increased operational costs in the case of components that require energy resources such as chillers and fans [23].

6.3.2. *Maintainability of HVAC Systems*

Inadequate performance of HVAC systems has been associated with occupants' health and productivity, biological growth, moisture problems and high-energy cost. Inaccessibility for maintenance is one of the most common causes of low maintainability HVAC system. Some examples of failures to make equipments accessible are high ceilings with HVAC equipments hidden within; restricted space in plant rooms that prevents maintenance personals from accessing the equipments. HVAC system is a mechanical system of which periodical system failure is inevitable, the ability for it to be quickly restored is thus crucial [24].

For HVAC systems, design changes are very costly to implement during the operation stage. It was also reflected that maintainability should be considered during the original design, during renovations and during equipment/system replacement. And when maintainability is designed or considered during the design stage, the benefits reaped includes:

- longer lifespan of systems and components,
- lower operating costs by performing at high efficiency,

- lesser unscheduled downtime, by preventing failures,
- lesser scheduled downtime, by decreasing the time required to perform a particular maintenance task.

Wani and Gandhi [2] used three design variables, namely, standardization, accessibility and condition monitoring as basis for evaluating the maintainability index for mechanical systems, in terms of mean time to repair (MTTR) and maintenance activity time.

Although the BAS system of buildings has improved over time, many buildings are not properly commissioned, operated and maintained. Lack of commissioning, improper operating practices and lack of proper maintenance practices lead to inefficient operation, excess expenditures on energy, poor indoor conditions and reduced lifetime for equipment. It was found that a large proportion of buildings suffered from problems with HVAC equipment [25].

6.3.3. *Common Defects*

HVAC systems waste energy and defects are prevalent [26, 27]. Various problems are found to occur in the long-term running of HVAC systems such as mold growth in supply air duct affecting occupants' health [28, 29], the loss of comfort and an approximate 15–30% wastage of energy [30]. The common problem of poor humidity control is actually indicator of many major defects. Air leakage on the return duct, wrong placement or omission of sensors and short-circuiting are only a few of the reasons [31]. It is hence crucial that before commissioning, the system specification should be checked from time to time with the designer and manufacturer. Rydzewski [32] mentioned about intermediate checking but his main concern was accessibility for maintenance. Poorly maintained cooling towers especially located in the direction of air intake may induce fatal Legionnaires' disease [33]. Micro-biologically induced corrosion (MIC) can cause a pitting and very rapid loss of metallic parts [34]. Apart from this, formicary corrosion and galvanic corrosion are also found to a large extent [35].

Shortage in air supply accompanied by internal finishes with high total volatile organic compounds (TVOC) can cause sick building syndrome [36]. Overflowing of condensate drain pan causes water stagnation that facilitates biological growth [37]. Heat exchanger fouling and particulate deposition diminish heat transfer performance, increase pressure drop and can cause contamination of working fluids [38]. Detailed analysis of system components have been conducted for: (1) sensors [39, 40] and (2) air handling units [41–43].

Figure 6.2 and Table 6.2 summarize some of the common problems and their causes observed in central chilled water HVAC of commercial buildings.

6.4. Elevators/Lifts

6.4.1. *General*

Mechanized transportation especially the lift system and its advancements have enabled the creation of high-rise buildings. Design of mechanized transportation, from the simple rope to the chain hoists, has tremendously improved and evolved significantly since 1853 with the invention of safety gears by E. G. Otis which made skyscrapers a reality.

Among many important functional decisions which must be taken by the designer of a multi-storey building, probably none is more than or as important as the proper selection of the mechanized transportation systems including elevators, escalators and even moving walkways. Not only these systems are critical to the designer, the quality of these systems also affects the tenant's choice of space against competing buildings [44]. Rising construction costs and persistent demand for quality space increase the value of designing a mechanized transportation system to deliver the most effective way of moving traffic within the building. For maximum serviceability and economy, mechanized transportation must be planned to meet the requirements of each building and its users.

In the last few decades, there have been significant developments in the field of mechanized transport systems [45]. A building and its mechanized transportation system are reciprocally interrelated; each, to a marked extent, affects the design of the other, therefore both should be planned concurrently to achieve the necessary integration of building and system. Without proper design considerations and planning executed

Fig. 6.2. Common defects in HVAC system.

Table 6.2. Common defects in HVAC system.

Defects and Coding	Possible Causes of Defects	Cause gr. D*	C*	M*	E*
Air handling unit (AHU) and fan coil unit (FCU)					
Fan belt emitting burnt smell/noise (F1)	Loose fan belt rubs against the pulley and gets very hot.			▲	
	Slippage of loose belts.			▲	
	Fan belts are broken/cracked with wear and tear from ageing. It produces a squeak upon contact with the pulley. Not replaced as per schedule.			▲	
	Wrong alignment of pulleys.		▲	▲	
	Excessive vibration of loose belts passing over the pulleys. The intensity will increase or decrease with the speed of the drive. The whole set should be tested for pressure, noise level, etc.		▲		
Noisy fan motor (F2)	Insufficient lubrication of bearings.			▲	
	Worn out bearings from mechanical wear and tear. Lack of lubrication may cause accelerated aging.				▲
	Impeller out of balance from manufacturing fault or careless servicing.		▲	▲	
	Loose mounting from poor installation or maintenance.		▲	▲	
	Vibration isolators are inadequate, worn out or rusted.	▲		▲	
Corrosion/scaling of cooling coil, filter etc. (F3)	AHU/FCU housing not sealed — Constant contact with moist air and dampness from environment.	▲		▲	
	Over dosage of chemical during chemical cleaning of coil.			▲	
	Formicary corrosion (local micro-pitting) from unsuitable cleaning agents. Steam cleaning may help (Field, 2002).			▲	

(*Continued*)

Table 6.2. (*Continued*)

Defects and Coding	Possible Causes of Defects	Cause gr. D*	C*	M*	E*
	Poor maintenance of the evaporator side of the chiller plant.			▲	
	Large volume of fresh water is added to replenish major leakage (F5). The undue amount of additional mineral forms scale on the coil.		▲		▲
Slimy bio-film or dust on coils: Reduces airflow (F4)	Cleaning of the surface row only but not the inner rows, where dust and water collects.			▲	
	Insufficient access or >8 rows of coil hindering maintenance.	▲			
	Due to mishandling filters are damaged (e.g. gaps between panels). It allows dust or particles to enter and settle on cooling coil. During maintenance, filters should be combed or replaced if required.		▲	▲	
Chilled water leakage (F5)	Weak welded joints — Should be pressure tested.		▲		
	Wear and tear after a normal lifespan of 10–15 yrs.				▲
	Condensate pan material if not corrosion resistant, leaks in long run.	▲			
Damaged parts: Fins, filters etc. (F6)	Accidental bending of parts especially fragile fins during installation/maintenance. Should be combed to rectify.		▲	▲	
	Unsuitable chemical cleaning makes aluminum fins brittle.			▲	
Excessive airflow and moisture carry over (F7)	Fan is oversized compared to the capacity of the cooling coil.	▲			
	Spacing between coils is not proper to limit the flow to specified level.	▲			
Water retention and biofilm on drain pan (F8)	Drain pipe or pan not laid into right gradient.		▲	▲	
	Inadequate performance of undersized drain trap.	▲			
	Inadequate cleaning of drain pipe choked with biological agents.			▲	
	Sagging of drain-off pipe from insufficient hanging brackets.	▲			

(*Continued*)

Table 6.2. *(Continued)*

Defects and Coding	Possible Causes of Defects	Cause gr.			
		D*	**C***	**M***	**E***
Condensation of drain-off piping (F9)	Not insulated at all due to cost cutting.	▲			
	Damaged insulation with time especially for pipes located near floor which are prone to be stepped upon during maintenance.	▲		▲	▲
Clogged air filter and drop in air pressure (F10)	Intake of dirty/polluted air if inlets are located near exhaust outlet or pollution source such as car parks and cooling towers.	▲			
	Inadequate maintenance or replacement of filters.			▲	
	Insufficient access space for cleaning or easy replacement.	▲			
	Problem remains unnoticed if pressure gauge is at inaccessible location or above eye level or not calibrated.	▲	▲	▲	
Condensation on exterior of components (F11)	Insufficient insulation — Poor specification.	▲			
	Air tightness of room or door is not checked for gasket around door is intact and door closing tightly.				
	Seals, tapes insulation lose efficiency with time making the housing leaky, e.g. door gaps, wall cracks, etc.			▲	
Leakage from actuator (F12)	Internal valve seals broken or damaged due to wear and tear process. Should be checked from time to time			▲	▲
Poor access for maintenance (F13)	Cluttered and haphazard layout of piping and ducting.	▲			
	Components located at very high level or obstructed by structure or services.	▲			
	Insufficient/wrong access panels to ceiling components.	▲			
	AHU room used as store, unsafe working condition, poor lighting.			▲	

(Continued)

Table 6.2. (*Continued*)

Defects and Coding	Possible Causes of Defects	Cause gr. D*	C*	M*	E*
Chiller					
Scale on condenser: Low chiller efficiency (F14)	Open loop design of condenser water.	▲			
	Inadequate maintenance and cleaning of condenser tubes or cooling tower.			▲	
Corroded parts of condenser — Face plate, tubes (F15)	Tubing material is not resistant to corrosion/pitting.	▲			
	Paint protection lost during careless installation or maintenance.		▲	▲	
	Harsh cleaning (hard brush, strong acid, etc.) cause thinning of copper tube walls, leading to of pitting or formicary corrosion.			▲	
Refrigerant leakage (F16)	For positively pressured chillers, refrigerant leaks through joints (at gasket, piping, monitoring socket, etc.), which might have been opened for maintenance and later not tightened.			▲	
	No checking for system leakage and remedial measures taken.	▲		▲	
Refrigerant contamination (F17)	Leakages of condenser tubes induce moisture to refrigerant.	▲			
	Incondensable gas (air, moisture) diffuse, leak or evaporate into chiller through joints, cracks and gaps. Requires purging.	▲		▲	
	Mixing of fines from metallic moving parts of compressor motor or lubricant. Selection of refrigerant is improper.	▲			
	Excessive oil added to refrigerant by faulty pump or its valve. As an indicator, oil level should be checked from time to time.			▲	

(*Continued*)

Table 6.2. (*Continued*)

Defects and Coding	Possible Causes of Defects	Cause gr. D*	C*	M*	E*
Surging of the chiller (F18)	Insufficient loading of the chiller, mismatch of capacity, operation under capacity of less than 25% of its handling capacity.	▲			
	Scaling in condenser tubes causing higher condenser pressure (F14).	▲		▲	
	Excessive amount of noncondensable present in refrigerant either due to leakage in evaporator or irregular purging.	▲		▲	
Excessive noise and vibration (F19)	Inadequate or worn out spring isolators.	▲		▲	
	Inadequate acoustical insulation in the machine room.	▲			
	Mechanical failure of parts — Should be tested from time to time.			▲	
	Compressor worn. Accelerated aging occurs if the chiller is constantly running or very frequent start-stop.	▲			▲
Poor access for maintenance (F20)	No access rout planned for replacement of big components.	▲			
	Insufficient space in plant room for maintenance.	▲			
	Plant room used for storage hinders maintenance activity.			▲	
Condenser pipes leaking (F21)	Poor workmanship at joints. Should be tested before commissioning.		▲		
	Harsh cleaning method makes pipe thin. Excessively thinned pipes should be plugged or replaced.			▲	
Condensation on chilled water pipes (F22)	Pipe surface become cold if insulation is inadequate.	▲			
	Insulation is damaged during careless maintenance of the equipments.			▲	

(*Continued*)

Table 6.2. *(Continued)*

Defects and Coding	Possible Causes of Defects	Cause gr. D*	C*	M*	E*
Water gushes/leaks out thru' monitoring device (F23)	Damaged mechanical seal of valve from wear and tear.				▲
	Not fitted properly after servicing, usually if access is difficult.	▲		▲	
	Poor workmanship at the socket joint.		▲		
Cooling tower					
Biological fouling (F24)	Design is not suitable to prevent dirt accumulation, stagnation.	▲			
	Inadequate cleaning of tower fill, basin, drift eliminators, etc.			▲	
	Inefficient water treatment method used.			▲	
Corroded parts — Louver, collection basin (F25)	Material is not corrosion resistant to counteract constant exposure to the condenser water and external environment.	▲			
	Damaged protective layer on metallic surface during installation.		▲		
	Parts are not cleaned, lubricated or painted in time to prevent corrosion and scaling.			▲	
Foaming on condenser water (F26)	Too much chemical treatment added.			▲	
	Bacteria boom. Cleaning and application of biocide is not done.			▲	
	Inadequate access to reach all parts for cleaning.	▲			
Inadequate heat rejection (F27)	Obstructing structures above the cooling tower.	▲			
	Short-circuiting: cooling tower located near heat exhausts.	▲			
	Condenser water pump is over supplying — Flow rate too high to reject heat properly and waterfall effect.	▲			
Noise and vibration (F28)	Mounting on isolator is not leveled.		▲		
	Inadequate checking for condition of fan, motor or lack of lubrication.			▲	

(Continued)

Table 6.2. *(Continued)*

Defects and Coding	Possible Causes of Defects	Cause gr. D*	C*	M*	E*
Air distribution and terminal systems					
Condensation on duct or supply air diffuser (F29)	Insufficient or damaged insulation.	▲	▲	▲	
	Leakage from air-duct can be trapped in the gaps between fiber insulation and aluminum foil and condensate. Chances are higher if there remains air gap between duct and insulation.		▲		
	Diffuser is located near infiltration of external air or near entrance.	▲			
	Settings of the sensors such as dampers, operators and velocity controllers have drifted off and needs recalibration.			▲	
	Inaccessible/access blocked for regular recalibration of sensors.	▲		▲	
Air Leakage (F30)	Flanged joints are prone to air leakages.	▲			
	The aluminum foil surface is not cleaned before sticking it to the duct tape. The dust works as bond breaker and the joint gets exposed as soon as the foil gets detached.		▲		
Corrosion (F31)	Galvanized protection layer cracked or scratched during installation, thus exposing the surface to air and moisture.		▲		
	Rainwater entry through fresh air inlet into the supply air duct and this moisture is transported into the ducts.	▲			
	Wet exhaust unable to drain out the moisture.	▲			
	Moisture carry-over from AHU transported into supply air ducts.	▲			
Dirty ducts (F32)	Poor jointing of ducts or rough finishes of plenum traps dust.		▲		
	Lack of cleaning of ducts — Poor access/maintenance.	▲		▲	
	Dirty or dusty air flow from damaged or poorly cleaned air filters in AHU.		▲	▲	

(Continued)

Table 6.2. (*Continued*)

Defects and Coding	Possible Causes of Defects	Cause gr. D*	C*	M*	E*
Noisy air flow (F33)	Turbulence inside under-sized ducts.	▲			
	Air not balanced in the ducting system. An opposed blade type damper provides a better option.	▲	▲		
	Insufficient or undersized diffusers resulting in high-velocity air.	▲			
	Malfunctioning of controls of air diffuser dampers.	▲		▲	
Poor airflow: Blocked diffusers/dumping of cold air (F34)	Diffusers located near return air grille cause short-circuiting. Diffused air returns before it is circulated in room.	▲			
	Diffusers blocked by other installations, furniture layout, wrongly chosen false ceiling.	▲		▲	
	Diffuser very dirty, especially those with difficult access.	▲		▲	
	Wrong choice of diffusers throws cold air directly at the user.	▲			
Faulty pneumatic control (F35)	Pneumatic hoses damaged or leaking out pressurized air.		▲		
	Failure of air compressor unit — esp. open type.	▲		▲	
	Dampers are jammed.			▲	

(*Continued*)

Table 6.2. (*Continued*)

Defects and Coding	Possible Causes of Defects	Cause gr. D*	C*	M*	E*
Internal environment					
Insensitive thermostat (F36)	Wrong location does not allow it to sense the room temperature.	▲			
	Sensors are not recalibrated at regular interval. May be due to poor accessibility or poor maintenance practices.	▲		▲	
Fungal growth/stains on walls (F37)	Moisture entry from fresh air inlet.	▲			
	Infiltration of external warm air — Short circuiting.	▲			
	Moisture carry-over from cooling coil.	▲			
Excessive noise from machinery spaces (F38)	Location of AHU plant room within occupancy space.	▲			
	Poor acoustical insulation of the mechanical room.	▲			
	The installation on isolator is not leveled.		▲		
	Worn out isolator or lack of lubrication of AHU, FCU or chiller plant equipment.			▲	
Users sick/SBS, dissatisfied (F39)	Air from cooling tower gets sucked by air intakes. It may cause legionella infection.	▲		▲	
	Mould growth in return air plenum. Using ducts, grilles and regular cleaning can prevent such problem.	▲		▲	
	Dumping of cold air or malfunction of VAV terminal/FCU (often accompanied by poor access for maintenance).	▲		▲	

D* — Design; C* — Construction; M* — Maintenance; E* — Environment.

during the design stage, it would be difficult to predict the outcome and result of the mechanized transportation system, resulting in difficulties in maintenance and inefficient operation.

6.4.2. *Mechanized Transportation System*

6.4.2.1. *Elevators*

An elevator is defined as a permanent lifting equipment serving two or more landing levels, including a car for transportation of passengers and/or other loads, running at least partially between rigid guide rails, either vertical or inclined to the vertical by less than 15° [46]. The three common types of elevators are:

(1) *Hydraulic*: Uses an electrically driven pump to force oil into a cylinder, thereby extending the piston and raising the cab, when hydraulic oil is diverted to a by-pass, the elevator will seek downward travel.
(2) *Gearless traction*: Relies on a large slow-speed electric motor to drive a pulley on which hoist cables have been installed to pull the elevator upward and lower it when required.
(3) *Geared traction*: Employs a high-speed motor to drive a worm and gear reduction unit to turn the hoisting sheave allowing cables on the sheave to raise or lower the cab [47].

6.4.2.2. *Elevator Operation and Control*

The control system of the elevator, as an overview of the operating system discussed, is defined as the designation for the equipment that determines individual performance characteristics of an elevator. For example, its speed of traveling from floor to floor includes speed of door opening and closing, built-in time factors for passenger transfer, its ability to travel swiftly and accurately, accurate display of hall lantern signals and all functions of the control applied to an elevator.

Another key performance of the elevator is door control. It was highlighted for its adjustment requirements recommended for a separate car to stop and hall stop door hold-open times to be specified for any building to match expected passenger transfer time with the operation of the elevator.

Strakosch [48] emphasized in detail the combination of all the times discussed to this point, adding up to a performance time for an elevator on a one-floor run or trip.

6.4.2.3. *Elevator Service Requirements*

There are different needs for different building uses for the vertical transport system to be effective. Certain facilities require speed while others may put stability or comfort as the top priority in performance. Therefore the speed of travel and handling capacity are emphasized for commercial buildings, where large numbers of passengers travel within the building expecting short lift time.

Even between different commercial buildings, individual building purposes such as commercial hotels and retail facilities have unique specifications for usage of lifts; hence it is needed to design and apply understanding to the particular requirements of users of the mechanized transport system.

6.4.3. *Codes and Standards*

The mechanical, electrical and operational design of elevators, escalators and moving walkways are specifically controlled according to the various standards or codes, depending on the country. It may typically be dictated by international, national, state, regional or city-based standards. Previously, the many standards that were introduced were highly prescriptive; however, there has been a paradigm shift towards performance-based standards. Therefore, it is the role of the designer to ensure that the design of the mechanized transport systems meets the project requirements.

In many cases and standards, the mechanized transport system, especially the elevators are to comply with standards that are related to earthquake resilience, fire standards, electrical wiring rules and so forth. Some of the national vertical transport system standards that are followed by the respective countries are:

(1) Australia — AS1735 Lifts, escalators and moving walkways: General requirements [49].

(2) Canada — CAN/CSA B44 M94 Safety Code for Elevators, Escalators, Dumbwaiters, Moving walks and Freight Platform Lifts [50]
(3) Europe — EN 81 series (EN 81-1, EN 81-2, EN 81-28, EN 81-70, EN 12015, EN 12016, EN 13015, etc.) Safety rules for the construction and installation of lifts [51].
(4) Japan — JIS A 4302 Inspection standard of elevator, escalator and dumbwaiter [52].
(5) Singapore — CP 2 Code of Practice for Installation, Operation and Maintenance of Electric Passenger and Goods Lifts/CP15 Code of Practice on the Installation, Operation and Maintenance of Escalators [53, 54].
(6) United States of America — ASME A17 Safety Code for Elevators and Escalators [55].
(7) Other international standards [56–71].

Besides, there are also additional codes and standards that enable access for disabled persons, for example, the International Standards, *ISO 9386 Power-operated lifting platforms for persons with impaired mobility: Rules for safety, dimensions and functional operation* [62, 63], or in several countries, may be made mandatory by laws or regulations such as the Disabilities Act in the US.

6.4.4. *Maintainability of Mechanized Transportation System*

Although modern vertical transportation systems are well-designed and manufactured under tight quality control, once in service, like any other electro-mechanical device, it requires periodic maintenance. A systematic maintenance program is essential for efficient performance throughout useful service life. It is important to bear in mind the importance for maintainability when designing the system, since the mechanized transport system is an essential building service for most high-rise buildings and often difficult to replace.

The mechanized transport systems characteristically have a relatively long service life of 15–30 years. Therefore it is important for designers and building owners to ensure whether the design and quality of the product enables continuous and easy maintenance activities for safety and continued performance, cleanliness and aesthetically pleasing appearance of the product.

There is a need to identify shortcomings in a system's design or a material's function before it is placed into use, as it is crucial to construct a system component that is maintainable and functions as it is intended to be. Such a practice will be able to sustain its optimum desired performance while at the same time minimize cost of maintenance and operation during the life cycle of building as the system built would be able to. The degree to which the design of a building embraces maintenance considerations has a major impact on its performance.

6.4.5. *Common Defects*

Elevators are complex and hi-tech equipment to form a safe and reliable vertical transportation spine of modern high-rise buildings. Since the invention of safety gear by E.G. Otis in 1953, significant improvement has taken place for elevators in traffic handling, controls, traction systems and vibration dampers. Despite high safety factor ($\approx$ 16) and ISO certification, elevator related accidents kill about 30 and seriously injure about 17,100 people each year in US alone. 50% of the victims are trained mechanics [72].

Safety rules of ISO are focused at various types of accidents: shearing, crushing, falling, impact, trapping, fire, electric shock, damaged material, accidents due to wear or corrosion [73]. Among the prevalent problems such as over speeding of car, uncontrolled low speed, car moving before door is closed, etc; the most common hazard is fatal fall through elevator shaft due to opening of lobby door on pressing of call button before the car arrives [74]. Moreover equipment shutdown incurs in expensive over time "call-backs" and inconvenience to the users. Such malfunctioning can be prevented by efficient maintenance practices [75].

Defects identified from field investigations and the knowledge gained through the literature review are elicited [76–79] and summarized in Fig. 6.3 and Table 6.3.

Machine room elements

Dirty machine room

Dirt, rust, poor access of traction machine

Oil seal leaking in traction machine

Governor machine – Dirt, oil, rust

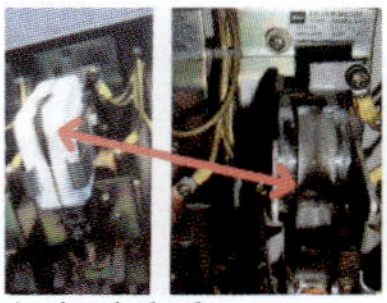
Arcing in brake contactor

Resistor with burnt mark

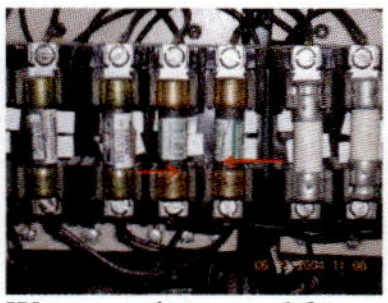
Wrong resistor used for replacement

Burnt mark in controller cable

Car and lobby door

Roller worn and rusty

Door closing on users

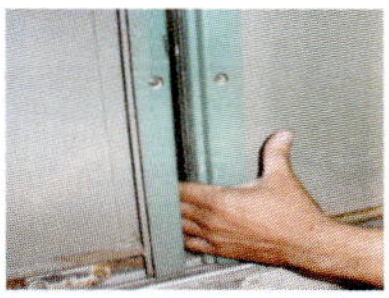
Large gap in door panel

Burnt door lock switch

Car (interior, top, bottom)

Trip hazard

Damaged flooring

Walls dirty with graffiti

Rubbish in the car

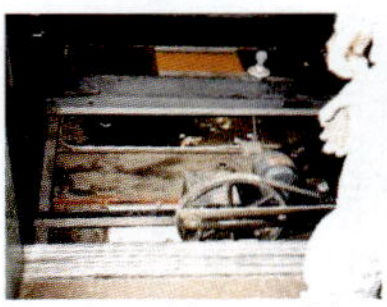
Unsafe working condition at car top

Corroded guide rail

Useless emergency call button

Cardboard packing to stop excess movement in roller shoe

Hoistway and hoisting rope

Dirty hoistway

Worn rope

Rusty sheave groove

Less folds in hoist

Fig. 6.3. Common defects in elevator [76–79].

Table 6.3. Common defects in elevator.

Defects and Coding	Possible Causes of Defects	Cause gr. D*	C*	M*	E*
General travel performance and hazards					
Long waiting time (G1)	Poor design consideration of expected traffic load.	▲			
	Sluggish movement of lift car due to dirt or wear on acceleration switches. Such problem is usually identified by computer tag.			▲	
Uncontrolled movement — Too fast, too slow, moving with open doors (G2)	Speeding caused by undue power surge or slippage of worn-out pulley. Should be stopped by "safeties" attached to car bottom.			▲	▲
	Dirt/wear on acceleration switch slow down the car.			▲	
	Wrong wiring in controller supplies power to lift when the door is still open (OSHA, 2004).		▲	▲	
Trip hazard — Leveling >5mm (G3)	Wrong tension in main rope due to wrong design or requires re-levelling after regular interval.	▲		▲	
	Landing floor or flooring and car floor not accurately adjusted.		▲		
Jerk, vibration, noisy movement, hitting at top/bottom (G4)	Overloading of car, especially if counterweight has just two ropes or shows deflection.	▲			
	Traction machine not well-lubricated.			▲	
	Uneven tension in main rope if less number of folds.		▲		
	Traveling cable and/or compensation cable not firmly secured.			▲	
	Poor selection of safety gear type.	▲			
	Poor brake adjustment.		▲		
	Machine room location is poorly planned — All noise and vibration are propagated to the other parts of building.	▲			

(*Continued*)

Table 6.3. *(Continued)*

Defects and Coding	Possible Causes of Defects	Cause gr. D*	C*	M*	E*
Vibration and jerky landing (G5)	Brake rubber coupling is worn/shaky.			▲	
	Brake rod not secured to plate.		▲		
	Brake level or drum is rusty, dirty, not lubricated.			▲	
Lift jam or stalled car (G6)	Safety switches not activated.			▲	
	Damaged restrictor leaves large gap in doors or repeated opening and closing (G49).			▲	▲
	Faulty circuit in controller.		▲	▲	
	Hall button unable to register call (G32).	▲		▲	▲
	Limit switch too close to governor rope and halts car from time to time. This is due to wrong selection of design factor.	▲			
Fall through hoistway (G7)	Opening of door when lift is not present (G53).	▲		▲	
	People forcefully open the door of stalled car and try to jump.				▲
	No safety precaution for the mechanic working on the elevator.			▲	
Falling on/hit by adjacent machines during maintenance (G8)	Inadequate space in machine room, hoist way or pit poses hazard of electrocution or hit by moving parts.	▲			
	The working surface is slippery with water or lubricant.	▲		▲	
	No partition in the well/pit accommodating more than one car.	▲			
	No safety precaution taken, e.g. tag off the group of elevator sharing the same pit. A mechanic can be injured in the machine room due to the same reason if the electromechanical parts of other elevators are in operation and there is no safe spacing.			▲	

(Continued)

Table 6.3. *(Continued)*

Defects and Coding	Possible Causes of Defects	Cause gr. D*	C*	M*	E*
Machine room					
Dirty with rubbish, carbon dust, lubricant (G9)	Infrequent cleaning of room. Worst if finishes generate dust.	▲		▲	
	Careless maintenance of machine parts, e.g. spillage of lubricant.			▲	
	Inadequate clearance around machines and control panel for proper cleaning. Either poor planning or storing material in the room.	▲		▲	
Poor lighting (G10)	Inadequate provisions of lux level or lights have short lamp life.	▲			
	Poor upkeep of luminaries.			▲	
Hot, damp, stuffy (G11)	Inadequate opening or the opening is closed.	▲		▲	
	Poor air circulation by dirty/dusty inverter cooling fan.			▲	
	Seepage through wall or ceiling.	▲	▲		
Controller (in machine room)					
Transformer noisy, dirty (G12)	Excess sparking and chattering.			▲	
	Oil leakage — Foreign make (unsuitable for tropical temperature).	▲			
Overheated parts (G13)	Inverter cooling fan is dirty or dusty.			▲	
	Arc, spark, high resistance due to loose connection.			▲	
Resistor with burnt mark — May cause intermittent fault or fire (G14)	Noncompatible resistor causing burnt. Especially happens due to patchwork type repair job.			▲	
	Careless installation or maintenance which may lead to partial discharges between the various parts, ground failure, etc.		▲	▲	
	Dirt collected on the windings may damage insulation, reduce air flow and result in burn out, shorts or grounds (NEII, 2005).			▲	

(Continued)

Table 6.3. (*Continued*)

Defects and Coding	Possible Causes of Defects	Cause gr. D*	C*	M*	E*
Traction machine and motor (in machine room)					
Oily/dirty machine (G15)	Joints at cover are weak due to bad welding or not corrosion proof.	▲	▲		
	Inadequate cleaning or spillage during lubrication. Often caused by poor accessibility.	▲		▲	
	Cover is misfit (usually after replacement) or improperly fitted.			▲	
	Oil-level gauge blurred or faulty due to mishandling.			▲	
Wearing marks on rope, sheave groove etc. (G16)	Unanticipated excessive friction — Under capacity design, uneven tension or inadequate lubrication.	▲		▲	
	Aging especially for ropes with less strands or resistance to corrosion or kink.	▲			
Overheated motor (G17)	Dirty cooling fan from inadequate cleaning.			▲	
	Problem with electrical circuit.		▲		
Excessive noise or vibration (G18)	Bearings are loosely fitted (poor workmanship).		▲		
	Worn out, rusty or insufficiently lubricated bearing. Wearing is faster if the specification is insufficient than the required strength.	▲		▲	▲
	Impeller out-of-balance — Needs re-alignment.			▲	
	Machine bed isolation rubber pad is absent or worn.	▲			▲
Brake assembly (in machine room)					
Unable to stop car (G19)	Brake plunger/felt/rubber cushion worn.				▲
	Brake rod not securely fixed to plate — Faulty installation.		▲		
	Brake contactor lips excessively — Parts worn, over lubrication.			▲	▲
	Arcing mainly due to loose connections.			▲	

(*Continued*)

Table 6.3. (*Continued*)

Defects and Coding	Possible Causes of Defects	Cause gr. D*	C*	M*	E*
Parts are not in proper working condition (G20)	Burnt lining, loose screws, uneven lever, insufficient clearance, scratching, etc. needs rectification.			▲	
	Careless installation or mishandling during maintenance.		▲	▲	
	Over or under lubrication.			▲	
Over-speed governor machine (in machine room)					
Noisy, oily/dirty (G21)	Wear and tear of parts, accelerated by poor lubrication.			▲	
	Accelerated wearing due to high friction from inappropriate pulley geometry of unsuitable material.	▲			
	Oil spillage or inadequate cleaning after lubrication. Inaccessible.	▲		▲	
Uneven tension in gov. rope (G22)	Wear and tear with time or due to poor specification.	▲			▲
	Irregular condition checking as never used except emergency.			▲	
Lift car					
Guide/roller shoe slipping (G23)	Loosely fitted due to poor workmanship or wrong replacement.		▲	▲	
Unsafe working condition (G24)	Low overhead — Poor provision of clearance in design.	▲			
	No warning sign — Negligence of safety codes.			▲	
	Inadequate lighting — Poor provision or spoiled bulbs not replaced.	▲		▲	
Dirty safety gear, linkage, etc. (G25)	Position makes it poorly accessible for cleaning.	▲			
	Inadequate cleaning after lubrication or other maintenance work.			▲	

(*Continued*)

Table 6.3. *(Continued)*

Defects and Coding	Possible Causes of Defects	D*	C*	M*	E*
Traveling cable or compensation cable noisy (G26)	Cables not firmly secured — Careless installation or mishandling during lubrication.		▲	▲	
	Lack of lubrication.			▲	
	Travel cable not covered with sheath.	▲			
Faulty safety gear (G27)	Jammed due to inadequate clearance (poor installation) or dirty.		▲	▲	
	Missing/worn off parts.			▲	
Dirty with lint, cobwebs (G28)	Inadequate cleaning.			▲	
	Not enough space/provision for cleaning in the lift pit.	▲			
Guide shoe/roller slipping (G29)	Dimension mismatch usually during replacement.			▲	
	Components not securely fitted.		▲	▲	
Dark and stuffy interior (G30)	Irregular cleaning or replacement — Worsened by graffiti, littering.			▲	▲
	Lights are damaged (usually if not guarded).	▲			
	Inadequate ventilation, fan not working.	▲		▲	
Noisy interior (G31)	Poor acoustical design unable to prevent noise from the hoistway.	▲			
	Loose false ceiling due to poor installation.		▲		
	Ventilation fan is noisy due to poor make or inadequate maintenance (e.g. damaged blades during careless cleaning).	▲		▲	

(Continued)

Table 6.3. *(Continued)*

Defects and Coding	Possible Causes of Defects	Cause gr. D*	C*	M*	E*
COP button cannot register call (G32)	Lack of co-ordination with controller. Such defect usually occurs if servicing is not done after designated interval.			▲	
	Jammed or loose buttons — Wear and tear with time.				▲
	Damaged buttons if nondurable design or suffers from vandalism (banging, hitting, repeated pressing). Having an integral type of COP is helpful in such case.	▲			▲
Spoilt emergency/trap door (G33)	Lack of security provided by design which is easily operable.	▲			
	Mishandling during maintenance of other parts.			▲	
Certificate plate lost/spoilt (G34)	Careless maintenance — Nonadherence to safety rules.			▲	
	Vandalism.				▲
Damaged/worn flooring (G35)	Material is not durable as per usage or vandalism.	▲			
	Wear and tear from unpredicted heavy usage.	▲			
Hoistway					
Guide rail dirty, slippery (G36)	No cleaning after regular maintenance, oil spillage. It hinders smooth movement of the car or counterweight.			▲	
Excessive noise and vibration in hoistway (G37)	Excess movement in parts — Usually due to size mismatch after replacement.			▲	
	Denting of guide rail — Not strong or not securely fitted.	▲	▲		
	Inadequate lubrication or dirt accumulation in the parts.	▲		▲	

(Continued)

Table 6.3. *(Continued)*

Defects and Coding	Possible Causes of Defects	Cause gr. D*	C*	M*	E*
Dirty, dusty, damp, stuffy (G38)	Ventilation provision inadequate, not functioning, churns dust.	▲		▲	
	Inner surface with projection or finish that accumulates dust.	▲			
	Water seepage through the structure.	▲			
	After maintenance (esp. lubrication) not cleaned.			▲	
	Less running clearance between car and hoistway hampers cleaning.	▲			
Various defects in elec. installation (G39)	A proper co-ordination with electrical team should be planned for all stages. Patchwork maintenance by inexperienced person is most common in such case, e.g. noncalibrated circuit breaker, haphazard wiring, wrong fuse wire, removal of trunking cover, etc.	▲			
	Dust/dirt deposit into the hoist way jams. Switch roller.	▲		▲	
Suspension and governor rope					
Wrong tension (G40)	Overloading of car w.r.t. rope — Rope not strong/less winding.	▲			
	Rope socket slanted/hitting.		▲	▲	
	Wrongly positioned wedge rope socket clip — Slipping due to faulty installation, wear and tear with age and careless maintenance.		▲	▲	▲
	Rope wedge is of wrong size usually from careless replacement.			▲	
Corrosion, broken wire (G41)	Inadequate/improper lubrication or throwing up of lubricant (for suspension rope only).			▲	
	If material not rust resistant, it corrodes on exposure to dust, dirt and water. A proper lubricant may deter or prevents corrosion.	▲		▲	
	The wire undergoes frequent bending or undesirable friction due to unsuitable pulley geometry or groove.	▲			

(Continued)

Table 6.3. *(Continued)*

Defects and Coding	Possible Causes of Defects	Cause gr. D*	C*	M*	E*
Lobby and car door					
Closing on moving users (G42)	Hall and COP (car operating panel) button unable to register call (G32).	▲		▲	▲
	Photo electric (PE) eyes or multi beams not aligned if hitting or banging of door takes place due to rough use/vandalism			▲	▲
	Faulty door interlock electrical contact/door lock switch.			▲	
	Door re-opening device has become faulty.			▲	
No self-closing (G43)	Button or COP unable to register the message (G32).	▲		▲	▲
	Debris between the doors/header track obstruct PE beam of the detector eye.			▲	
Slow/fast (G44)	Faulty door operator.			▲	
Different speed of two doors (G45)	Door clutch, linkage or catch device roller — Any of the parts are worn out, misaligned and thus become faulty.			▲	
Noisy/erratic operation, i.e. door is not moving freely (G46)	Roller for door or hanger is worn, rusted due to damp environment or poor make or excessively dirty.		▲	▲	
	Roller shoe is worn out especially if not fitted with rubber lining.	▲			
	Inadequate clearance of shoe — Faulty installation or rough usage.		▲		▲
	Door operator becomes faulty. The reason may be various: loose/worn parts such as belt, sprocket wheel, inadequate cleaning and lubrication or mechanical damage if the car hits the hoistway ceiling due to uncontrolled movement (G2).		▲	▲	▲

(Continued)

Table 6.3. (*Continued*)

Defects and Coding	Possible Causes of Defects	Cause gr.			
		D*	C*	M*	E*
Rubbing or large gap (>10 mm) in door panel (G47)	Poor adjustment during installation.		▲		
	Damaged restrictor by vandalism (forceful opening of the door).				▲
	Lock contact worn or clearance out due to hitting — Needs readjustment.			▲	
	Catch device is hitting/clearance out due to dirt accumulation.			▲	
Banging (G48)	Tuning problems due to bad installation or careless maintenance.		▲		
Repeated opening and closing (G49)	Door interlock electrical contacts (door lock switch) jammed/burnt, obstructed detector.			▲	
Jamb is loose, rusty, missing or damaged stopper (G50)	Door components not strong or corrosion resistant.	▲			
	Not durable w.r.t usage.	▲			▲
	Vandalism, e.g. banging on jamb to forcefully open the door.				▲
Bulging sill (G51)	Door unable to withstand rough use or vandalism, e.g. banging.	▲			
Opening of lobby door when car is not present (G52)	Malfunctioning of car safety shoes — Worn out, loose or out of balance due to rusting or careless maintenance.			▲	
	Electro-mechanical interlock installed on the lobby door is not synchronized with car door or has become faulty with time.	▲		▲	
Opening of car door in unlocking zone (G53)	Mishandling during maintenance damages micro-switch faulty.			▲	
	Constant hitting or banging on door to open it forcefully damages the micro-switch.				▲

(*Continued*)

Table 6.3. (*Continued*)

Defects and Coding	Possible Causes of Defects	Cause gr. D*	C*	M*	E*
Lobby/landing and accessories					
Faulty button-jam/ loose (G54)	Damaged by heavy usage — Design mismatch of durability and usage. In few areas a vandal proof design is must.	▲			
Faulty detector — Lift may reach fire floor (G55)	Omission in design.	▲			
	Not replaced in time.			▲	
	Damage during maintenance — Hitting hard, painting over the surface.			▲	
Service switch jammed (G56)	Improper and forceful use of other keys in the slot of mechanic's or fireman's service switch — Not guarded, careless maintenance.	▲		▲	
Lift pit					
Dirty, wet often leads to rust (G57)	Inadequate cleaning (esp. if not safe to access or work), oil spillage	▲		▲	
	Poor water-tightness of structural member.	▲	▲		
	Dropping of construction waste if the pit is left unprotected.		▲		
Dimly lit (G58)	Damaged switch or light — Not changed with time, mishandling, breaks easily if not guarded.	▲		▲	
Compensation cable causing noise (G59)	Cable and sheaves are misaligned due to careless installation or mishandling during maintenance.		▲	▲	
	Wrong tension due to unpredicted heavy use.	▲			
	Lack of lubrication.			▲	

D* — Design; C* — Construction; M* — Maintenance; E* — Environment.

6.5. Electrical System

6.5.1. *General*

Electrical system plays an important role particularly for modern commercial buildings, which demand for complex and sophisticated electrical system to meet and support new innovation and technology in this fast changing environment.

An electrical system consists of various components and subcomponents, for instance, transformer, cabling and wiring, plant and equipment, control gear, lighting, emergency power supply as well as earthing and lightning protection. As such, initial installation cost for electrical system is always high. Lack of proper maintenance will result in system unreliability, high hazard, short-life span of plants and equipment, requiring expensive repair and replacements. In the worst case, fire, system shutdowns, injuries and deaths may occur and this should always be prevented from happening.

Conventionally, corrective and preventive maintenance had always been adopted in maintaining building system and services. Regular inspection, testing, repair and replacement formed major part of them. However, in order to maximize the efficiency and effectiveness of electrical system and avoid any circumstances that would affect its performance throughout a building life cycle, predictive maintenance incorporate maintainability issues that should be taken into account as from the design stage.

6.5.2. *Maintainability of Electrical System*

A typical electrical system in a commercial building will start at the main services entrance where incoming from the service provider is located; the electrical energy will then go through a series of voltage step down processes before distributing to the end utilization devices.

For electrical systems consist of so many different components and sub-components (Fig. 6.4), it definitely carries an expensive initial capital cost for installation. However, any design changes or retrofitting during the operational stages would be even higher, due to the coverage of many other building services that are running on electricity.

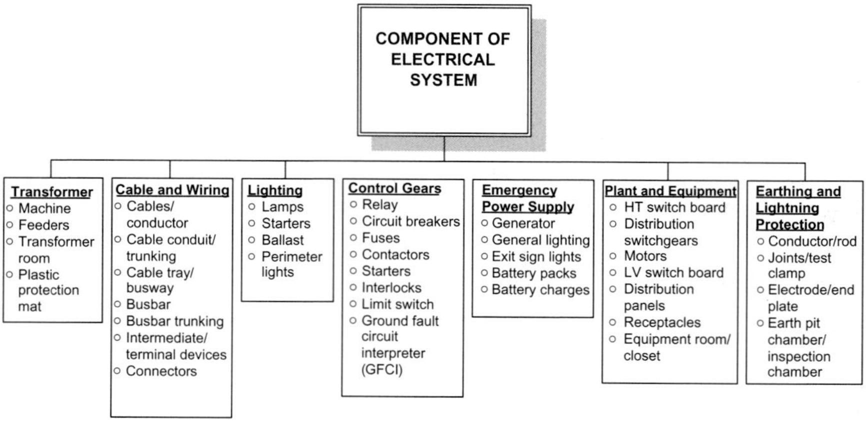

Fig. 6.4. Components and sub-components of electrical systems.

6.5.3. *Common Defects*

Electricity is notorious for claiming life and causing injuries. In US alone, 2287 workers died and 32,807 workers were injured due to electrical shock or burn between 1992 and 1998 [80]. However Floyd *et al.* [81] argue that such databases fail to include "near miss accident data" which may be 30,000 times more than the fatality statistics (Fig. 6.5).

Electricity is also closely related to fire hazards as it supplies the spark or arc — one of the components of fire triangle is overheated devices, loose connections or damaged insulation [82]. Accidental bridging of a gap between two conductors or a charred insulation can generate an arc [83]. Once generated, it can maintain itself for a mere 20 V per cm and become devastating with temperature reaching up to 4000°C [84]. Current with voltage higher than 600 V can cause burn and death. Unfortunately death can occur at a voltage as low as 120 V if current passes through chest. Even a 0.5 mA feeble current undetected by GFCI (ground fault circuit interrupter) can induce a startle shock and lead to fatal fall or contact with dangerous equipment [85]. Explosion in oil insulated transformer and over-burnt dry transformer have been reported by many [86–89]. Apart from hazards, system performance is another area of concern. Dry type transformers often suffer from harmonic-induced heating and severe frequency-dependent

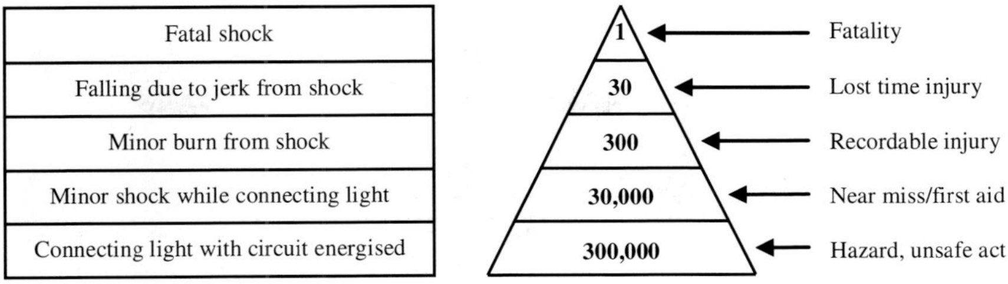

Fig. 6.5. Relationship of accidents of increasing severity of consequences.

losses [90]. Reliability of circuit breakers has been questioned in Ref. 91. Surges can reduce working life of the components by 50% [92].

Defects identified in this study are summarized in Fig. 6.6 and Table 6.4.

6.6. Fire Protection System

6.6.1. *General*

A fire protection system forms one of the minute yet most important building services. The system encompasses both passive and active fire protection measures such as fire extinguishers, detectors, sprinklers and piping with the ability to prevent, protect, suppress and control of the fire.

Having an appropriate and suitable fire protection system in a commercial building is extremely important as it concerns with the safety of the occupants in the building. Any outbreak of fire in a building will cost not only damage to the assets and properties, but also threats to lives and even results in catastrophic state.

Therefore, such system will come in useful in an emergency condition. The two prime objectives of building fire protection systems are to save lives and protect assets. Besides the two main objectives, it serves another function, which is to prevent, suppress or isolate the spread of fire. All in all, fire protection system prevents fire right from the planning stage with proper layout; protects the occupants and property of a building; detects and alerts the occupants when there is fire as well as to suppress the spread of fire through suppression systems.

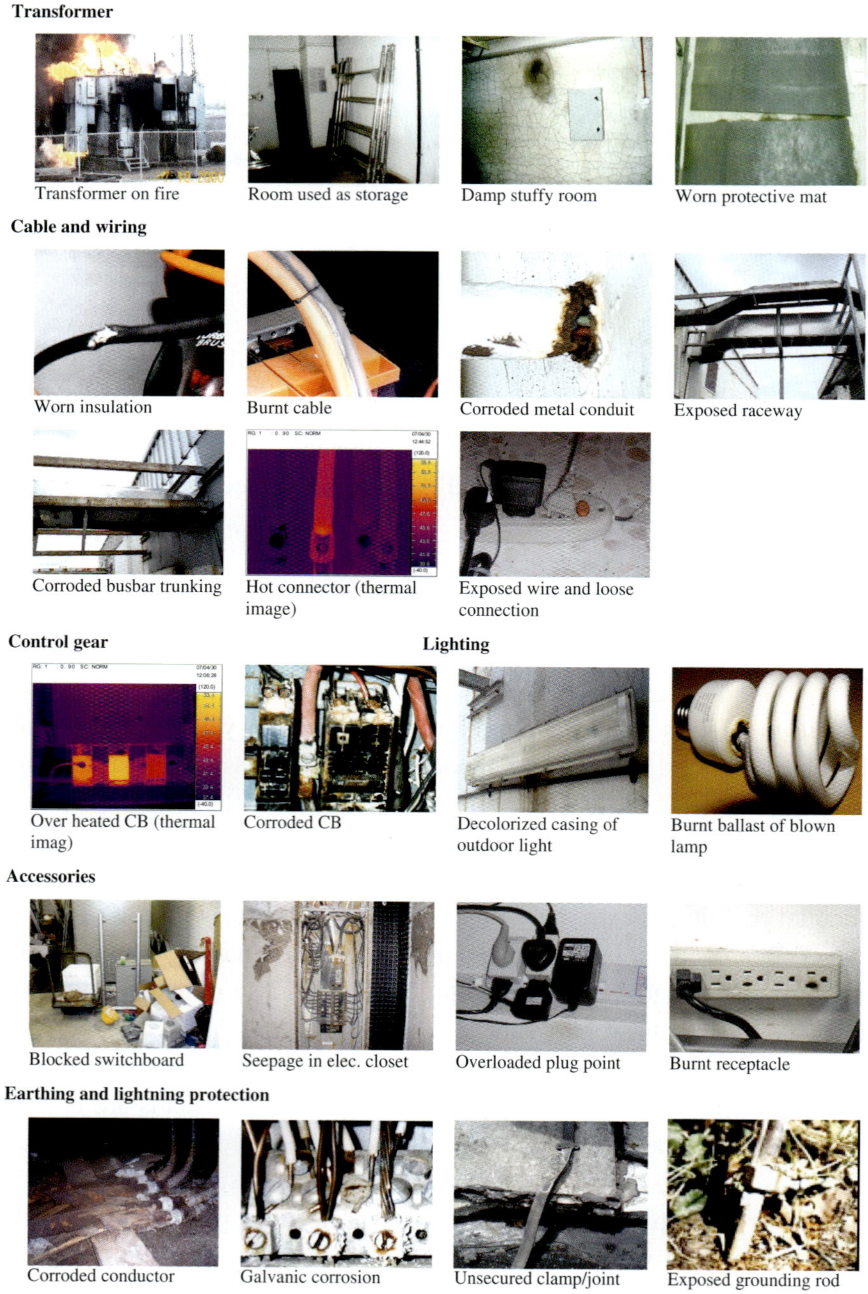

Fig. 6.6. Common defects in electrical system.

Table 6.4. Common defects in electrical system.

Defects and Coding	Possible Causes of Defects	Cause gr. D*	C*	M*	E*
General performance and hazards					
Short circuit (H1)	Poor design details of busbars or cables in terms of center line spacing, size, strength and provision of mechanical supports.	▲	▲		
	No/faulty protective device.	▲		▲	
Poor performance of equipment (H2)	Poor insulation or inadequate conductor spacing causes impedance.	▲	▲		
	Current supplied at lower than rated is example of poor planning or wrong tap adjustment of transformer.	▲	▲	▲	
	Voltage-drop during starting of big motors causes momentary poor performance. Supply is of under capacity or not divided into sections.	▲			
	Voltage-drop in undersized conductors.	▲			
	Wrong wiring connection at receptacles causes different voltage.		▲		
	Rating mismatch of equipment and supply.	▲	▲		
Nuisance tripping (H3)	Supply is far away from load center of building and hence overloading at one part of the building occurs.	▲			
	Fuse of wrong rating or noncalibrated circuit breaker (CB) — Usually from careless maintenance (Love, 2001).			▲	
	Transformer tap setting at lower standard may hike load.	▲	▲	▲	
	Loose connections causing spark, overheating or burning.		▲	▲	
	Single phase circuit catering three phase equipment.	▲			
	Under-capacity of circuit to cater appliances with higher load factor.	▲			

(Continued)

Table 6.4. *(Continued)*

Defects and Coding	Possible Causes of Defects	Cause gr.			
		D*	C*	M*	E*
Shock and electrocution (H4)	Wrong polarity (e.g. switch gear connected to neutral). This may give shock even when the equipment is switched off.		▲		
	Deterioration of protective coverings/insulation with time or carbonization due to high voltage difference across insulation.	▲			▲
	Condensation on exposed live part. This happens if other services cross through electrical room or closet.	▲			
	Improper grounding or "mis-wiring".		▲		
	Faulty/no GFCI. Accidentally if any metal tool or ladder bridges the gap between two conductors, current flows through an unintended part of the circuit.	▲		▲	
	No/poor connectivity of grounding system to clear the fault (H38).	▲	▲	▲	▲
Arc, spark (H5)	Damaged/charred insulation by careless splicing, mechanical impact, use of wrong cleaning chemical and exposure to high sources or aging.	▲	▲	▲	▲
	Bridging two exposed conductors by metallic tool. Places under regular O&M should not have exposed cabling.	▲			
	Connectors are loose or have inadequate tip pressure, nonremoval of oxide, rusting, etc.		▲	▲	
	Very high resistance at connections if conductors are clamped together without sufficient contact pressure, corrosion of contactor or use of dissimilar material with different thermal expansion.	▲	▲		
	Mechanical failure (creep) with time.				▲
	Water enters through poorly detailed/finished transition points and corrodes conductor.	▲	▲		
	Short circuit heating in undersized conductor.	▲			

(Continued)

Table 6.4. *(Continued)*

Defects and Coding	Possible Causes of Defects	Cause gr. D*	C*	M*	E*
Difficult identification/ rectification of defect (H6)	Haphazard wiring and wrong color coding. Cramped wires in undersized junction box or outlet box also results in haphazard wiring.	▲	▲		
	No "as built" drawing available, esp. for changes made by tenants. Distribution boards (DB) should also be preferably labeled.	▲			
	Inadequate number of intermediate pull or junction box.	▲			
	Difficult pulling over sharp bends in conduit. Long distance between junction boxes makes it difficult to replace any damaged wire.	▲			
Total power cut from one fault (H7)	Wrong/no provision of circuit breaker at desired isolation.	▲			
	Ballasts if not fitted with in-line fuse-holder and time delay fuse, one failed unit causes black-out in entire area.	▲			
Transformer					
Vibration and noise (H8)	Overloading and fluctuation density of core.	▲			
	Loose connection of bolt on clamping structure. Use of wrong sized crimp or improper tool.			▲	
	No precaution of sound isolating pad/vibration damper for mounting.	▲			
Overheated or burnt body, feeder, etc. (H9)	Insufficient air ventilation in the room for heat dissipation.	▲	▲		
	Loose or wrong sized parts increase resistance.		▲	▲	
	Over greasing, lack of proper tool and training cause loose connections.			▲	
	Eddy current, excessive overloading.	▲			
	No thorough investigation to detect the problem in inner parts.			▲	
	Dirt or grime from insulation and windings prevent heat dissipation.			▲	

(Continued)

Table 6.4. (*Continued*)

Defects and Coding	Possible Causes of Defects	Cause gr. D*	C*	M*	E*
Oil leakage (H10)	Foreign make — Not suitable for higher ambient temperature of tropics.	▲			
	Careless lubrication.			▲	
	Excessive corrosion overlooked for a very long time.			▲	
Damaged insulation (H11)	Damage during transportation is not checked before installation.		▲		
	Poor cleaning — Often hindered by poor access.	▲		▲	
	High acidity of cooling liquid. Should be checked and changed.			▲	
	Breakdown of insulation from overheating (H9).	▲	▲	▲	
Unsuitable/unsafe working condition (H12)	Insufficient space design or transformer room is used as storage.	▲		▲	
	Protection mat deteriorated with time or spillage of chemicals.			▲	▲
	Inadequate lighting and ventilation.	▲			
	Slippery and dirty floor surface from oil spillage or scattered parts.	▲		▲	
	Omission or missing warning sign or locks.	▲			▲
	Door mechanism of transformer room is not set to de-energize transformers upon opening the door.		▲		
Cable and wiring					
Overheated/burnt (H13)	Narrow cross section or excessive length causes resistance heating.	▲			
	Exposed to direct sunlight or other heat sources.	▲			
	Short circuit or ground fault.	▲	▲	▲	
	Overloaded busbar. K-rated transformer can reduce harmonics.	▲			

(*Continued*)

Table 6.4. (*Continued*)

Defects and Coding	Possible Causes of Defects	Cause gr. D*	C*	M*	E*
Mechanical damage (H14)	Movement or dangling if support is not secured or insufficient.	▲	▲		
	Broken/loose supports from excessive corrosion.			▲	
	Excessive tension — Careless workmanship.		▲		
	Too close to structural frame or no hangers to allow movement. Damage is more prominent in absence of expansion joint or open run.	▲			
	Inadequate protection from crushing, impact or abrasion.	▲	▲		
	Sharp or too many bends, pulled over excessive length. Intermediate junction/pull box can solve such problem to a great extent.	▲			
	Nicks during careless cutting results in broken wire beneath the cover.		▲		
	Damaged during maintenance of other services running too close/interfering (result of poor planning).	▲			
	Conductor not firmly secured — Pulsating current causes sway.	▲	▲		
	Missing cover of junction/pull box (H17) encourages vandalism, esp. if the cables can be pulled easily due to gap between cable and knockouts.	▲	▲	▲	▲
Corroded metal parts (H15)	Water seepage due to poorly designed/finished joint.	▲	▲		
	Use of unsuitable lubricant, pulling or cleaning agent.		▲	▲	
	Lashing water jet for cleaning enters through ends or joints which are not properly designed or sealed.	▲	▲		
	Adjacent structure is not water resistant.	▲			
	Galvanic corrosion.	▲			

(*Continued*)

Table 6.4. (*Continued*)

Defects and Coding	Possible Causes of Defects	Cause gr.			
		D*	C*	M*	E*
Damaged insulation (H16)	Softened from prolonged exposure to high temperature (>70°C).	▲			
	Use of unsuitable lubricant, pulling or cleaning agent.		▲	▲	
	Narrow cross section or excessive length causes resistance and heating.	▲			
	High inrush current from short circuit (H1) or ground fault.	▲	▲	▲	
Damaged elec. box (H17)	Intermediate (junction or pull) box — Not made of rust proof material.	▲			
	Knockouts are made using improper tools or methods. Should be done before mounting.		▲		
	Knockouts are made forcefully big to accommodate too many cables.	▲			
	Screw/cover missing during careless installation or maintenance.		▲	▲	
Distribution equipment					
Unsafe switch-board or panel-board (H18)	Insufficient space or inadequate access for proper operation and maintenance.	▲		▲	
	Warning notice/label — Not provided, peeled off or not visible.	▲			▲
	Insulation is damaged (crack, blister, warping) by overheating, physical impact or from spillage of cleaning chemicals.	▲		▲	
	Moisture ingress in the metal closet or it is unsuitable for exposure condition.	▲			
	Shock hazard (H4).	▲	▲		▲
Burnt mark or smell in switch-board or panel-board (H19)	Rust, dirt causing high resistance and heating at cable contacts.	▲		▲	
	Burnt or charred insulation by exposure to external heat source or constant arc produced inside. Inner components should be checked with special means.	▲		▲	
	Catering to too many protection devices or connections.	▲			

(*Continued*)

Table 6.4. (*Continued*)

Defects and Coding	Possible Causes of Defects	Cause gr. D*	C*	M*	E*
Outlet box unsafe (H20)	Adapter catering to too many devices — Poor planning to accommodate future extension.	▲			
	Poor design or furniture layout, stacking makes it inaccessible.	▲		▲	
Noise/spark (H21)	Resonant to the magnetic field generated by higher-frequency harmonic current. Use of *k*-rated transformer is one of the solutions.	▲			
No power at receptacle (H22)	Internal contacts burnt or worn out from constant sparking.	▲			
	Careless usage, e.g. forceful insertion of flat pins in round slots.				▲
Jam/loose switchgear/ switch (H23)	Rough use or vandalism (esp. switches in outlet box).				▲
	Irregular cleaning and lubrication.			▲	
	May require replacement after certain time.			▲	
	Corrosion from poor make or water seepage in electrical closet.	▲			
Inaccurate meter (H24)	Faulty device needs resetting.			▲	
	Broken wires, dirt or dust impeding the movement.			▲	
Protective device					
Frozen/unable to trip (H25)	Damaged by dirt, corrosion or mishandling.		▲	▲	
	Contacts are burnt if they are loosely fitted.		▲		
	Improper design rating or wrong installation — Should be tested.	▲	▲		
	Fuse of wrong rating or noncalibrated CB — Usually from careless maintenance (Love, 2001).			▲	

(*Continued*)

Table 6.4. (*Continued*)

Defects and Coding	Possible Causes of Defects	Cause gr. D*	C*	M*	E*
	GFCI-deformed or missing pins, damaged insulation and incorrect wiring		▲	▲	
	Interlock is worn with time or key is missing for mechanical type. As a result CBs cannot function properly.			▲	▲
Noisy contactor (H26)	Poor contact in control circuit.		▲	▲	
	Low voltage causes chatter — Rating mismatch with supply.	▲			
Lighting					
Lamps blink, flicker, freq. blow off (H27)	Voltage dip if cable between lamp and fused splitter is too long or a heavy motor without reduced voltage starter is in same supply line.	▲			
	Exhausted lamp life. Replacement needed.			▲	
	No over current protection device (fuse/circuit breaker) provided.	▲			
Less light (H28)	Shadow casting by over-decorative luminaire.	▲			
	Lumen output of lamps is inadequate.	▲			
	Luminaire spacing is too high. Usually such mismatch occurs when the luminaire cannot be shifted according to new interior layout.	▲			
	Covered by dirt, dust and grime. Poor maintenance or design of luminaire is prone to collect dust.	▲		▲	
	Reduced lamp life with time — No regular replacement.			▲	

(*Continued*)

Table 6.4. *(Continued)*

Defects and Coding	Possible Causes of Defects	Cause gr. D*	C*	M*	E*
No light (H29)	Connection of lamp and corresponding switch is not proper. Lights controlled by 2, 3 or 4 way switch mainly suffer from such coordination issues.		▲		
	Blown lamp is not replaced.			▲	
	Metal halide lamps upon voltage drop take few minutes to restart.	▲			
Damaged casing (H30)	Wear and tear with time.				▲
	Exposure to direct weather condition.	▲			
Overheated/burnt ballast (H31)	Flickering lamps (H27).	▲		▲	
	One of the paired lamps is removed or blown. If it is not replaced for a long time, ballasts get overheated from spark.			▲	
Noisy ballast (H32)	Loose fitting in the ballast causes vibration.		▲	▲	
	Vibrations get amplified by large radiating surface of luminaires. Good quality products are tested to pass such design defects.	▲			
Standby/emergency power supply					
Generator: Noise, vibe and overheating (H33)	Insufficient ventilation by radiator — Water level is low or it has inadequate opening in the generator room.	▲		▲	
	Overloading or unbalanced load.	▲			
	No acoustical control is provided.	▲			
	Vibration isolator is ineffective due to improper mounting.		▲		
	Vee belt is loose or too tight.		▲	▲	

(Continued)

Table 6.4. (*Continued*)

Defects and Coding	Possible Causes of Defects	Cause gr. D*	C*	M*	E*
Dirty, leaky generator body (H34)	Filters are clogged and dirty.			▲	
	Poor servicing — Rusts should be removed and after that, parts should be painted or lubricated.			▲	
Damaged insulation (H35)	Damaged during transportation and not checked before installation.		▲		
	Frequent/extended operation at stand by rating.	▲		▲	
	Overheating is not detected through testing for a long time.			▲	
No/delayed/less power supply (H36)	Power source is not reliable or under capacity.	▲			
	Wrong circuit design or wiring — It is unable to activate the generator or battery when the normal power supply is cut. Such control should be tested and marking on DBs are recommended to avoid ambiguity.	▲	▲		
	Generator fails to start automatically due to insufficient air intake for combustion, excessive corrosion or faulty limit switch.	▲		▲	
	Battery goes flat — Over discharged or not charged properly, tampered, electrolyte condition not satisfactory.	▲		▲	
	Leakage in battery due to excessive rusting, poor cleaning or painting.			▲	
	Poor contact at any junction or battery terminals — Careless maintenance or rusting esp. in damp and dirty storage place.	▲	▲	▲	
	Voltage-dip during transfer of power to alternate source if the transfer mechanism is faulty. UPS (uninterruptible power supply) should be used for critical circuits.	▲		▲	

(*Continued*)

Table 6.4. *(Continued)*

Defects and Coding	Possible Causes of Defects	Cause gr. D*	C*	M*	E*
Grounding and lightning protection system (LPS)					
Corroded parts (H37)	Dissimilar metals are connected without special precaution. As a result galvanic corrosion takes place.	▲	▲		
	Protective coating is damaged. Should be checked regularly.			▲	
No/poor conductivity (H38)	Earth electrodes/conductors are not resistant to weathering or deteriorate over a long period of time.	▲			▲
	Inadequate contact with surrounding soil.	▲			
	Incomplete network formed for all conductors in the building.	▲			
	Surrounding soil has poor conductivity and not treated for the same.		▲		
	Inadequate burial depth or improper driving damage conductors.	▲	▲		
	Fasteners are not corrosion resistant. They become rusted and loosen contact over time. Should be checked regularly.	▲		▲	
	Vandalism if cover of earth pit/inspecting chamber is not secured.	▲			▲
	Thorough tests not conducted after installation and at specified interval to detect such faults.		▲	▲	
Damaged electrode (H39)	Inadequate burial depth or careless excavation in vicinity.	▲	▲		
	Vandalism if cover of earth pit/inspecting chamber not secured.	▲		▲	

D* — Design; C* — Construction; M* — Maintenance; E* — Environment.

While the construction cost of a fire protection system may be as low as 1% and seldom more than 5% [93], a safe building begins with carefully analyzed architectural layout and planned maintenance schedules. This includes the appropriate locations and construction of the various subsystems and frequently ongoing maintenance and repairs. Incontestably, the importance of an efficient and efficient fire protection system and management cannot be overlooked.

Most designs and operations of the fire protection system is strictly governed by fire local protection codes such as the Code of Practices for various fire equipments and subsystems and international standards such as National Fire Protection Association. The designers and facilities managers (FM) need to be conscious of these codes during the design and post occupancy phases. Indeed, the scope of FMs in maintaining the system efficiently is obligatory as it deals with the lives of the occupants.

In order to achieve an efficient performance of fire protection system, it is crucial to maintain the fire protection system well. As it is not put into regular use, the working condition of the system is often overlooked. With proper inspection and maintenance, the system can perform at its optimum function. When the system is kept in proper condition, it also helps to eliminate the extra costs and inconvenience generating from instances such as false alarms.

6.6.2. *Maintainability of Fire Protection System*

In many countries, there is a range of mandatory inspections of the components of the fire protection system on weekly to annual basis. However, carrying out the same to prevent defects is only a near ideal situation. Poor design, bad workmanship, mishandling and wear and tear are almost impossible to avoid. Hence, it is important for the fire managers to understand and recognize the shortcomings of the system that can affect the system performance.

Fire protection system can be subdivided into four major components, namely, (1) detection and communication system; (2) fire fighting system; (3) control and monitoring devices and (4) fire escape. Each component can be further structured and divided into major sub components (Fig. 6.7), which are commonly found in commercial buildings.

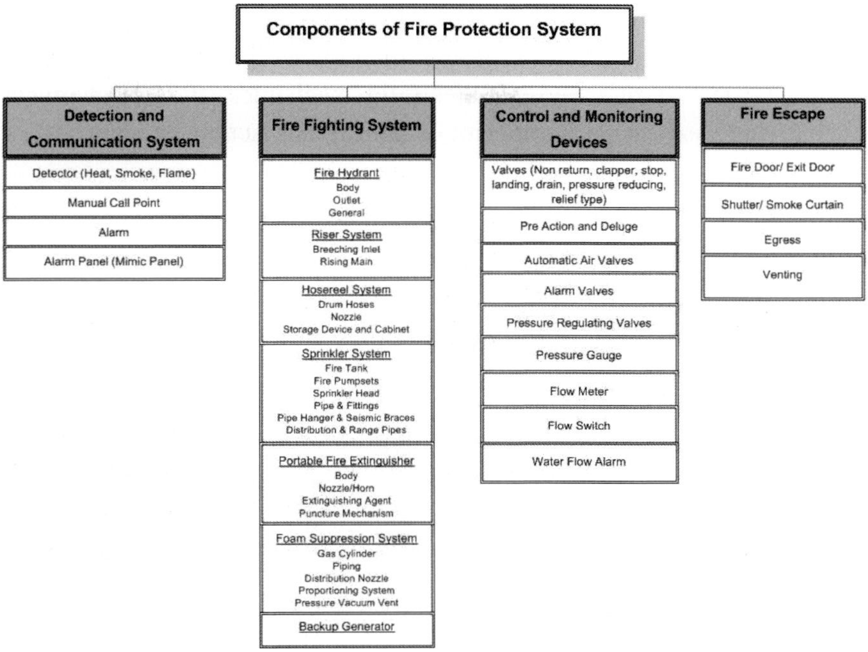

Fig. 6.7. Outline of fire protection system in a typical commercial building.

6.6.3. *Common Defects*

Fire hazards have been recorded since early dates of history starting from arson fire of Rome in 390BC [94]. Fire in buildings costs about 1% of GDP each year for developed nations [95]. Contrary to its destructive nature of loss of life and property, such incidents have contributed to development of fire safety engineering. Instead of traditional fire triangle of oxygen, fuel and heat, concept of fire quadrangle with the fourth element of chemical chain reaction has been introduced in development of extinguishers. Similarly planning aspects such as efficient occupant evacuation and early detection or suppression of fire are being dealt with probability theory, risk assessment or decision tree [96–98].

Fire protection engineering is multidisciplinary in true sense [99]. Architecturally it is the layout and planning of escape route, while fire rating of the construction is covered by civil engineering. Mechanical aspects involve flow of water through the pipe work, discharge of extinguishing agent through nozzle and fluid mechanics. Finally chemical engineering deals with hazards of chemical interaction and processes.

Fig. 6.8. Common defects in fire proctection system.

Table 6.5. Common defects in fire protection system.

Defects and Coding	Possible Causes of Defects	Cause gr. D*	C*	M*	E*
Detection and communication					
Under performance or false alarm of detector (J1)	Flame detector is misaligned, damaged by shock if loosely fitted.		▲		
	Wrong selection (exposed to usual but high heat, smoke or light) .	▲			
	Function not checked through simulation test after installation and during regular maintenance.		▲	▲	
	Obstructed or covered by paint — Poor housekeeping.			▲	
	Dirty — Especially for poorly accessible points.	▲		▲	
Manual call point faulty (J2)	No power supply (unreliable power source, faulty wiring or battery goes flat).	▲	▲	▲	
Inaudible or unidentifiable alarm (J3)	Wrong selection of bells — Not distinguishable from general clutter.	▲			
	Loose gong bolt or damaged alarm — Blocked, corroded, spoilt by extreme temperature fluctuation.	▲		▲	
	Faulty alarm panel to indicate the exact alarm — No lucid diagram, corresponding bulb blown off or battery goes flat.	▲		▲	
Alarm panel — No signal, spoilt exterior (J4)	Proper connectivity to response center is not provided.	▲	▲		
	Panel is tampered — Lack of concern for safety or unsafe location.	▲			
	Bulbs, buzzer etc cannot work if not reliable or battery goes flat.	▲		▲	
	Blown bulbs are not replaced.			▲	
	Excessive damp location or poor make leads to corrosion of cabinet.	▲		▲	

(*Continued*)

Table 6.5. (*Continued*)

Defects and Coding	Possible Causes of Defects	Cause gr. D*	C*	M*	E*
Fire hydrant and accessories					
Difficult access/ detection (J5)	Obstructed by parking, landscaping — Poor project coordination.	▲			
	Not installed at adequate height as per specification.		▲		
Faulty — Leaky, jammed (J6)	Damaged by corrosion, physical damage. e.g. knocked by vehicles.	▲			
	Parts (stem, cap, plug, thread, etc.) damaged during installation.		▲		
	Corroded parts if not lubricated or painted at regular interval.			▲	
	Outlet is not properly tightened after testing.		▲	▲	
	Missing operating nozzle — Careless use, not secured.		▲	▲	
Unacceptable wkg. condn. (J7)	Such examples are: Dry riser having water or less pressure in wet riser. It happens mainly if valves are faulty or leaky.		▲		
Rusted riser (J8)	Poor make.	▲			
	Cyclic wetting and drying if the tests are conducted carelessly.		▲		
	Use of raw/dirty water. Tanks may not be cleaned for long.	▲		▲	
Inlet jam (J9)	Dirt deposition in the rubber cap or breeching inlet.			▲	
Fire hose					
Damaged hose — Cut, kink, leak, missing part, abrasion (J10)	Misuse — Washing floors or vehicles etc., mishandling.			▲	
	Wear and tear with time. Accelerated aging if of poor make.	▲			▲
	Improper winding due to careless handling. Velcro strap damaged.	▲	▲	▲	

(*Continued*)

Table 6.5. *(Continued)*

Defects and Coding	Possible Causes of Defects	Cause gr.			
		D*	C*	M*	E*
Damaged nozzle (J11)	Mishandling, careless maintenance, e.g. dislodged from stem or not connected to hose rack nipple or valve.			▲	
	Jammed, obstructed if not cleaned properly or flushed out.		▲	▲	
Cabinet difficult to open (J12)	Hinges not lubricated, corroded from poor make especially if in extreme damp location.	▲		▲	
Damaged cabinet (J13)	Located in a place vulnerable to mechanical damage.	▲			
	Glazing broken or glass break device missing due to vandalism.				▲
Sprinkler system					
No more functional (J14)	Unsuitable choice of type and function of sprinkler makes it nonoperative within a short period of time.	▲			
	Covered by paint or other foreign material.			▲	
	Poor maintenance — Dirty, damaged by use of wrong wrench.			▲	
	Empty gas bulb — Needs replacement with time.			▲	
Inadequate spray/ discharge (J15)	Poor consideration for potential obstructions in planning.	▲			
	Wrong orientation/hindrance by supports (careless installation).		▲		
	Obstructed by poor housekeeping — Stacking of material.			▲	
	Replaced with different type (usually if spare unit not in stock), grease/dirt in spray nozzle.			▲	

(Continued)

Table 6.5. (*Continued*)

Defects and Coding	Possible Causes of Defects	Cause gr. D*	C*	M*	E*
Leakage in pipe (J16)	Excessive corrosion and scaling if pipe is not rust resistant.	▲			
	No remedial measure taken for initial corrosion and pitting.			▲	
	Microbiologically influenced corrosion (MIC) due to use of raw/untreated water, irregular flushing.	▲		▲	
Blocked pipe (J17)	Excessive scale and sedimentation from raw water, inferior material or irregular flashing.	▲		▲	
	Misaligned, careless workmanship — Tools, loose fittings left inside.		▲		
Sagging pipe (J18)	Effect of vibration, water flow, etc. not taken into account in design or mounting location is not enough strong.	▲			
	Strength of supports were not cross checked during installation.		▲		
	Pipe hanger and seismic brace become loose and worn with time.				▲
Pump-excess vibration, noise overheating (J19)	No safe working condition or enough space in pump room for regular inspection, maintenance or replacement of pumps.	▲			
	Insulating pad not provided to reduce noise and vibration.	▲			
	Defects during installation remain unnoticed.		▲		
	Pumps are not maintained regularly, e.g. cleaning, lubrication, inspection and testing.			▲	
Overflowing/leaky tank(J20)	Tank body not made of corrosion proof material.	▲			
	Tanks not maintained regularly — Sludge removal, overflow pipes, etc.			▲	

(*Continued*)

Table 6.5. *(Continued)*

Defects and Coding	Possible Causes of Defects	Cause gr.			
		D*	C*	M*	E*
Leakage in vales (J21)	Valves are not closed/opened fully if jammed from dirt or rust from poor material, raw water or irregular lubrication.	▲		▲	
	Valves damaged by mishandling, e.g. use of wrong wrench.			▲	
	Valves not properly closed or opened during maintenance/after testing.			▲	
	Valves are misfit — Usually due to wrong replacement.			▲	
	Gasket/washer not aligned or wear and tear with time.		▲	▲	▲
Missing (J22)	Theft of valve/parts such as hand wheel, padlock, strap, etc.			▲	▲
Auto valve is no more auto (J23)	Water pressure that regulates the valve is too low. Should be detected by pressure gauge.			▲	
	Leakage through rubber washer which is missing or groves dirty.			▲	
False alarms (J24)	Retarding chamber or alarm drain leaking.			▲	
	Supervisory switch tampered by vandalism or mishandling.			▲	▲
	Pressure gauge has loose spring or worn out dial. Accuracy of reading should be tested.			▲	
	Pressure regulating valve or switch not calibrated after testing.		▲	▲	
	Rusting or wear/tear of flow switch tongue giving wrong signal.			▲	
	Misfit flow switch — Usually during replacement.			▲	
Preact. and deluge valve — Low discharge (J25)	Drop in supervisory air pressure due to leakage at jointing.			▲	
	Diaphragm not opening full if rusted, dirty (esp. use of raw water).	▲		▲	

(Continued)

Table 6.5. (*Continued*)

Defects and Coding	Possible Causes of Defects	Cause gr. D*	C*	M*	E*
Portable fire extinguisher					
Missing/damaged parts or the whole unit (J26)	Corroded if not rust resistant or exposed to very damp environment.	▲		▲	
	Denting or abrasion due to mishandling.			▲	
	Mishandling may damage parts (seal trap, thread, gasket, handle, etc.)			▲	
	Stolen — Vandalism. Mounting should be able to prevent that.		▲		▲
Wrongly placed (J27)	No concern for safety rules, misuse, e.g. as door stopper.			▲	
	Hangers are broken due to corrosion, not secured to structure or used for hanging heavier extinguisher.		▲		
Less pressure in nozzle or poor discharge (J28)	Irregular servicing — Leaky pressure valve, poor content of extinguishing agent or clogged by the dated agent.			▲	
	Damaged nozzle such as cracked, deformed, spoilt thread, etc. from mishandling or vandalism. Mounting not vandal proof.		▲	▲	▲
	Illegible weight marking may leave gas cartridge partially empty.			▲	
	Extinguisher if not hydrostatically tested, may not be suitable to cater to the full pressure.			▲	
	Leakage due to corrosion of inferior quality pipe material.	▲			

(*Continued*)

Table 6.5. (*Continued*)

Defects and Coding	Possible Causes of Defects	Cause gr. D*	C*	M*	E*
Unacceptable extinguishing agent — Needs change or unsuitable discharge (J29)	Hardening of powder in damp environment.				▲
	Water/foam has incorrect weight or of expired date.			▲	
	Liquid becomes gummy if not changed in time and clogs the pressure vacuum vent or the discharge nozzle.			▲	
	Malfunction of components with time — Suppression activator, proportioning device, etc. should be replaced as recommended.			▲	
	Components of nozzle cap, plug, etc. corroded, missing or misaligned.			▲	
	A change in type is required to suit the anticipated type of fire and hazard class. Usually happens due to change in building use.	▲			
Faulty puncture mechanism (J30)	Poor make.	▲			
	Thread is blunt or jammed — Damaged during installation/servicing.		▲	▲	
Fire escape and related services					
Door not fully closing/ opening (J31)	No/faulty self closure (usually damaged if no cushioned back check).	▲			
	Not set to operate with minimum force.		▲		
	Jammed/sagging from poor workmanship (e.g. tilted hinge), rough use.		▲		▲
	Stopper provided or obstructed due to structural member or stacking of material nearby.	▲		▲	
	Accuracy of gap should be maintained during all stages.	▲	▲	▲	

(*Continued*)

Table 6.5. (*Continued*)

Defects and Coding	Possible Causes of Defects	Cause gr. D*	C*	M*	E*
Damaged door-panel, label, glass broken etc. (J32)	Accelerated aging due to poor specification or installed door differs from tested prototype. Usually such doors have lower fire rating than expected.	▲			
	Panels stripped off with time, rough usage or mishandling during installation. A regular checking or minor repair is recommended.		▲	▲	▲
	Label is peeled off (not vandal-proof metal plate type) or covered during careless painting. Labels are required for confusion free operation and maintenance.	▲		▲	
	Glass is not vandal proof or of larger area than recommended.	▲			
Faulty door release (J33)	Faulty sensor — Paint/coat on its surface or aging.			▲	▲
	Corrosion of the link or hinges. Lubrication helps.			▲	
No/less light for exit and sign (J34)	Inadequate provision of lux level.	▲			
	Bulbs dirty, blown or faulty starter/ballast. Better to use LED.			▲	
	Power source is not reliable — Use of standby generator which fails mostly as controlled by mains.	▲		▲	

(*Continued*)

Table 6.5. (*Continued*)

Defects and Coding	Possible Causes of Defects	D*	C*	M*	E*
	Wrong circuit design or wiring — It is unable to switch on escape lighting when the normal power is switched off. Regular testing and marking on DBs are recommended to avoid ambiguity.	▲	▲	▲	
	Battery goes flat — Over discharged or not charged properly, tampered, electrolyte condition not satisfactory or leaking.	▲	▲	▲	
	Battery or light not working at all — If mounted in wrong position.		▲		
Smoke curtain/shutter half closed (J35)	Careless maintenance. After testing not reset into operating condition.			▲	
	Parts are corroded/not lubricated — Not operating smoothly.			▲	
Insufficient ventilation (J36)	Poor intake by fans which suffers from faulty installation, poor servicing or clogged ventilating screen.		▲	▲	
	Not enough provision of standby fan to compensate poor performance of duty fans.	▲			

D* — Design; C* — Construction; M* — Maintenance; E* — Environment.

Many of these common defects could be avoided if maintainability is considered right from the design stage [100–102]. In most of the cases the mistakes reported are (1) exits not properly marked or blocked or locked; (2) no/inoperative alarm system; (3) fire hose, standpipe or extinguishers missing; (4) no sprinkler installed or sprinkler did not respond and (5) no fire training provided to occupants or maintenance team.

Fire protection systems are required only during fire hazard and such hazard is a rare incident. Hence unless a maintainability plan is adopted, there is a high possibility that during emergency it may remain inoperative. Especially for nonsprinkled buildings fire brigade relies upon fire hydrants. As these are usually under supervision of local fire department, it has a higher probability of being forgotten or being camouflaged by landscaping, parking or any other outdoor structure.

The common defects often faced by maintenance team are summarized in Fig. 6.8 and Table 6.5.

References

[1] P. S. Dunston and C. E. Williamson, "Incorporating maintainability in constructability review process", *Journal of Management Engineering*, **15**(5), 56–60, 1999.

[2] M. F. Wani and O. P. Gandhi, "Development of maintainability index for mechanical systems", *Reliability Engineering and System Safety*, **65**, 259–270, 1999.

[3] PUB, *Code of Practice on Sewerage and Sanitary Works*, PUB (Public Utilities Board), Singapore, 2004.

[4] M. A. Hassanain, "Guidelines for the design of water and sewage systems in buildings", *Journal Architectural Engineering*, **11**(4), 117–121, 2005.

[5] V. T. Manas, *National Plumbing Code Handbook: Standards and Design Information*, 1st Edition, McGraw-Hill, NY, 1957.

[6] SPRING Singapore, SS *CP 48: Code of Practice for Water Services*, SPRING, Singapore, 2005.

[7] R. D. Woodson, *Int. Plumbing Codes Handbook*, McGraw-Hill, NY, 2006.

[8] NAHB Research Center Inc., *Design Guideline: Residential PEX Water Supply Plumbing Systems*, NAHB (National Association of Home Builders), Upper Marlboro, MD, 2006.

[9] V. Lauer, "Plumbing in the 21st century", *Heat./Piping/Air Con. HPAC Engineering*, **62**(10), 43–47, 1990.

[10] I. J. Karassik, *Pump Handbook*, 4th Edition, McGraw-Hill Education, 2007.

[11] C. H. Cheng, "A physical study of plumbing life cycle in apartment houses", *Building and Environment*, **36**(9), 1049–1056, 2001.

[12] ENV and PUB (23 August 2000), *Water Contamination at Bukit Timah Plaza: Update*, from http://app.mewr.gov.sg/press.asp?id=SAS788. Retrieved 7 September 2007.

[13] W. Mathys, J. Stanke, M. Harmuth and E. J. Mathys, "Occurrence of Legionella in hot water systems of single-family residences in suburbs of two German cities with special reference to solar and district heating", *International Journal Hygiene and Environmental Health*, 2007.

[14] A. Plottu-Pecheux, C. Democrate, B. Houssais, D. Gatel and J. Cavard, "Controlling the corrosiveness of blended waters", *Desalination*, **138**(1), 237–249, 2001.

[15] R. Fertmann, S. Hentschel, D. Dengler, U. Janßen and A. Lommel, "Lead exposure in drinking water: An epidemiological study in Hamburg, Germany", *International Journal Hygiene and Environmental Health*, **207**(3), 235–244, 2004.

[16] S. I. Percival, J. S. Knapp, R. G. J. Edyvean and D. S. Wales, "Biofilms, mains water and stainless steel", *Water Resources*, **32**(7), 2187–2201, 1998.

[17] M. Y. L. Chew, S. Das, N. DeSilva and F. Y. Foon, "Grading maintainability parameters for sanitary-plumbing system for high-rise residential buildings", *Proc BEAR2008: Int. Conf. in Bldg. Edu. Res.*, Sri Lanka, 2008.

[18] (E3) Energy Efficient Engineered Designs, Heating, Ventilating and Air Conditioning (HVAC) System Survey, retrieved from http://www.e3designs.com/articles/LCA_summary_5_17_04.html, 2002.

[19] L. Lu, W. Cai, Y. S. Chai and L. Xie, "Global optimization for overall HVAC systems — Part I problem formulation and analysis", *Energy Conversion and Management*, **46**(7–8), 999–1014, 2005.

[20] S. B. Austin, "Chilled water system optimization", *ASHRAE Journals*, **35**(7), 50–56, 1993.

[21] T. B. Hartman, "Global optimization strategies for high-performance controls", *ASHRAE Transactions*, **10**(2), 679–687, 1995.

[22] O. Ahmed, "DDC applications in variable-water-volume systems", *ASHRAE Transactions*, **97**(1), 751–758, 1991.

[23] C. D. Heidler, "Life cycle costing: Getting approval for the budget you need", *APPA Proceedings*, Appendix VI-D-1, 1–19, 1994.

[24] L. Chen and J. Cai, "Using vector projection method to evaluate maintainability of mechanical system in design review", *Reliability Engineering and System Safety*, **81**, 147–154, 2003.

[25] K. Srinivas, R. B. Michael and L. Larry, "Automated proactive techniques for commissioning air-handling units", *Transactions of the ASME*, **125**, 282–291, 2003.

[26] D. Nathan, "$175 million — That's the energy bill for offices", *The Straits Times*, **62**, 9 December 1999.

[27] M. Y. L. Chew and S. Das, "Maintainability aspects of central chilled water HVAC system", paper presented at the *4th. Int. Conf. on Const. in the 21st Century (CITC-IV)*, Gold Coast, Australia, 2007.

[28] B. Yu and A. H. C. van Paassen, "State-of-the-art of energy fault", paper presented at the *Int. Symp. on Air Conditioning in High Rise Building*, Shanghai, China, 2000.

[29] R. Kitter, "Mechanical dehumidification control strategies and psychometrics", *ASHRAE Transactions*, **102**, 613–617, 1996.

[30] S. Katipamula and M. R. Brambley, "Methods for fault detection, diagnostics and prognostics for building systems — A review", Part I. *International Journal of Heating, Ventilating, Air Conditioning and Refrigerating Research*, **11**(2), 3–25, 2005.

[31] B. K. Browning, "Case studies of commissioning HVAC systems", *Proc. ICEBO — Int. Conf. for Enhanced Bldg. Operations*, Texas, 2001.

[32] J. Rydzewski, "Stupid things we do to mess up maintainability", *A Seminar on Designing for Maintainability*, ASHRAE, Minneapolis, 2000.

[33] MEWR, *Code of Practice for the Control of Legionella Bacteria in Cooling Towers*, MEWR, Singapore, 2001.

[34] ASHRAE, *2004 ASHRAE Handbook. Heating, Ventilating and Air-Conditioning Systems and Equipment* (SI Ed.), ASHRARE, Atlanta, GA, 2004.

[35] T. Notoya, "Localized corrosion in copper tubes and the effect of anti-tarnishing pre-treatment", *Journal of Materials Science Letters*, **10**(7), 389–391, 1991.

[36] Y. Y. Hammad, "The problem of the "sick" building — Facts and implications: Identifying and measuring indoor nonbiologic agents", *Journal of Allergy and Clinical Immunology*, **94**(2), 389–393, 1994.

[37] D. W. Bearg, *Indoor Air Quality and HVAC Systems*, CRC Press, Boca Raton, FL, 1993.

[38] J. A. Siegel and W. W. Nazaroff, "Predicting particle deposition on HVAC heat exchangers", *Atmospheric Environment*, **37**(39), 5587–5596, 2003.

[39] Z. J. Hou, Z. W. Lian, Y. Yao and X. J. Yuan, "Data mining based sensor fault diagnosis and validation for building air conditioning system", *Energy Conversion and Management*, **47**(15–16), 2479–2490, 2006.

[40] S. Wang and F. Xiao, "Detection and diagnosis of AHU sensor faults using principal component analysis method", *Energy Conversion and Management*, **45**(17), 2667–2686, 2004.

[41] J. E. Pakanen and T. Sundquist, "Automation-assisted fault detection of an air-handling unit, implementing the method in a real building", *Energy and Buildings*, **35**(2), 193–202, 2003.

[42] H. Yoshida and S. Kumar, "ARX and AFMM model-based on-line real-time data base diagnosis of sudden fault in AHU of VAV system", *Energy Conversion and Management*, **40**(11), 1191–1206, 1999.

[43] H. Yoshida, S. Kumar and Y. Morita, "Online fault detection and diagnosis in VAV air handling unit by RARX modelling", *Energy and Buildings*, **33**(4), 391–401, 2001.

[44] W. J. McGuinness and B. Stein, *Building Technology: Mechanical and Electrical Systems*, John Wiley and Sons, New York, 1997.

[45] N. Shibata, K. Hasegawa, K. Hashimoto and S. Akasaka, "Recent elevator design supported by computer technology", *Elevator World*, 1999.

[46] L. Janovsky, *Elevator Mechanical Design*, 3rd Edition, Elevator World, Inc, 2001.

[47] S. M. Levy, *Construction Databook*, McGraw-Hill, New York, 1998.

[48] G. R. Strakosch, "A new dimension in elevator monitoring", *Elevator World*, **XXXXII**(8), 126–129, 1994.

[49] AS1735 Australian Standard Lift Code, 2001.

[50] CAN/CSA B44 M94 Safety code for elevators, escalators, dumbwaiters, moving walks and freight platform lifts, Canada, 1994.

[51] EN 81 series (EN 81-1, EN 81-2, EN 81-28, EN 81-70, EN 12015, EN 12016, EN 13015, etc.), Safety rules for the construction and installation of lifts, European Standard, 2001.

[52] JIS A 4302 Inspection standard of elevator, escalator and dumbwaiter, Japan, 2006.

[53] CP2: 2000 code of practice on the installation, operation and maintenance of electric passenger and goods lifts, Singapore, 2000.

[54] CP15: 2004 code of practice for installation, operation and maintenance of escalators and passenger conveyors, Singapore, 2004.

[55] ASTM A17-1: 2007 Safety code for elevators and escalators.

[56] ISO 4190-1: 1999 lift (US: Elevator) installation — Part 1: Class I, II, III and IV lifts.

[57] ISO 4190-2: 2001 lift (US: Elevator) installation — Part 2: Class IV lifts.

[58] ISO 4190-3: 1982 Passenger lift installations — Part 3: Service lifts class V.

[59] ISO 4190-5: 1987 lifts and service lifts (USA: Elevators and dumbwaiters) — Part 5: Control devices, signals and additional fittings.

[60] ISO 4190-6: 1984 lifts and service lifts (USA: Elevators and dumbwaiters) — Part 6: Passenger lifts to be installed in residential buildings: Planning and selection.

[61] ISO 7465: 2001 passenger lifts and service lifts: Guide rails for lift cars and counterweights: T type.

[62] ISO 9386-1: 2000 power-operated lifting platforms for persons with impaired mobility: Rules for safety, dimensions and functional operation — Part 1: Vertical lifting platforms.

[63] ISO 9386-2: 2000 power-operated lifting platforms for persons with impaired mobility: Rules for safety, dimensions and functional operation — Part 2: Powered stairlifts for seated, standing and wheelchair users moving in an inclined plane.

[64] ISO/TR 11071-1: 2004 comparison of worldwide lift safety standards — Part 1: Electric lifts (elevators).

[65] ISO/TR 11071-2: 1996 comparison of worldwide lift safety standards — Part 2: Hydraulic lifts (elevators).

[66] ISO/TS 14798: 2000 lifts (elevators), escalators and passenger conveyors — Risk analysis methodology.

[67] ISO 16368: 2003 mobile elevating work platforms — Design calculations, safety requirements and test methods.

[68] ISO 16369: 2000 elevating work platforms — Mast-climbing platforms.

[69] ISO/TR 16765: 2003 comparison of worldwide safety standards on lift for firefighters.

[70] ISO 18738: 2003 lifts (elevators) — Measurement of lift ride quality.

[71] ISO/TR 22559-1: 2004 safety requirements for lifts (elevators) — Part 1: Global essential safety requirements.

[72] M. McCann, *Deaths and Injuries Involving Elevators or Escalators, Research paper*, The Center to Protect Workers' Rights, Silver Spring, March 2004.

[73] J. Staal and J. Quackenbush, "Elevators, escalators and hoists", J. M. Stellman (ed.), *Encyclopaedia of Occupational Health and Safety*, Vol. III, International Labour Office, Geneva, 1998, pp. 40–44.

[74] C. E. Vlahovic, "Rationale for new rules in CSA-B44 safety code for Eeevators", Elevator World Educational Division (ed.), *Elevator World Educational Package and Reference Library*, Vol. III, Elevator World Inc, Mobile, AL, 1990, pp. 66–72.

[75] NEII, *Building Transportation Standards and Guidelines, Part 7: Maintenance Guidelines*, NEII (National Elevator Industry Inc), Salem, NY, 2007.

[76] J. McKain, *Elevator Maintenance Manual*, Elevator World Inc, Mobile, AL. 1999.

[77] J. O'Donoghue and T. L. Jarboe, *Elevator and Escalator Rescue: A Comprehensive Guide*, Penwell, Tulsa, Okla, 2007.

[78] G. R. Strakosch, *The Vertical Transportation Handbook*, 3rd Edition, Wiley, NY, 1998.

[79] M. L. Chew, S. Das and N. H. B. Sulaiman, "Quantifying maintainability parameters for vertical transport system", *Proc. 11dbmc: 11th Int. Conf. on Durability of Bldg. Material and Components*, Istanbul, Turkey, 2008.

[80] J. C. Cawley and G. T. Homce, "Occupational electrical injuries in the United States, 1992–1998 and recommendations for safety research", *Journal of Safety Research*, **34**(3), 241–248, 2003.

[81] H. L. Floyd, K. Eastwood and D. P. Liggett, "How we can better learn from electrical accidents?", *IEE Industry Applications Magazines*, 16–23, 2000.

[82] R. E. Loyd and National Fire Protection Association, *Hazardous Classified Locations: Electrical Design and Installation*, Delmar Publishers, Albany, NY, 1997.

[83] T. H. Gillman and I. L. M. "Mechanical and electrical failures leading to major fires", *Engineering Failure Analysis*, **14**(6), 995–1018, 2007.

[84] E. K. Greenwald, *Electrical Hazards and Accidents: Their Cause and Prevention*, Van Nostrand Reinhold, NY, 1991.

[85] ANSI, *ANSI/UL C101-1992 Leakage Current for Appliances*, ANSI (American National Standards Institute), NY, 1992.

[86] W. F. Cooper and D. A. Dolbey Jones, *Electrical Safety Engineering*, 3rd Edition, Butterworth-Heinemann, Oxford, 1993.

[87] FMEC, *Handbook of Industrial Loss Prevention*, 2nd Edition, FMEC (Factory Mutual Engg. Corpn.), NY, 1967.

[88] A. A. Hattangadi, *Electrical Fires and Failures*, McGraw Hill, NY, 2000.

[89] C. R. Lux, D. S. Crames, K. A. Wellmaker and S. R. Salaymeh, *WSRC-RP-95-915: Savannah River Site Hazard Analysis Generic Initiator Database*, Westinghouse Savannah River Company, Aiken, SC, 1995.

[90] R. Dywer and R. D. Mueller, "Selection of transformers for commercial buildings", *Industry Applications Society Annual Meeting*, 1335–1342, 1992.

[91] C. R. Heising, Comparison of results from several power circuit breaker reliability surveys in the United States and the CIGRE 13-06 Worldwide Survey, *Proc. 10th Annual Engg. Conf. on Reliability, Availability and Maintainability for the Electric Power Industry*, Montreal, Canada, 1983.

[92] N. Sclater and J. E. Traister, *Handbook of Electrical Design Details*, 2nd Edition, McGraw-Hill, NY, 2003.

[93] R. R. Janis and W. K. Y. Tao, *Mechanical and Electrical Systems in Buildings*, Pearson/Prentice Hall, Upper Saddle River, NJ, 2005.

[94] D. P. Nolan, *Encyclopedia of Fire Protection*, 2nd Edition, Thomson Delmar Learning, Clifton Park, NY, 2006.

[95] G. Cox, "Fire research in the 21st century", *Fire Safety Journal*, **32**(3), 203–219, 1999.

[96] B. Ashe and T. J. Shield, "Analysis and modelling of the unannounced evacuation of a large retail store", *Fire and Materials*, **23**(6), 333–336, 1999.

[97] D. Charters, D. McGrail, N. Fajemirokun, Y. Wang, N. Townsend and P. Holborn, Preliminary analysis of the number of occupants, fire growth, detection times and pre-movement times for probabilistic risk assessment, *Proc. 7th Int. Symp. on Fire Safety Science*, Worcester, MA, USA, 2002.

[98] A. M. Hasofer and D. Bruck, "Statistical analysis of response to fire cues", *Fire Safety Journal*, **39**(8), 663–688, 2004.

[99] J. Lataille, *Fire Protection Engineering in Building Design*, Butterworth-Heinemann, Oxford, 2003.

[100] R. Burke, *Fire Protection: Systems and Response*, CRC Press, Boca Raton, FL, 2008.

[101] M. Fontana, J. P. Favre and C. Fetz, "A survey of 40,000 building fires in Switzerland", *Fire Safety Journal*, **32**(3), 137–158, 1999.

[102] NFPA, *Major Causes of Fire*, from http://www.nfpa.org/display Content.asp?categoryID=952. Retrieved 15 June 2008.

CHAPTER 7
MATERIAL SELECTION

7.1. General

In addition to structural and aesthetical properties, functionality, durability, maintainability and sustainability are important considerations to be taken into account during the selection of materials. However, due to the wide range of materials available for use, selecting the correct material is seldom an easy task. As a result, the material selected may not fulfill its intended function.

With sufficient knowledge on the use, design, installation and maintenance of a wide range of materials provided from project onset to post completion, suitable materials may be selected and costs associated with over-specification may be greatly reduced. The frequency of repair and replacement and their associated budgeting may also be planned and managed more effectively throughout the life cycle of a building.

This chapter discusses durability and performance of common construction materials and the Material Manual (MM) in the Maintainability Website (Fig. 7.1) (http://www.hpbc.bdg.nus.edu.sg/). The MM provides a prototype to facilitate selection of suitable materials for long-term performance of a facility throughout its lifetime. It illustrates a framework with critical factors which users may adjust accordingly for each material with regards to its durability, property and maintainability. Defects common to these materials are also discussed.

7.2. Interface of the Material Manual

With an interface similar to that of the Defect Library, the MM comprises of various building elements. Each building element is in turn studied for their "Durability", "Property", "Maintenance" and "Defect".

Fig. 7.1. Main page of material manual.

As an example, under "Façade", materials commonly used in façade construction are categorized as follows:

(1) *Structural frame* (included hot rolled steel, cold formed steel, concrete encased steel, reinforced dense concrete and prestressed precast concrete).
(2) *Masonry* (included brick, concrete block and mortar as a jointing material).
(3) *Concrete wall* (included reinforced concrete wall and precast wall).
(4) *Finishes* (included plaster, paint and tiles).
(5) *Cladding* (included materials used for framing such as stainless steel, aluminum, mild steel; infill panels, i.e. stainless steel, aluminum, mild steel, GRC, natural stone; and jointing materials such as gasket and sealant).
(6) *Curtain wall* (included materials used for framing such as stainless steel, aluminum, mild steel; infill panels such as stainless steel,

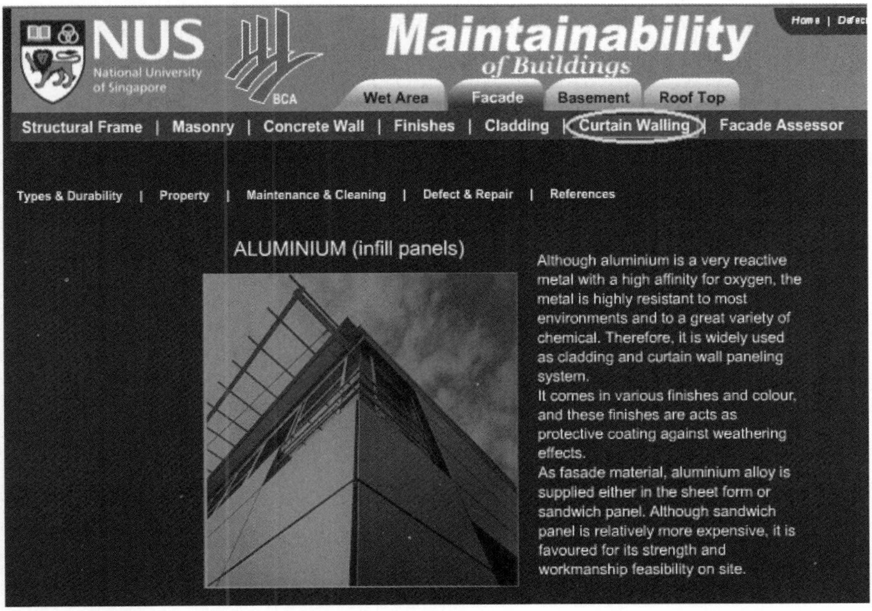

Fig. 7.2. Main page of "aluminum panel" (Material Manual → Façade → Curtain Walling → Infill Panel/aluminum).

aluminum, mild steel, glass; and jointing materials such as gasket and sealant).

Figure 7.2 shows the opening page for aluminum infill panels within the wider heading of "Curtain Walling". For each material within the MM such as the aluminum infill panel, a brief description of the material is given and further information on its types, durability, properties, maintenance and cleaning methods; and the types of failures and repair methods are discussed.

Another example is "Wet Area" which is categorized as follows:

(1) *Floor component* (included floor slab, i.e. reinforced concrete and precast concrete slab; screed and waterproofing).
(2) *Floor and wall finishes* (included manufactured tiles, natural stone, plaster and paint).
(3) *Jointing material* (included tile bedding material, grout and sealant).

(4) *Ceiling component* (included supporting member, i.e. mild steel and aluminum; false ceiling panel, i.e. mineral board, gypsum and calcium silicate board).
(5) *Pipe or plumbing* (included cast iron, plastics and copper pipes).
(6) *Sanitary ware and fittings* (included wash basin, i.e. glass, plastics, steel and vitreous china; mixers and tap ware, i.e. ceramic, metal and plastics; shower tray, i.e. cast iron, plastics, steel, polystyrene and vitreous china; WC, i.e. fireclay and vitreous china).

Figure 7.3 shows the opening page for ceramic tile of "Floor and Wall Finishes" from the "Wet Area" section of the Material Manual.

7.3. Durability

Durability may be defined as the expected lifespan of a material during which no excessive expenditure on operation, maintenance and

Fig. 7.3. Main page of "ceramic tile" (Material Manual → Wet Area → Floor and Wall Finishes → Ceramic Tile).

replacement are required. The lifespan of a material may be defined as "service life". Any failure or major defect occurring before the end of the material's expected lifespan is considered as "premature failure".

Ideally, materials are expected to perform up to their expected lifespan. However, when assembled into a system, their performance may be affected by interactions and compatibility issues between the materials. Design detailing, construction quality (workmanship) and maintenance are crucial factors to observe in upholding the performance of the whole system. By selecting suitable and high-quality materials, premature failure may be minimized or prevented, hence reducing the need for replacements. The durability of each selected material is crucial in determining the lifespan of the overall building system. Durability is in turn found to be dependent on the properties of the material. Through advancements in technology, the physical and chemical properties of materials are continuously improved to reduce the incidence of failures [1–5].

The durability of a material is dependent on various factors:

- installation in accordance with manufacturer's specifications;
- appropriate design details;
- testing to international and local standards;
- proper maintenance in accordance with manufacturer's specifications.

7.3.1. *Adjustment Factors for Durability of Material*

In the Materials Manual, materials are subdivided into "Primary materials" (i.e. cladding panel and finishes) and "Secondary materials" (jointing material and framing members).

Materials perform differently under different exposure conditions, e.g. under tropical exposure conditions, materials are subjected to thermal shocks and harsh weathering agents and hence their performance differs greatly as that under temperate conditions. Adjustment needs to be made by the user accordingly to reflect actual durability under the respective conditions [6–25]. The MM provides a framework for the adjustment using a weighted percentage which takes account of factors including location and orientation differences. Across materials, adjustment percentages differed. An example of adjustment provision for glass under the element "Façade" and grout under the element "Wet Area" are shown in Figs. 7.4 and 7.5 [26–29].

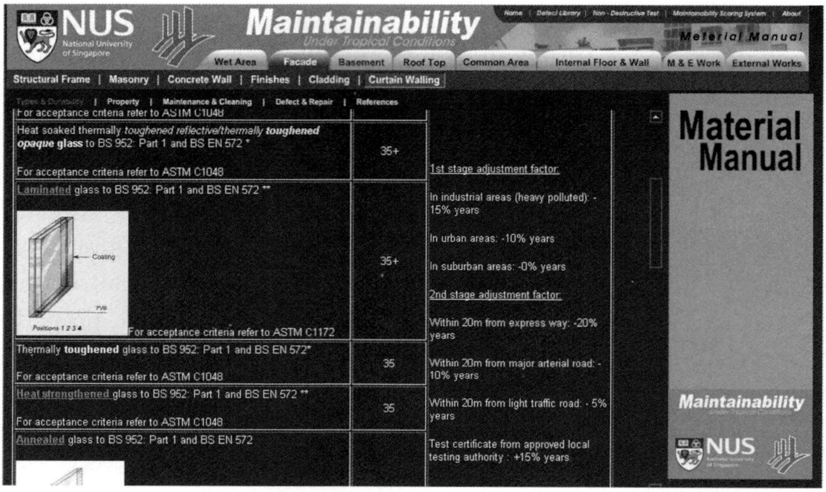

Fig. 7.4. Adjustment percentages for façade materials (Material Manual → Façade → Curtain Walling → Infill Panels/Glass → Type and Durability).

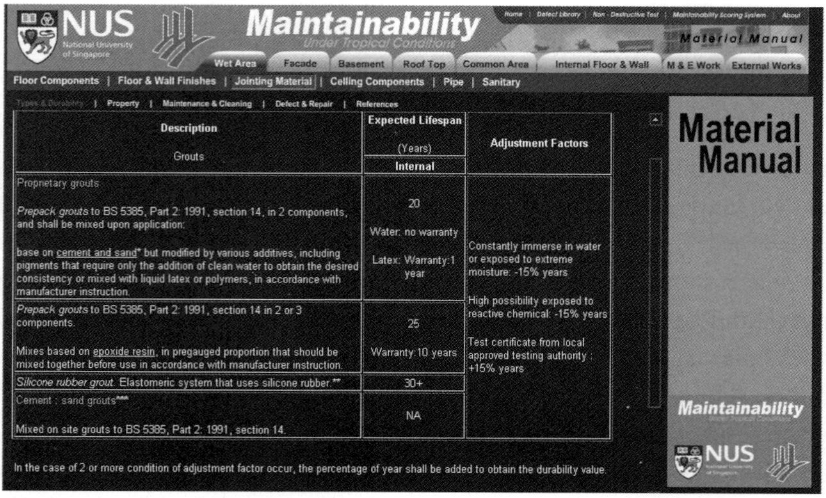

Fig. 7.5. Adjustment percentages for wet area materials (Material Manual → Wet Area → Jointing Material → Grout → Type and Durability).

In addition, similar adjustments can be made by the user on the following:

- atmospheric condition;
- land-use zone — Industrial, urban, suburban;
- orientation — Highway, vegetation, UV.

Figures 7.6–7.9 show an example of the durability of some primary materials with multiple adjustments.

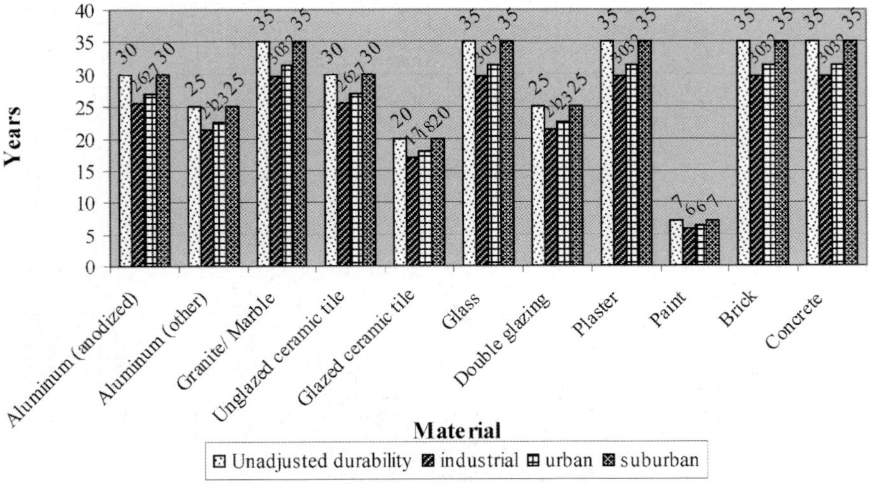

Fig. 7.6. Durability under different zones.

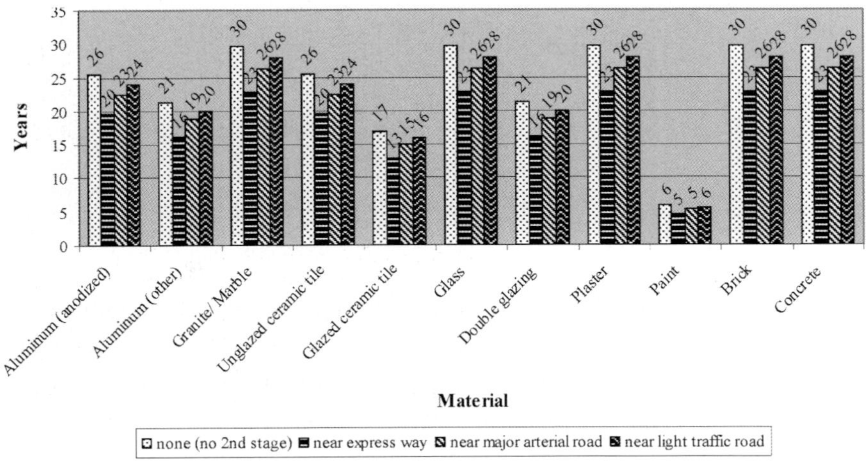

Fig. 7.7. Durability under industrial zone.

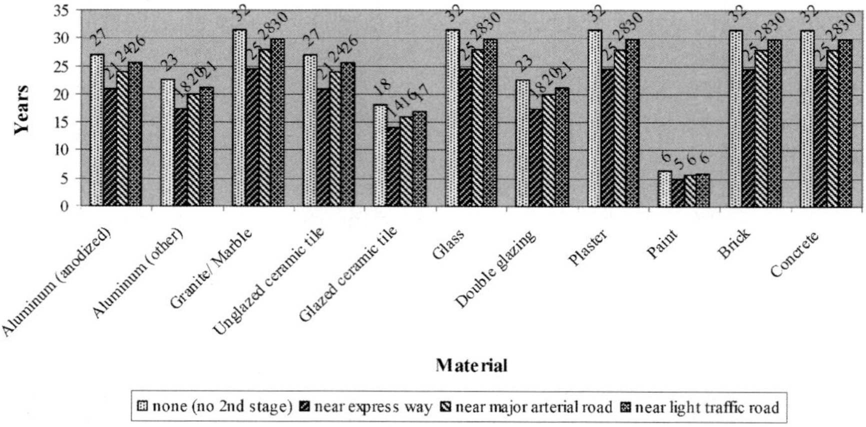

Fig. 7.8. Durability under urban zone.

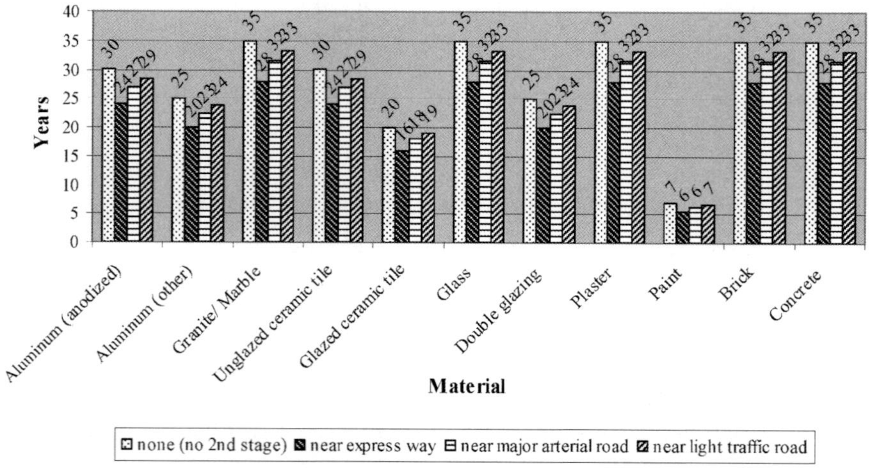

Fig. 7.9. Durability under suburban zone.

It should be noted that the above serve only as general examples. As the significance of each adjustment factor varies from one place to another, due to the influence of material composition, weather, exposure condition, usage etc., a material from different sources and/or used in different locations may perform significantly different. It is important that users consult their suppliers for data obtained from tests conducted under the local conditions.

Tables 7.1–7.15 show examples of "Types and Durability" of some common construction materials extracted from MM (http://www.hpbc.bdg.nus.edu.sg/). The information here is only meant as a guide. Users should consult their local suppliers for the material test data specific to their local conditions and usage.

7.4. Properties

The physical and chemical properties of basic materials are compiled. Tables 7.16–7.20 show examples of "Properties" of some common construction materials extracted from MM (http://www.hpbc.bdg.nus.edu.sg/). The information here is only meant as a guide. Users should consult their local suppliers for the material test data specific to their local conditions and usage.

7.5. Maintenance and Cleaning

As modern buildings are designed to meet higher standards and expectations, the influence of design on the maintenance of buildings is greater than ever before. The degree to which the design of a building embraces maintenance considerations has a major impact on its performance. Its consideration at the design stage is of great importance on the future performance of the building.

Different kinds of material require different maintenance methods. Besides that, different finishes of the same material would require different maintenance methods; e.g. anodized and liquid organic coated aluminum panel required different maintenance method. In addition, different cleaning frequencies has to be set, based on different environmental condition and exposure, e.g. aluminum cladding in industrial areas may require cleaning every three months, whereas those in residential areas may require cleaning only every six months.

Figures 7.10 and 7.11 show examples of maintenance and cleaning for façade and wet area, respectively.

Table 7.1. Durability of hot rolled steel [30–32].

Description (Hot Rolled Steel)	Expected Lifespan (Years) External	Expected Lifespan (Years) Internal	Adjustment Factors
Mild steel, hot dipped galvanized after any cutting, welding or drilling to BS 5493 specification SB3 (equivalent in either case to a zinc coating weight of 1500 g/m² or 210 micron thickness).	30	35+	**First stage adjustment factor:** In industrial areas (heavy polluted): –15% of no. of years.
Mild steel, hot dipped galvanized after any cutting, welding or drilling to SB2 (equivalent in either case to a zinc coating weight of 1000 g/m² or 140 micron thickness).	25	35+	In urban areas: –10% of no. of years. In suburban areas: No adjustment.
Mild steel, molten zinc sprayed after any cutting, welding or drilling to BS 5493 specification SC2Z (equivalent to a zinc coating weight of about 1050 g/m² or 150 micron thickness).	25	35+	**Second stage adjustment factor:** Structural members subjected to "very severe" exposure conditions to Table 3.2 of BS 8110: Part 1,
Mild steel, molten zinc sprayed after any cutting, welding or drilling to BS 5493 specification SC1Z (equivalent to a zinc coating weight of about 700 g/m² or 100 micron thickness).	20	35+	without additional surface protection: –30% of no. of years. Within 20 m from express way:
Mild steel, hot dipped galvanized after any cutting, welding or drilling to BS EN 729 (to steel at least 5 mm thick) or to BS 5493 specification SB1 (any thickness steel) equivalent in either case to a zinc coating weight of 610 g/m² or 85 micron thickness.	20	35	–20% of no. of years. Within 20 m from major arterial road: –10% of no. of years. Within 20 m from light traffic road: –5% of no. of years.

(*Continued*)

Table 7.1. (*Continued*)

Description (Hot Rolled Steel)	Expected Lifespan (Years)		Adjustment Factors
	External	Internal	
Mild steel, hot dipped galvanized after any cutting, welding or drilling to BS EN 729 to steel 2–4.9 mm thick (equivalent to a zinc coating weight of 460 g/m² or 64 micron thickness).	15	30	Test certificate from approved local testing authority: +15% of no. of years.
Mild steel, hot dipped galvanized after any cutting, welding or drilling to BS EN 729 to steel 1–1.9 mm thick (equivalent to a zinc coating weight of 335 g/m² or 43 micron thickness).	10	25	
Mild steel, blast cleaned after any cutting, welding or drilling and protected with an appropriate Micaceous Iron Oxide, Chlorinated Rubber or similar high performance finish to give a minimum dry film thickness of 250 microns.	10	25	

Table 7.2. Durability of reinforced dense concrete [33–42].

Description Reinforced Dense Concrete	Expected Lifespan (Years) Internal	Adjustment Factors
In situ columns/beams complying with BS 8110: Parts 1 and 2, with cover to steel complying with Table 3.3 and specified to match the appropriate exposure category selected from Table 3.2. Minimum nominal cover to all reinforcement of 20 mm. Minimum content 275 kg/m³.	35+	**First stage adjustment factor:** In industrial areas (heavy polluted): –15% years. In urban areas: –10% years. In suburban areas: –0% years.
Unclassified, i.e. beams/columns not complying with BS 8110: Parts 1 and 2.	<35+	**Second stage adjustment factor:** Structural members subjected to "very severe" exposure conditions to Table 3.2 of BS 8110: Part 1, without additional surface protection: –30% years. Within 20 m from express way: –20% years. Within 20 m from major arterial road: –10% years. Within 20 m from light traffic road: –5% years. Within 20 m from sea water/prone to salt spray: –15%. Test certificate from approved local testing authority: +15% years.

Table 7.3. Durability of clay bricks.

Description Clay Brick	Expected Lifespan (Years) External Masonry Walls	Adjustment Factors
First grade designated bricks to SS 103: 1974. Includes facing brick.	35+	**First stage adjustment factor:** In industrial areas (heavy polluted): −15% of no. of years. In urban areas: −10% of no. of years. In suburban areas: No adjustment. **Second stage adjustment factor:** Within 20 m from express way: −20% of no. of years. Within 20 m from major arterial road: −10% of no. of years. Within 20 m from light traffic road: −5% of no. of years. Test certificate from approved local testing authority: +15% of no. of years.
Second grade designated bricks to SS 103: 1974.	35+	
Third grade designated bricks to SS 103: 1974.	35+	
*Non*designated bricks as to SS 103: 1974.	<35+	

Table 7.4. Durability of homogeneous tiles.

Description Homogeneous Tiles	Expected Lifespan (Years) External	Adjustment Factors
Polished homogeneous tiles tested to ISO 10545.	35+	**First stage adjustment factor:** In industrial areas (heavy polluted): −15% years. In urban areas: −10% years. In suburban areas: −0% years. **Second stage adjustment factor:** Within 20 m from express way: −20% years. Within 20 m from major arterial road: −10% years. Within 20 m from light traffic road: −5% years. Within 20 m from sea water/area prone to salt spray: −15% years. Test certificate from approved local testing authority: +15% years.
Matt finished homogeneous tiles tested to ISO 10545.	35+	
Rock finished homogeneous tiles tested to ISO 10545.	35	
Structured finished homogeneous tiles tested ISO 10545.	35	
Homogeneous tile *not* tested to ISO 10545.	5–10	
Unclassified, i.e. any material. Quality/duty not specified.	<5	

Table 7.5. Durability of ceramic tiles.

Description Ceramic Tiles	Expected Lifespan (Years) External	Adjustment Factors
Unglazed ceramic tiles or *quarry tiles* to BS 6431 Part 1 or SS CP 68: 1997, with test certificate to ISO 10545.	30	**First stage adjustment factor:** In industrial areas (heavy polluted): −15% years. In urban areas: −10% years. In suburban areas: –0% years.
Glazed ceramic tiles to BS 6431 Part 1 or SS CP 68: 1997, with test certificate to ISO 10545.	20	**Second stage adjustment factor:** Within 20 m from express way: −20% years. Within 20 m from major arterial road: −10% years. Within 20 m from light traffic road: −5% years. Within 20 m from sea water/area prone to salt spray: −15% years. Test certificate from approved local testing authority: +15% years.
Unglazed ceramic tiles or *quarry tiles* not to BS 6431 Part 1 or SS CP 68: 1997.	10	
Glazed ceramic tiles not to BS 6431 Part 1 or SS CP 68: 1997.	5	
Unclassified, i.e. any material. Quality/duty not specified.	<5	

Table 7.6. Durability of concrete tiles.

Description Concrete Tile	Expected Lifespan (Years)	Adjustment Factors
Single lap concrete interlocking tile classified as group B product as to BS 473/BS 550 (or BS EN 490) or ASTM C 1492-01 of body thickness not less than 9.5 mm and at interlocking portion not less than 6.3 mm.	35+	**First stage adjustment factor:** In industrial areas (heavy polluted): −15% years. In urban areas: −10% years. In suburban areas: –0% years.

(*Continued*)

Table 7.6. (*Continued*)

Description Concrete Tile	Expected Lifespan (Years)	Adjustment Factors
Double lap concrete plain tiles classified as group A products as to BS 473/BS 550 (or BS EN 490) or ASTM C 1492-01 of thickness not less than 9.5 mm at any cross section. Either smooth or granular finish.	35+	**Second stage adjustment factor:** Within 20 m from express way: –20% years. Within 20 m from major arterial road: –10% years. Within 20 m from light traffic road: –5% years. Within 20 m from sea water/ area prone to salt spray: –15% years. Test certificate from approved local testing authority: +15% years.
Double or single lap concrete imitation slate or stone slate manufactured and tested in accordance with BS 473/ BS 550 (or BS EN 490) or ASTM C 1492-01 of thickness not less than 9.5 mm at any cross section. Manufactured by high-pressure extrusion and compaction process. Slates to be tightly fixed to battens, stone slates to be hung on battens.	30	
Bitumen-based reinforced sarking felt not manufactured to BS 747 Type 1F.	<10	None

Table 7.7. Durability of granite [43–51].

Description Granite and Other Igneous Rocks	Expected Lifespan (Years) Internal	Adjustment Factors
Natural granite and other igneous rocks specified and selected to BS 5385-1 or ASTM C1528-02. Performance and specification to be confirmed with supplier.	35+	Quality relies on preservative or water repellent: –30% years. Contact with soluble salt: –30% years. Constantly in contact/immersed in water or exposed to extreme moisture: –15% years. High possibility of exposure to reactive chemical: –15% years.
Natural granite and other igneous rocks specified and selected to BS 5385-1 or ASTM C1528-02. Performance and specification to be confirmed with supplier.	35	Test certificate from local approved testing authority (including porosity, crystallization, etc.) for specific batch delivered to site: +15% years.
Unclassified, i.e. natural granite and other igneous rocks *neither* specified nor selected to BS 5390.	<35	None

Table 7.8. Durability of marble [52, 53].

Description Marble (Metamorphic Stones)	Expected Lifespan (Years) General	Adjustment Factors
Natural marble specified under Group A. Performance and specification to be confirmed with supplier.	35+	Quality relies on preservative or water repellent: –30% years. Contact with soluble salt: –30% years. Constantly in contact/immersed in water or exposed to extreme moisture: –15% years.

(*Continued*)

Table 7.8. (*Continued*)

Description Marble (Metamorphic Stones)	Expected Lifespan (Years) General	Adjustment Factors
Natural marble specified under Group B. Performance and specification to be confirmed with supplier.	35+	High possibility of exposure to reactive chemical: −15% years. Test certificate from local approved testing authority (including porosity, crystallization, etc.) for specific batch delivered to site: +15% years.
Natural marble specified under Group C. Performance and specification to be confirmed with supplier.	35+	
Natural marble specified under Group D. Performance and specification to be confirmed with supplier.	35+	
Unclassified, i.e. natural marble neither specified nor in above groups.	<35	None

Table 7.9. Durability of tile bedding.

Description Tile Bedding	Expected Lifespan (Years) Internal	Adjustment Factors
Cementitious adhesive (mix upon application with water or liquid admix to BS 5980:1980, i.e. Liquid Latex) or EN 12004:2001. Complying with requirements of type 1, class AA or class A in BS 5980:1980, with test certification for shear, tensile adhesion strength and mold resistance.	+ water ≤10 (no warranty given) + liquid latex = 10+ (local warranty: 10 years)	Constantly immersed in water or exposed to extreme moisture: −15% years. High possibility of exposure to reactive chemical: −15% years. Test certificate from local approved testing authority: +15% years.

(*Continued*)

Table 7.9. (*Continued*)

Description Tile Bedding	Expected Lifespan (Years) Internal	Adjustment Factors
Cementitious adhesive not to BS 5980:1980 or EN 12004:2001 and/or without certification for shear and tensile adhesion strength test.	<8	None
Organic adhesive should comply with the requirements for class AA or class A in BS 5980:1980 or EN 12004:2001, for *type 3 (dispersion/cement adhesive)* type 5 (reaction resin adhesive) with test certification for shear, tensile adhesion strength and mold resistance.	5+	Constantly immersed in water or exposed to extreme moisture: −15% years. High possibility of exposure to reactive chemical: −15% years. Test certificate from local approved testing authority: +15% years.
Organic adhesive not comply with the requirements for class AA or class A in BS 5980:1980 or specified in EN 12004:2001.	<5	None
*Epoxy-based adhesive**, to 5980:1980 or EN 12004:2001 with test certification for shear, tensile adhesion strength, and mold resistance.	35+ local warranty: 10 years	Constantly immersed in water or exposed to extreme moisture: −15% years.
Polyurethane-based adhesive, to 5980:1980 or EN 12004:2001 with test certification for shear, tensile adhesion strength and mold resistance.	30+ local warranty: 10 years	High possibility of exposure to reactive chemical: −15% years. Test certificate from local approved testing authority: +15% years.

Table 7.10. Durability of glass [54–69].

Description Glass (Curtain Walling)	**Expected Lifespan (Years)** General	**Adjustment Factors**
Heat soaked thermally *toughened laminated* glass to BS 952: Part 1 and BS EN 572*. For acceptance criteria refer to ASTM C1172 and ANSI Z97.1 CPSC Standard 16 CFR 1201.	35+	**First stage adjustment factor:** In industrial areas (heavy polluted): −15% of no. of years. In urban areas: −10% of no. of years. In suburban areas: No adjustment.
Heat soaked thermally *toughened/thermally toughened body-tinted* glass to BS 952: Part 1 and BS EN 572*. For acceptance criteria refer to ASTM C1048 and more stringent requirements.	35+	**Second stage adjustment factor:** Within 20 m from express way: −20% of no. of years. Within 20 m from major arterial road: −10% of no. of years. Within 20 m from light traffic road: −5% of no. of years. Test certificate from approved local testing authority: +15% of no. of years.
Heat soaked thermally *toughened reflective/thermally toughened opaque glass* to BS 952: Part 1 and BS EN 572*. For acceptance criteria refer to ASTM C1048.	35+	
Laminated glass to BS 952: Part 1 and BS EN 572**. For acceptance criteria refer to ASTM C1172 and ANSI Z97.1 CPSC Standard 16 CFR 1201.	35+	
Thermally *toughened* glass to BS 952: Part 1 and BS EN 572*. For acceptance criteria refer to ASTM C1048.	35	

(*Continued*)

Table 7.10. (*Continued*)

Description Glass (Curtain Walling)	Expected Lifespan (Years) General	Adjustment Factors
Heat strengthened glass to BS 952: Part 1 and BS EN 572**.		
For acceptance criteria refer to ASTM C1048.		
Annealed glass to BS 952: Part 1 and BS EN 572.	35	
For acceptance criteria refer to ASTM C1036.	35	
Glass *not* to BS 952: Part 1 and/or BS EN 572.	<35	

Table 7.11. Durability of solvent-based sealants [70, 71].

Description Sealants (Solvent-Based)	Expected Lifespan (Years) Internal	Adjustment Factors
Oil/solvent base		Constantly immersed in water or exposed to extreme moisture: −15% years. High possibility of exposure to reactive chemical: −15% years. Test certificate from local approved testing authority: +15% years.
Oil-based mastic, *butyl* or bitumen rubber, plastic but no elastic properties.	5	
Acrylic		
Acrylic (solvent type). Plastic and elastic properties.	15	
Polysulfide		
2 part polysulfide to BS 4254 (mainly elastic properties).		
1 part polysulfide to BS 5215 (mainly elastic properties).	20	
1 or 2 part polysulfide *not* to BS 5215 or BS 4254.	<5	

(*Continued*)

Table 7.11. (*Continued*)

Description Sealants (Solvent-Based)	**Expected Lifespan (Years)** Internal	**Adjustment Factors**
Polyurethane		
2 part polyurethane (mainly elastic properties). 1 part polyurethane (mainly elastic properties).	20	
Silicone		
1 part low modulus silicone to BS 5889 Type A; elastic properties.	20	
Silicone sealant *not* to BS 5889 Type A. Also Type B to BS 5889.	<5	

Table 7.12. Durability of gasket.

Description Gasket	**Expected Lifespan (Years)** External	**Adjustment Factors**
Thermoplastics (PP, PE, PS, PVC)*. Flexible PVC specified in ASTM C 509 and C 864, with test certification for compression set and stress relaxation, weatherability and UV resistance, heat aging, fluid resistance.	10	**First stage adjustment factor:** In industrial areas (heavy polluted): –15% years. In urban areas: –10% years. In suburban areas: –0% years. **Second stage adjustment factor:** West or south orientation: –10% years. Applied to reseal existing joint: –15% years. Test certificate from approved local testing authority (as specified in SS 212:2000 for glazing material): +15% years.
Uninsured. Thermoplastics gaskets *not* specified in ASTM C 509 and C 864 and without test certification.	<10	

(*Continued*)

Table 7.12. (*Continued*)

Description Gasket	Expected Lifespan (Years) External	Adjustment Factors
Thermoset rubber/elastomer (silicone, EPDM, neoprene, NBR).** Neoprene (chloroprene rubber) as specified for glazing material and to ASTM C 509 and C 864, with test certification for compression set and stress relaxation, weatherability and UV resistance, heat aging, fluid resistance.	20+	
*EPDM Rubber*** as specified for glazing material and to ASTM C 509 and C 864, with test certification for compression set and stress relaxation, weatherability and UV resistance, heat aging, fluid resistance.	20+	
Silicone Rubber and to ASTM C 509 and C 864, with test certification for compression set and stress relaxation, weatherability and UV resistance, heat aging, fluid resistance.	20+	
Uninsured. Thermoset gaskets *not* specified in ASTM C 509 and C 864 and without test certification.	<20	

(*Continued*)

Table 7.12. (*Continued*)

Description Gasket	Expected Lifespan (Years) External	Adjustment Factors
Thermoplastic Elastomer — TPE (Santoprene, Kriburge, Evoprene, Forprene, Alcryn). TPE as specified in ASTM C 509 and C 864, with test certification for compression set and stress relaxation, weatherability and UV resistance, heat aging and fluid resistance.	20+	
Unclassified. Thermoplastic Elastomer gaskets *not* specified in ASTM C 509 and C 864 and without test certification.	<20	

Table 7.13. Durability of waterproofing [72–88].

Description Basement Waterproofing	Expected Lifespan (Years)	Adjustment Factors
Preformed membrane. Sheet/roll applied waterproofing with base materials to CP 82:1999, with method of fixing includes self adhesive, adhesive bonding, chemical welding, hot air welding and mechanical fixing, in accordance to manufacturer instruction. • *Bituminous*: Asphalt, bitumen, modified bitumen. • *Nonbituminous*: Ethylene-copolymer bitumen (ECB), polyvinyl chloride (PVC), polyethylene (PE), chlorinated	10	Existence of aggressive ground water*: −15% years. Hydrostatic head** exceeding system design (except crystalline system): −10% years. Test certificate from approved local testing authority: +15% years.

(*Continued*)

Table 7.13. (*Continued*)

Description Basement Waterproofing	Expected Lifespan (Years)	Adjustment Factors
polyethylene (PE), chlorinated polyethylene (CPE), butyl rubber, ethylene propylene diene (EPDM), neoprene, polyisobutylene (PIB) and bentonite clays.		
Liquid applied system	10	
Liquid (cold/hot) applied with base material to CP 82: 1999, with method of fixing includes by means of squeegee, roller, brush, trowel or spray, in accordance to manufacturer instruction. • *Bituminous.* • *Nonbituminous*: acrylic, cementitious, epoxy resin, hypalon, neoprene, polyurethane, polysulfide, rubber and silicone.		
Liquid (cold/hot) applied with base material *not* in accordance to CP 82: 1999, with method of fixing includes by means of squeegee, roller, brush, trowel or spray, in accordance to manufacturer instruction.	<10	
Integral system*		
Integral waterproofing system with method of application includes by means of masonry brush or compressed air-spray equipment or by dry shake or as admixture, in accordance to manufacturer instruction. • crystalline waterproofing, • chemical admixture, • waterproofing screed.	10+	

Table 7.14 Durability of plastic pipes.

Description Plastic Pipes	**Expected Lifespan (Years)**	**Adjustment Factors**
uPVC to SS 141: 1976, SS 213: 1998 for soil, waste and vent applications, SS 272: 1983 for drainlines and sewers. Joints and fittings for use with unplasticized PVC pressure pipes shall comply with SS 174: 1977.	30	Pipes embedded in screed but not plastic sheated or fully wrapped: Limited to five years. Pipes embedded in screed and plastic sheated or fully wrapped: −15% of no. of years. Mains water below 7pH: −15% of no. of years.
Plastic pipes or tubes *not* specified as above.	<30	Test certificate from approved local testing authorities: +15% of no. of years.

Table 7.15. Durability of paints (1)

Types	Expected Lifespan (Years)	Adjustment Factors
	Exterior	
Water-based, tested to ASTM D5324:1998 or equivalent with resin component: • Silicone acrylic • VA/acrylic copolymer • PVA copolymer	Normal: 7 acrylic-textured: up to 8	**First stage adjustment factor:** In industrial areas (heavy polluted): –15% of no. of years. In urban areas: –10% of no. of years. In suburban areas: No adjustment. **Second stage adjustment factor:** Within 20 m from express way: –20% of no. of years. Within 20 m from major arterial road: –10% of no. of years. Within 20 m from light traffic road: –5% of no. of years. Test certificate from approved local testing authority (i.e. tested for fungus resistance to SS 345): +15% of no. of years. Silicone-based is preferred.

(*Continued*)

Table 7.15. (*Continued*)

Types	Expected Lifespan (Years)	Adjustment Factors
	Exterior	
Solvent-based, tested to ASTM D5146:1998 or equivalent with resin component: • fluorocarbon, • epoxy, • silicone, • polyurethane, • alkyd, • acrylic.	Normal: 7 PU-textured: up to 10	**First stage adjustment factor:** In industrial areas (heavy polluted): –15% of no. of years. In urban areas: –10% of no. of years. In suburban areas: No adjustment. **Second stage adjustment factor:** Within 20 m from express way: –20% of no. of years. Within 20 m from major arterial road: –10% of no. of years. Within 20 m from light traffic road: –5% of no. of years. Test certificate from approved local testing authority (i.e. tested for fungus resistance to SS 345): +15% of no. of years. Polyurethane-based is preferred.

(*Continued*)

Table 7.15. *(Continued)*

Types	Expected Lifespan (Years) Exterior	Adjustment Factors
Baking fluoropolymer • Polyvinylidene fluoride (PVDF) Complying with BS 2569 for iron and steel, BS 3987 and BS 4842: 1992 for aluminum.	Normal: 10	**First stage adjustment factor:** In industrial areas (heavy polluted): –15% of no. of years. In urban areas: –10% of no. of years. In suburban areas: No adjustment. **Second stage adjustment factor:** Within 20 m from express way: –20% of no. of years. Within 20 m from major arterial road: –10% of no. of years. Within 20 m from light traffic road: –5% of no. of years. Test certificate from approved local testing authority (i.e. tested for fungus resistance to SS 345): +15% of no. of years.

Notes:

The type of resin component is listed descendent according to its durability.

Oil-based paint is normally refer to solvent-based paint with alkyd resin component.

PVA resin component normally used for interior application.

Emulsion-based paints fall under water-based category.

Enamel-based paints fall under solvent-based category.

Table 7.16. Properties of hot rolled steel.

HOT ROLLED STEELCONSTITUENTS

Constituents

The term "steel" usually refers to plain carbon steel, defined as alloys of iron and carbon that do not contain more than 2% carbon and are malleable in block and or ingot form.

Steel will have chemical composition containing:

- carbon: 0.16–0.25%;
- manganese: 1.5–16%;
- silicon: 0.1–0.5%.

Grade

Grade	Ultimate Tensile Strength (MPa)
40	340–500
43	430–580
50	490–640
55	550–700

Source: BS 4360: 1990.

Carbon Content and Properties

Carbon amounts up to 0.8% increases strength with each increment of carbon added and decreases ductility in the same proportion. Hardness also increases with greater carbon content.

For structural purposes, steel needs carbon to give strength but not in such quantity as to affect ductility.

Type of Steel	Carbon Percentage	Ultimate Tensile Strength (MPa)	Yield Tensile Strength (MPa)
Very mild	0.05–0.15%	275.80–379.23	165.48–206.85
Mild	0.15–0.25%	303.96–448.18	206.85–248.22
Low carbon	0.25–0.40%	413.70–482.65	248.22–275.80
Medium carbon	0.40–0.65%	482.65–551.60	275.80–303.96
Higher carbon	0.60–0.70%	551.60–648.13	303.96–386.12
Spring	0.70–0.80%	648.13–813.61	386.12–441.28
Pearlitic	0.75–0.85%	827.4	482.65

(*Continued*)

Table 7.16. (*Continued*)

Classification

Both cast and wrought carbon steel have been generally classified into approximate grades base on the carbon content:

Ingot Iron	Lowest possible content of carbon and other alloying elements, made in basic open-heart furnaces, corrosion resistant, used largely for sheet.
Extra soft or dead soft steel	0.08–0.18% carbon, this type of steel is characterized by ductility, toughness, weldability and is used where cold-workability is desirable; generally it is used where strength or stiffness is not important.
Mild, structural steel	0.15–0.29% carbon; characterized by strength combined with easy machinability, used for buildings, bridges, bolts, boilers, railroadrolling stock.
Medium grade	0.25–0.35% carbon; harder and stronger than mild structural grade yet can be hot-forged; used for shipbuilding and machinery.
Medium hard grade	0.35–0.65% carbon.
Spring grades	0.85–1.05% carbon.
High carbon tool steel	1.05–1.20% carbon.

Manufacturing Process

1. Casting
   Most steel compositions can be cast directly into complex shaped components. The casting is usually heat treated to relieve any residual stresses caused by uneven cooling and to refine the grain structure of the material.
2. Rolling
   Rolling is the most widely used process; it can be done hot or cold. A wide range of cross sectional shapes can be produced.
3. Forging
   Hot forging is used to make shaped, usually blocky, artefacts which need to have properties better than those of casting.
4. Seamless tube forming
   Seamless tubing is produced by hot deformation of initially cylindrical billets. Rolls which generate a spinning motion about the billet axis as well as motion along the billet axis are used to force the billet around a mandrel which defines the internal diameter of the tube.

(*Continued*)

Table 7.16. (*Continued*)

5. Drawing
   In wire drawing, a feedstock is pulled through a die by the product. If the material does not work harden, the process would not work. Successive reductions through many dies can be used to impart very high strength by work hardening. The price paid is reduction in ductility and toughness.

Finishes
Hot dipping, electro galvanizing, spraying and sherardizing are all processes whereby a protective coat of zinc is applied to steel and iron to seal them against corrosion.

Galvanized sheets that are heat-treated after coating produce a zinc–iron alloy surface free of spangles and is suitable for painting after cleaning. Where severe corrosive conditions exist, it is advisable to protect galvanized steel with paint.

Galvanizing
Galvanizing is the application of a coating of zinc to steel products for protection against many types of corrosion. The coatings are measured in g/m² and vary from 275 to 915.45 g/m² for sheet products; and 91.55 to 475.73 g/m² for surface for wire and prefabricated products.

The advantage of coating with zinc is that, should the iron and steel become exposed through wear, aging or discontinuities such as pinhole in the coating, galvanic reaction between coating and base metal causes the zinc to corrode and form compounds that cover and continue to protect the iron and steel for as long as any zinc remains.

Density and Thermal Properties

	Density at 20°C (kg/m3)	**Coefficient of thermal expansion × 10-6/°C**	**Thermal Conductivity (W/m K)**	**Specific Heat (J/kg K)**
Pure iron	7870	12.1	78	455
Wrought low carbon steel	7860	12.2	65	480
Cast low carbon steel	7860	12.2	49	480
Medium carbon steel	7850	12.2	51	485
High carbon steel	7850	11.1	48	490

Elasticity

Modulus elasticity of mild steel:

- 0.75% C: 210 GPa;
- 1.0% C, 1.0% Mn, 0.65% Cr, 1.0% W (hardened): 200–205 GPa.

Table 7.17. Properties of marbles.

MARBLE

General Description

Marble is a metamorphic stone. It was at one time a limestone or dolomite that has been structurally altered by recrystallization. Pure marble is white. The abstract variety of colors, mottling, clouds, swirls, shadings, streaks and veining are derived from the accessory minerals that were present in the limestone or dolomite.

Group	Metamorphic
Rock-forming	Calcareous
Composition	Calcite, dolomite or a combination of both and accessory minerals acting as coloring agents. Veining minerals range from calcite, quartz and a variety of other minerals.

Constituent	%
CaO	53.70–55.90
CO_2	42.17–53.54
MgO	0.06–0.92
Al_2O_3	0.06–0.94
Fe_2O_3	0.005–0.42

Hardness	Hard
Porosity %	0.06–2.3 m [2]
Density	2.4–2.8 gr/cm^3 [2]
Abrasion resistance	0.25–2.9 mm
Compressive strength	62.46–122.64 MN/m^2 [1]
Absorption %	0.1–0.6
Coefficient of linear expansion	2.7–5.1/°C^{-1} × 10^{-6} [2]
Specific gravity	2.64–23.73 kg/m^3 [1]
Impact strength	21.20–60 cm
Weight	2651.55–2731.90 kg/m^3 [1]

(Continued)

Table 7.17. (*Continued*)

Weathering characteristics	Moisture sensitive as a rule. All polished marble will lose their polish in exterior applications.
Tendencies	Absorbs oils and other liquids, easily scratched, acid sensitive
Colors	Every color is possible due to the variability of accessory minerals.
Finishes	Polished, honed.
Thickness	10–15 mm for internal wall; up to 20 mm for floor.

Table 7.18. Properties of glass.

Glass

1. Classification

1.1. ***Single Glazing/Monolithic***

This term refers to the basic flat glass which is either annealed (un-tempered), tempered, cleared, opaque, tinted, coated, etc. and can be one or two component of laminated and double glazing unit.

1.2. ***Multi-Glazing***

Consisting of two or three parallel sheets of glass separated by a thin dehydrated air space maintained by an airtight glass seal between the sheets and located around their edges. Because of the seal and the dehydrated air, there is no condensation or accumulation of dirt within the air space. The outer sheet is usually a heat absorbing glass. Various other combinations, which may include tempered and/or laminated glass are used. Multi-glazing unit also reduces sound transmission and is known as insulating glass.

The *u* value for this type of glass with the air space ranges from 6.35 to 12.70 mm is between 2.28 and 6.47 is W/m^{2}°C, whilst for single glazing is around 2.8 W/m^{2}°C.

1.3. ***Heat Absorbing Glass*** **(*Tinted/Reflective*)**

This type of glass is a kind of glass which because of its special composition, excludes a high percentage of UV rays to which it is exposed, transmit less solar heat into building and has less bleaching effort on colored fabric. The advantage of using this glass in one can reduce the air conditioning loads and cost.

(*Continued*)

Table 7.18. (*Continued*)

Weight (kg/m²)	16.01–19–92
Transmittance (%)	16–50
Reflectance (%)	18–35
Relative heat gain (%)	258.71–425.93

Various processes have been developed for the application of thin coatings to large areas of flat glass. Coating is applied to alter the performance of the basic float glass, i.e. light and radiant heat transmission, light and radiant heat reflection and surface emissivity. Coating which has been developed to have high transparency but high reflectivity of long wave radiation is used to improve the thermal insulation properties of the treated glass. Some coatings provide distinctive colors, whilst others are neutral in appearance.

Two groups of coatings:

- Metallic coating/metallic alloy coating: Applied by means of a vacuum deposition.
- Metallic oxide coating: Can be applied in normal atmosphere.

1.4. ***Laminated Glass***

It is produced by bonding two glasses together with a plastic material or a resin. The interlayer, which is usually polyvinyl butyral (PVB) or fiberglass, can be either clear or tinted. The number of interlayer may vary from 1 layer to 6 layers. The bonding is achieved by heating the glass/interlayer sandwich and applying pressure. Laminated glass can incorporate multiple layers of interlayer and glass plus other material such as polycarbonate to achieve specific performance characteristics. Toughened glasses can also be used as components of a laminated product. Other than façade, laminated glass is also used as overhead glazing, as it can minimize the risk of dangerous broken glass fragments falling since it holds to the frame when it breaks. It is suitable for protection against vandalism, hampers or break-in.

1.5. ***Toughened/Tempered Glass***

Toughened glass is usually produced by reheating a piece of annealed glass to a temperature of approximately 700°C at which point it begins to soften.
The surfaces of the heated glass are then cooled rapidly.

As the strength of this glass is almost 4–5 times than annealed glass, it has a good impact performance and safety based on standard EN 12150 "Glass in Buildings — Thermally toughened safety glass" and BS 6206: 1982. When broken, it disintegrates into small fragments with dulled edges that are unlikely to cause serious injury.

(*Continued*)

Table 7.18. (*Continued*)

One disadvantage of these types of glass is that the component contains nickel sulfide which is prone to expand or shrink on drastic temperature difference, resulting on sudden breakage.

1.6. *Heat Strengthened Glass*

It is thermally treated, first to over 600°C, then cooled down at a carefully controlled speed. This process imbues the surface of the glass with permanent compression stress and gives it its special characteristic.

It is thermally treated, first to over 600°C, then cooled down at a carefully controlled speed. This process imbues the surface of the glass with permanent compression stress and gives it its special characteristic.

1.7. *Figured Glass*

Mainly for decorative purposes or diffusing light. Includes surface modified such as acid etched glass and sand blasted glass.

1.8. *Safety Glass*

Safety glass is defined as the glass which must have passed an impact test (BS 6202: 1981 and *graded as per SS 341:2001*). The critical location in internal and external walls, where safety glass required to be used is specified in BS 6262: Part 4.

- bullet proof glass;
- wired glass (one kind of glass reinforcement which incorporate steel mesh into the glass body to ensure the strength and to have better fire resistance property);
- laminated and tempered glasses also fall under this category.

2. Component

A typical composition for architectural glass/"flat glass" is as follows:

- sand (silica) 72%;
- soda ash (sodium carbonate) 13%;
- limestone (calcium carbonate) 10%;
- dolomite (calcium magnesium carbonate) 4%;
- various other oxides such as alumina and magnesia, to improve the physical properties of the glass, including its resistance to atmospheric pollutant;
- for body tinted glass, metal oxide can be also be incorporated.

3. Density

Density of glass is 2.5 g/cm^3 [3]

(*Continued*)

Table 7.18. (*Continued*)

4. Sizes

Typical maximum sizes for toughened glass

Thickness (mm)	Size (m^2)
4	2300 × 1300
5	2600 × 2000
6	4200 × 2000
10	4200 × 2000
12	4200 × 2000
15	4000 × 1800
19	4000 × 1800

Source: Doran, DK, *Newnes Construction Material. Pocket Book;* Gutterworth- Heinemann Ltd, Oxford, 1994.

5. Heat Transmission

		Thermal Insulation for Degree of Site Exposure		
Glass Type	**Air Space Width (mm)**	**Sheltered (W/m^2 K)**	**Normal (W/m^2 K)**	**Severe (W/m^2 K)**
Single glazing		5.0	5.6	6.7
Double glazing	3.0	3.6	4.0	4.4
	6.0	3.2	3.4	3.8
	12.0	2.8	3.0	3.3
	25 or more	2.8	2.9	3.2
Triple glazing	3.0	2.8	3.0	3.3
	6.0	2.3	2.5	2.6
	12.0	2.0	2.1	2.2
	25 or more	1.9	2.0	2.1

(*Continued*)

Table 7.18. *(Continued)*

Recent glass technology manages to improve the *u*-value.

Low emissivity is coating on glass which has the ability to reflect the long wave radiation. In air-filled cavities with uncoated surfaces, the long wave radiation exchange between glass surfaces is high, amounting to about 60% of the total heat exchange across the cavity. With one of the glass surfaces having a coating with emissivity less than 0.2 (compared with 0.88 for uncoated glass surface), the radiation exchange is reduced by approximately 75% and consequently the *u*-value is reduced.

Argon and Krypton are known as ideal gases since they are inert gases that do not react chemically and have no smell. Although Krypton is more expensive, it is still preferred since it gives better *u*-value.

6. Light Transmission (Solar Radiation)

This may vary depending on the glass thickness, whether it is a tinted glass, coated or multiglazing unit with and without coating.

Coating	Light Transmission (%)	Total Solar Transmission (%)	Solar Reflectance (%)
Gold	42	22	40
Copper	42	25	39
Silver	42	27	47

Source: Doran, DK, *Newnes Construction Material Pocket Book*; Gutterworth-Heinemann Ltd, Oxford, 1994.

7. Compressive Strength

620.55 to 1241.10 MPa [3]

8. Tensile Strength

Annealed glass: 40 MPa
Toughened glass: 120–200 MPa (depending on thickness, edgework, holes, notches, etc.)

9. Elasticity

Glass is a perfectly elastic material. It does not exhibit permanent deformation, until breakage. However, it is fragile and will break without warning if subjected to excessive stress.

$$E = 7 \times 10^{10}\ \text{Pa} = 70\ \text{GPa}\ [3].$$

(Continued)

Table 7.18. *(Continued)*

10. Coefficient of Linear Expansion

The coefficient of linear expansion of glass is: 9×10^{-6} m/mk [3].

11. UV Radiation

UV radiation can be virtually eliminated by the use of PVB laminated glass. This type of glass can transmit only about 0.4% UV, compared to 44% for the normal type of glass.

12. Acid Resistance

Glass is only attacked by hydrofluoric acid.

13. Other Resistivities

Glass is not affected by UV (UV resistance), resistance to biological growth and water absorption.

14. Scratches

When visually inspected from a distance of 1.8 m, no scratch longer than 75 mm shall be visible within the normal viewing area (the central area of the glass that encompasses 80% of its height and width).

Scratch from 75 to 100 mm shall only be allowed outside the normal viewing area.

Table 7.19. Properties of Roof Deck Waterproofing.

ROOF DECK WATERPROOFING

Density

Depending on the type of waterproofing, the density is ranged from 1.2 to 1.6 g/cm³.

Preformed membrane	Generally quite dense and sometimes its self-weight may become a deterrent factor due to creep when installed on vertical surfaces such as basement walls.
Liquid-applied	Generally low density.
Cementitious	Can be relatively dense due to the cement and aggregate contents.

Thermal Conductivity

About 0.691 W/m² K.

As all waterproofing membranes are relatively thin (0.6–4 mm) thermal conductivity is normally insignificant and for roofs, this is usually provided by using a layer of insulation material such as extruded polystyrene, etc.

(Continued)

Table 7.19. (*Continued*)

Water Vapor Transmission

Herewith is the general comparison between different types of waterproofing in terms of water vapor transmission:

Preformed membrane	Depending on the generic types, i.e. bitumen-based generally have very low vapor permeability while PVC types usually are formulated with relative high vapor permeability.
Liquid-applied	As above except that vapor permeability is usually affected by the thickness of the membrane and liquid systems tend to have lower thickness therefore generally better vapor transmission.
Cementitious	Good vapor permeability due to its structure.

In SS 374: The WVT requirement is $<2 g/m^2$ 24 h. If the amount of water vapor in the substrate is high and it has to permeate the film of low WVT, there will be a pressure on the film which can cause debonding from the substrate, resulting in blistering which is common among liquid-applied roof coating.

For liquid-applied elastomeric wall coatings, the specification is $>80 g/m^2$ 24 h.

Elastomeric wall coating have higher WVT due to the lower applied film thickness and therefore reduce the risk of debonding from the substrate, caused by the high pressure of water vapor.

Tensile Strength

Range about 4–6 N/mm^2

Types	
Preformed membrane	Generally high especially if reinforced.
Liquid-applied	Medium to low depending on generic types and also whether reinforced.
Cementitious	Generally low tensile strength and normally not reinforced except at detailing.

Modulus Elasticity (Elongation at Break)

Types	
Preformed membrane	Generally good for nonreinforced. Elongation however is affected by types of reinforcements (range 50–1000%).

(*Continued*)

Table 7.19. (*Continued*)

Liquid-applied	Average from medium to high depending on generic types (range 50–700%), i.e. — bitumen based: Low. — rubber-based: High.
Cementitious	Low, normally at around 150% for polymer modified type.

Reinforcement

Base Reinforcements	Benefits	Limitations
Woven glass fiber tissues and mats	• High modulus (300–1000 N/mm) • Dimensional stability • Rot proof	• Low elongation (3%) • Fragile • Low tear resistance • Capillary action along strand
Glass/polyester (mixed nonwoven fibers)	• High modulus (400 N/mm) • High dimensional stability • Rot proof • Improved resistance to crack propagation	• Low elongation
Polyester	• Modulus (100–300 N/mm) • High elongation (over 20%) • Rot proof • High-puncture resistance • High-tear resistance	• Moderate dimensional stability • Less flexible (stiff) • Heavier weights

Source: CITI training note; *Certificate in Waterproofing Supervision*.
With courtesy from Hitchin Pte Ltd.

Property	**APP Modified**		**SBS Modified**		**SS 374: 1994 Requirements**
Nominal thickness in mm include surfacing	Glass[a]	Polyester[b]	Glass[a]	Polyester[b]	
Dimension stability (both longitudinal and transverse) %	4 mm	4 mm	4 mm	4 mm	—

(*Continued*)

Table 7.19. (*Continued*)

Property	APP Modified		SBS Modified		SS 374: 1994 Requirements
Tensile strength longitudinal (N/mm^2) transverse (N/mm^2)	0.1%	+ 0.14% − 0.10%	0.2%	0.3%	+ 0.5% −0.5%
Elongation at break					
longitudinal (%)	2%	45%	2%	45%	Both max
transverse (%)	2%	45%	2%	45%	min. 45%
Tear strength/shear					
longitudinal (N/mm)	80	175	110	180	Tensile shear at joints — Failure occurs outside (>500 N)
transverse (N/mm)	100	210	130	220	
Water absorption (%)	0.5%	1.3%	—	—	<1.0%
Vapor penetration resistance	40,000	46,000	70,000	50,000	Water vapor transmission in g/h m^2 (<0.5)
Hydrostatic test	2 bars	2 bars	2 bars	2 bars	>3 bars — No leakage observed
Low temperature flexibility, max. °C	−15	−18	−25	−25	NA
Heat resistance, max. °C	+130	+155	+100	+100	NA

[a]Glass fiber 50 g m/m².
[b]Polyester mat 180 g m/m².
Source: CITI training note; *Certificate in Waterproofing Supervision* [5].
With courtesy from Hitchin Pte Ltd.

Property	PVC[a]	EVA[b]	ECB[c]	SS 374: 1994
Membrane thickness (mm)	1.2	1.2	2.0	
Dimensional stability %				
longitudinal	0.0	<0.4	<1.0	
transverse	0.0	0	—	±2.0% without aging

(*Continued*)

Table 7.19. (*Continued*)

Property	PVC[a]	EVA[b]	ECB[c]	SS 374: 1994
Tensile strength (MPa)				
longitudinal	15.4	12.5	6	
transverse	15.6	12.5	4	>15.0 N/mm^2
Elongation at break %				
longitudinal	250%	320%	350%	>300%
transverse	270%	320%	490%	>240%
Vapor permeability coefficient	<15,000	<20,000	<80,000	
Water vapor transmission (g/m^2 day)	11.4	1.65	0.25	Vapor transmission <12 g/m^2 24 h
Water absorption after 24 h%	0.11	NA	0.31[a]	<1.0%
Resistance to water penetration	4 bars	4 bars	6 bars	>3 bars — No leakage observed
Low temperature flexibility (°C)	–30	–20	–20	
Root resistance	Tested	NA	Tested	

[a]PVC membrane reinforced with polyester scrim.
[b]ECB membrane glass fleece backing.
[c]EVA membrane nonreinforced.
[d] Water absorption after 7 days.
Source: CITI training note; *Certificate in Waterproofing Supervision* [5].
With courtesy from Hitchin Pte Ltd.

Property	EPDM	CSPE	SS 374: 1994
Membrane thickness in mm	1.2	0.9	
Dimensional stability %			
longitudinal	±2	<0.4	
transverse	—	0	±2.0% without aging
Tensile strength (MPa)			>15.0 N/mm^2
longitudinal	11.7	10.4	ASTM D412 requires
transverse	—	—	>9 N/mm^2
Elongation at break %			
longitudinal	475%	500%[a]	>300%
transverse	—	—	>240%

(*Continued*)

Table 7.19. (*Continued*)

Property	**EPDM**	**CSPE**	**SS 374: 1994**
Water vapor transmission. g/m^2 24 h	0.25	2.1	Vapor transmission <12 g/m^2 24 h
Water absorption after 7 days. %	2.0	1.25	<1.0% after 24 h
Low temperature flexibility (°C)	–65	–45	

[a]Elongation of rubber only, fabric elongation at 35%.
Source: CITI training note; *Certificate in Waterproofing Supervision* [5].
With courtesy from Hitchin Pte Ltd.

Recommended Testing Method
Applicable for cementitious membrane waterproofing:

Item	**Test Method**	**Specification**
Characterization of polymer	FTIR, DTA and TGA	FTIR, DTA and TGA graphs used as fingerprint records for future comparison of subsequent testing.
Initial hardness (Shore A) (*7 days curing*)	ASTM D2240	min 40
Crack bridging	ASTM C836	No cracking at 2 mm width. No crack after 10 cycles of stretching and closing at 1 mm width.
Adhesion to substrate (*Condition as cast*)	ASTM D4541	min 0.3 N/mm^2
Tensile strength		
(a) Condition as cast *Change in strength* (%)		min 1.5 N/mm^2
(b) After aging at 50°C for 14 days		min 1.3 N/mm^2 and negative change max. 40%

(*Continued*)

Table 7.19. (*Continued*)

Item	Test Method	Specification
(c) After 72 h chemical immersion at room temperature:	ASTM D412	
(i) 0.5% (v/v) NaOCl		}
(ii) 1.25% (v/v) NH_4OH		} No limit for positive change
(iii) 3.7% (v/v) HCl		}
Elongation		
(a) Condition as cast *Change in elongation (%)*		min 150%
(b) After aging at 50°C for 14 days		min 120% and negative change max 40%
(c) After 72 h chemical immersion at room temperature:	ASTM D412	
(i) 0.5% (v/v) NaOCl		}
(ii) 1.25% (v/v) NH_4OH		} No limit for positive change
(iii) 3.7% (v/v) HCl		}
Chloride content	ISEA Method	min 0.1%
Resistance to water penetration	Adopted DIN 1048: Pt 5	No water penetration allowed at 0.2 kgf/cm² for 6 h
Set to touch time RH 65 ± 5% and 30 ± 2°C	ASTM D1640	Touch dry within 2 h

Source: CITI training note; *Certificate in Waterproofing Supervision* [5].
With courtesy from Hitchin Pte Ltd.

Table 7.20. Properties of sealants.

SEALANT
Classification

1. *Caulking*: Installed as filler between dissimilar materials in an interior controlled environment no subject to thermal or structural movement.
2. *Sealant*: Exterior application, using high-performance material (e.g. silicones), which are typically colored.
3. *Glazing material*: Used in construction of window panels or curtain walls, where higher strength is more important than movement capability.
   - Water-based sealant is those used in the interior application where not in contact with water, e.g. latex sealant.
   - Solvent based/oil-based sealant is used on the external façade as it is able to resist weathering effect (rain and UV). Because of its load bearing capability, it is considered structural sealant; e.g. butyl, acrylic, silicone, polyurethane and polysulfide.

Grade

According to ASTM C 920:

- **Grade P:** (pourable or selflevelling) sealant shall have flow characteristics such that when tested in accordance with test method C 639 it shall exhibit a smooth, level surface.
- **Grade NS (nonsag)** or gunnable sealant shall have flow characteristics such that when tested in accordance with test method C 639 it does not sag more than 4.8 mm in vertical displacement. Also the sealant shall show no deformation in horizontal displacement.

Constituents

1. Polymer
2. Filler: Part of sealant that gives the "body"
3. Cross linker: To link the polymer
4. Adhesion promoter
5. Plasticizer (solution/oil part of sealant)
6. Pigment

Base Materials

1. Acrylic
   Factory mixed, one component materials polymerized from acrylic acid. These are not used on joints subject to high movement because of their relatively low movement capability. They are commonly used in remedial applications with acrylic-based waterproof coatings and small movement joints such as windows and door perimeters.

(Continued)

Table 7.20. (*Continued*)

2. Butyl
   Produced by copolymerization of isobutylene and isoprene rubbers. Butyl sealants are used in metal curtain wall construction because of their ability to function in very thin application. Although they have low-movement capability, they have excellent adhesion to most substrate and have excellent weathering resistance.
3. Latex
   Latex sealants are typically acrylic emulsion or polyvinyl derivatives. It seldom used for exterior application since it is sensitive to water immersion and traffic. Typically used for interior purposes when a fast cured is desired for painting.
4. Polysulfide
   Polysulfide materials are produced from synthetic polymers of polysulfides rubbers. It exceeds the movement capability of paints and therefore should not be painted. However, they are manufactured in both one and two component packaging in a wide range of color. It is considered as one of the good performance sealants and thus suitable for wide range of applications, includes curtain wall joints, cast *in situ* and precast panels. Of all commercially available sealants, polysulfides are best suited for total-immersion joints, such as swimming pool, water and waste water treatment structures, fountains and ponds.

   Polysufides should not be installed in joints that might have bituminous residue or contamination such as premold joint filler. It should also not be applied over oil or solvent-based joint sealants.
5. Polyurethane
   Urethane sealants are polymers produced by chemical reactions formed by mixing di-isocynate with a hydroxyl. It ranges from one-component, self-levelling materials in a pourable grade for horizontal joints in plaza decks to two-component nonsagging materials used for vertical expansion joints. Urethane is not recommended for continual immersion application.

   Two-component urethanes are low-modulus sealants and have high joint movement capability averaging 25%. Polyurethane cannot be used in joints containing a polysulfide sealant or residue. Urethane sealant should also not be used in glazing applications of high-performance glass, plastics, or acrylics. Joints contaminated with asphalts, tar or for release agents must be cleaned before sealing work.
6. Silicones
   Silicones-based sealants are derivatives of silicones polymers produced by combining silicone, oxygen and organic materials. As it has high thermal stability, silicone is the most common sealant used in Singapore. Silicones are available in a wide range of compositions that are extremely effective in high-movement joints, including precast panels and expansion joints.

(*Continued*)

Table 7.20. (*Continued*)

Silicone sealant cannot be used for below grade applications, horizontal applications subject to vehicular traffic and water immersion joints. It is extremely important not to install silicone materials that might bleed through a silicone. This includes oil, solvents or plasticizer, which will cause staining and possible silicone failure.

Silicone materials exceed the movement capability of paints and as most paints will not adhere to silicones, they should not be painted over. It has excellent adhesion to almost all building products including wood, ceramic, aluminum and natural stones. It may be used in curtain wall joints (typically the high tensile-strength silicones with lower movement capability), precast concrete panel, expansion and control joints.

7. Precompressed foam sealant
Foam sealants are manufactured by impregnating open-cell polyurethane foam with chemical sealant containing neoprene rubbers. An adhesive is applied to one side and covered with a release paper. The foam is then compressed and supplied in rolls with various widths of up to 12 inches. During application, it swells and expands to fit tightly against both joint sides, allowing for any irregularities in joint width. Splices in material are prepared by overlapping or butting joint ends. This material eliminates the need for joint backing, primers and tooling.

Foam sealant is compatible with other sealant materials and allows elastomeric sealant to be applied over the foam, providing good barrier in critical water proofing joints.

Property	Acrylic/ Latex	Butyl	Polysulfide	Polyurethane	Silicone	Pre-compressed Foam
Maximum joint movement capability %	±7	±5	±25	±25	±50	±25
Weathering resistance	Good, excellent	Excellent	Good	Excellent, good	Excellent	Excellent
Recovery %	25–75	Poor	80	90	100	100

(*Continued*)

Table 7.20. (*Continued*)

Property	Acrylic/ Latex	Butyl	Polysulfide	Polyurethane	Silicone	Pre-compressed Foam
Adhesion	Good	Excellent	Good	Good	Excellent	Excellent
Joint design (number of times expected movement)	12	20	6	4	4	a
Shrinkage %	12–20	18	10	5	3	NA
Tack-free time (h)	1–72	24	72	72	3	NA
Water immersion	No	No	Yes	Some	No	No
Paintable	Yes	Yes	No	No	No	No
Primer required	Some cases	No	Metal, masonry	Horizontal masonry	Metal, natural stone	No
Ultimate elongation %	Low-450	Low	1000	700	1600	Very low
Hrz. joints	No	No	Yes	Yes	No	Yes
Modulus of elasticity (MPa)	0.126–0.28	0.175	0.21	0.245	0.21	0.175

[a] The process of weathering is defined as the action of atmospheric elements in altering the color, texture, composition or form of exposed objects, ultimately leading to disintegration or failure to perform a function. The well-known elements of weather are radiation, moisture, thermal conditions and gases. Particularly for sealant material, the following weathering factors will affect the performance:

1. UV
2. Ozone + UV
3. Acid rain + ozone + UV
4. Heat (especially for west and south orientation)
5. Building movement = thermal expansion and contraction + building settlement movement + building flexural movement + thermal shock movement (drop of temperature during a rain storm resulting in extreme short time movement in joint areas
6. Miscellaneous chemicals in the air

(*Continued*)

Table 7.20. (*Continued*)

Adhesion Strength

(BS 5889:1989)

The ability of sealant to bond to a particular substrate, including adhesion during substrate movement, since materials differ substantially in their adhesive strength to a particular substrate, manufacturer should be consulted for adhesion test samples on proposed substrate.

The "Adhesion in peel" test according to BS 5889:1989, the material shall not fail in adhesion over more than 25% of the test area.

ASTM C 794 tests the adequacy of adhesion of sealant to substrates and other sealant types.

Cohesion Strength

The ability of a material's molecular structure to stay together internally during movement. Cohesive strength has a direct bearing on elongation ability. Cohesive failure is when a sealant tears or splits apart due to excessive joint movement or improper installation. According to BS 5889:1989, adhesion and cohesion shall meet:

- Initial test

Type	Condition
A	7 days at 50% extension
B	7 days at 25% extension

- Cyclic adhesion test
- Adhesion and cohesion after exposure to UV radiation through glass

Type	Condition
A	Withstand 100% extension for 24 h UV exposure and water immersion
B	Withstand 60% extension for 24 h UV exposure and water immersion

Elasticity

According to BS 5889:1989 the types of sealant according to the elastic recovery.

Type of Sealant	Recovery
A	≥ 60% from 100% extension
B	≥ 90% from 60% extension

(*Continued*)

Table 7.20. *(Continued)*

Shore Hardness

Shore hardness is resistance to impact by a durometer gauge. This property becomes important in selecting sealant subject to traffic or punctures, such as horizontal paver joints (ASTM D 2240).

Type of Sealant	Hardness[a]
Epoxide polysulohide and flexibilized epoxide (two-part)	70–95
Acrylic (solution type)	25–30
Polysulfide (one part)	15–40
Polysulfide (two-part)	
High Modulus	40–60
Low Modulus	15–20
Polyurethane (one part)	15–40
Silicone	
High Modulus	20–30
Low Modulus	10–20

[a]Shore A Hardness is too rigid to be used in curtain wall or cladding system.

According to ASTM C 920, the classification of hardness is as follows:

1. Use T (traffic); shall have a hardness reading after being properly cured, of not less than 25 or more than 50 when tested in accordance with test method C 661.
2. Use NT (nontraffic); shall have a hardness reading, after being properly cured, of not less than 15 or more than 50 when tested in accordance with test method C 661.

Service Temperature Limit

Type of Sealant	Service Temperature Limits (°C)
Epoxide polysulfide and flexibilized epoxide (two-part)	–20 to +80
Acrylic (solution type)	–35 to +90
Polysulfide (one part)	–20 to +80
Polysulfide (two-part)	
High Modulus	–20 to +80
Low Modulus	–20 to +80

(Continued)

Table 7.20. (*Continued*)

Type of Sealant	Service Temperature Limits (°C)
Polyurethane (one part)	–40 to +70
Silicone	
High Modulus	–60 to +180
Low Modulus	–50 to +120

According to ASTM C 920 on effect of heat aging; the sealant shall not lose more than 7% of its original weight or show any cracking or chalking when tested in accordance with test method C 1246.

Movement Capability

Sealant must have a cyclic movement capability of ±12%–±25% through its useful service life. Refer to test methods specified in ASTM C719 or equivalent national standard.

Modulus

High, medium and low modulus.
Modulus at 100% elongation: ±0.4 MPa.
(ASTM D412)

Tensile Strength

Tensile strength: 1.4–2 MPa (ASTM D412).

Substrate

1. Concrete and masonry
   Newly placed concrete or masonry must be allowed to cure before applying sealant. If wet, it shall allow to dry at least 24 h in good drying weather before application.
2. Stone
   These surfaces generally provide good sealant adhesion. However, some stone should be primed before sealant application. If the surface area of stones appears to be flaking or dusty, it must be cleaned by either water blasting, sandblasting or wire brushing before application.
3. Glass
   These excellent surfaces for sealants once their surfaces are cleaned of contaminants and oils. Methyl ethyl ketone or alcohol is an ideal cleaner.

(*Continued*)

Table 7.20. (*Continued*)

4. Painted and lacquered surfaces
   Depending on where these surfaces are located and their exposure to weather, sealant should not be applied to flaking painted or lacquered surfaces. Sound painted should be first cleaned by wiping with a solvent to remove oil and dust.
5. Aluminum with a mill finish
   A good degreasing solvent, such as trichloroethane or xylene, will clean these surfaces properly. A rub down with fine steel wool or fine emery cloth might permit better adhesion.
6. Aluminum with an anodized finish
   Generally provides excellent surface for sealant application. However, it should be wiped down with methyl ethyl ketone or xylene to remove any surface contaminants.
7. Galvanized Steel
   New galvanized surfaces present more difficult surfaces for adhesion than weathered galvanized surfaces. Consultation with sealant manufacturer is recommended.
8. Stainless Steel
   This is difficult surface for adhesion purposes. Primers are often recommended along with solvent cleaning of the surface.

Staining

Possibility of sealant causing staining to certain substrates is tested in accordance with ASTM C1248 or D2203-01.

Compatibility with Other Material (for façade)

Compatibility of sealant with other materials such as gasket, spacer, setting block, etc. is tested in accordance with ASTM C1087.

Adequacy of adhesion of sealant to substrates and other sealant types is tested in accordance with ASTM C794.

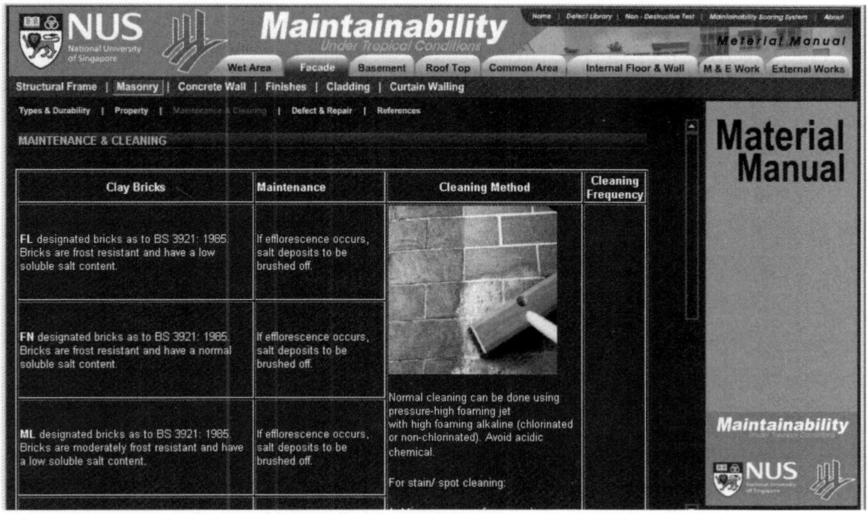

Fig. 7.10. Façade maintenance and cleaning.

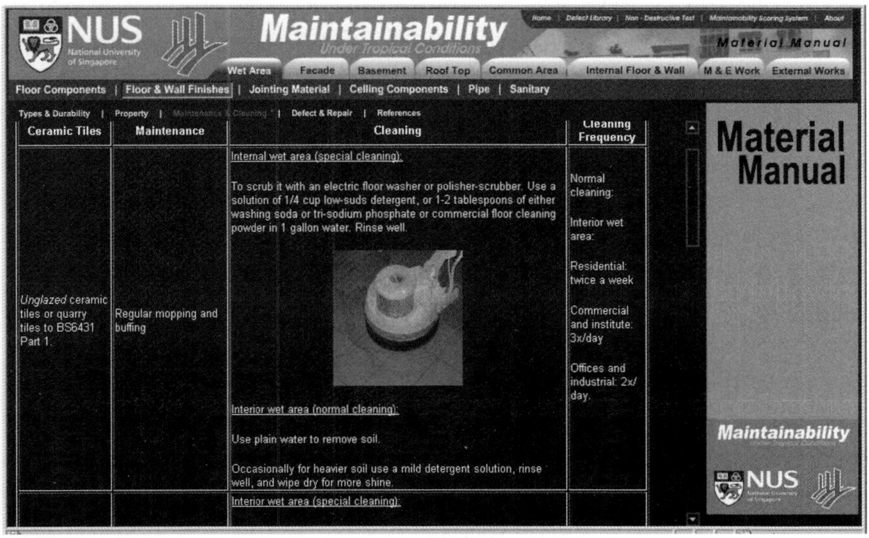

Fig. 7.11. Wet area maintenance and cleaning.

Tables 7.21 to 7.28 show the typical maintenance methods for a façade's main material as listed in this section.

Tables 7.29 to 7.31 show the maintenance method for materials used as wet area finishes.

Table 7.21. Aluminum panel.

Material	Maintenance	Cleaning Frequency
Aluminum (*anodized*)	Cleaning: • Regular cleaning with nonalkaline detergent. Under no circumstances shall rough scrubbing tools or materials causing grinding or scratch marks, such as sand paper, steel wool or wire brushes, be used. Replacement: Renewal of sealants/gaskets as necessary.	Industrial: 4 × /year Urban: 2 × /year Suburban: 1×/year 5–20 years
Aluminum (*liquid organic coating*)	Cleaning: • Regular cleaning with nonalkaline detergent. Do not use any powerful alkaline cleaning agents such as potassium hydroxide, sodium carbonate or caustic soda, or any powerful acidic products. • Under no circumstances shall rough scrubbing tools or materials causing grinding or scratch marks, such as sand paper, steel wool or wire brushes, be used. • Heavily abrasive scouring agent or lacquer-dissolving cleaning agents should not be used. Replacement: Renewal of sealants/ gaskets as necessary.	Industrial: 4 × /year Urban: 2 × /year Suburban: 1 × /year 5–20 years

(*Continued*)

Table 7.21. (*Continued*)

Material	Maintenance	Cleaning Frequency
Aluminum (*polyester powder coating*)	Cleaning: • Regular cleaning with nonalkaline detergent. Under no circumstances shall rough scrubbing tools or materials causing grinding or scratch marks, such as sand paper, steel wool or wire brushes, be used.	Industrial: 4×/year Urban: 2×/year Suburban: 1×/year
	Replacement: • Renewal of sealants/gaskets as necessary.	5–20 years

Table 7.22. Granite cladding panel.

Maintenance	Cleaning Frequency
Inspection: • Visual inspection every 5 years. If necessary, cleaning, repointing and surface repairs in accordance with BS 6270: Part 1: 1982 or CP 67 Part 1: 1997. • Thermographic tests should be done in every 5 years to detect tile delaminations of tiled walls.	1 × /year
Cleaning: • Generally, using high-foaming jet is recommended with neutral cleaner. Avoid acidic, high alkaline and chlorinated products. • Unpolished Granite o Pressure washing (though bush hammer and flame finished are more difficult to clean). o Use hydrofluoric acid together with orthophosphoric acid to remove stubborn soiling. o Alkaline cleaners followed by neutralization with a weak organic acid may also be successful. o Caustic soda cleaners should not be used.	1 × /year

(*Continued*)

Table 7.22. (*Continued*)

Maintenance	Cleaning Frequency

• Polished surfaces
  ○ Wash with clean water/pressure washing and a little mild liquid detergent and then clean off.
  ○ Chemical containing hydrofluoric acid and abrasive methods should not be used except in areas where repolishing is necessary. Alkaline agents followed by weak organic acid are suitable.
  ○ Atmospheric soiling on polished granite may be removed using nonionic soap in cold or hot water and associated with plastic pot scourers. Surfaces should then be thoroughly rinsed and wiped dry to prevent water spotting.	

Table 7.23. Ceramic tile wall.

Material	Maintenance	Cleaning Frequency
Unglazed ceramic tiles *Glazed* ceramic tiles	Inspection: • Inspection for delamination, crack, elongation: Test every 5 years	Durability refer Fig. 3.6
	Cleaning: • Pressurized hydro foaming jet to remove soil. For heavier soil, use a mild detergent solution, rinse well and wipe dry for more shine. • Alternatively, use a neutral to mild alkaline cleaner (chlorinated or nonchlorinate)	1×/year

Table 7.24. Glass curtain wall.

Maintenance Method	**Cleaning Frequency**
Inspection:	
• More than 25% of the glass wall should be visually inspected every six months for sealant failures and water leakages.	Industrial: 2 × /year Other area: 1 × /year
Replacement:	5–20 years
• Replace glazing gaskets, weather stripping, etc. as required.	
Cleaning:	
• For normal environmental conditions, basic washing with water and soap should be sufficient; unless there is excessive pollutant. All glass products must be cleaned periodically using neutral nonabrasive cleaning agents that contain no fluorine compounds. • Alkalis and acid should not be in contact with glass. • Immediately after washing, the entire surface of glass has to be rinsed and dried. In the event of persistent pollution or fuel oil contamination, stains can be removed by either a solution of clear water and cerium oxide in a concentration varying between 50 and 160 g/l. Use soft cloth dipped in the solution to clean the glass and only exert the slightest pressure on glass (strong pressure can deteriorate the glass since the cerium oxide then becomes abrasive). Rinse thoroughly. • Fingerprints, grease marks or traces of mastic can be removed with solvents such as acetone, methyl ethylcetone (MEC) or ammonia, since these products do not attack the seals or penetrate the rabbet. • These methods must not be applied to the coated side of glass. • Special care must be taken into account when cleaning the coated glass that is exposed, since it is easily scratched and the coating damaged permanently.	Industrial: 2 × /year Other area: 1 × /year

Table 7.25. Plaster wall.

Maintenance	Cleaning Frequency
Cleaning:	
• Can be washed by water jetting at mains pressure or using high-foaming jet (without pressure) with neutral to mild alkaline. Avoid acidic chemicals.	1 × /year
Replacement:	
• Cracks and damages are to be cut out and rectified according to CP 56: 1991.	Durability refer Fig. 3.6

Table 7.26. Painted wall.

Maintenance	Cleaning Frequency
Inspection:	
• Regular inspection to detect or identify defects immediately to take corrective maintenance at early stages to prevent further complexities. • Visual inspection techniques or instrumentation techniques used for regular inspections. Such fungi growth is relatively unsightly and quick response to repair the defect is essential to prevent the defect from spreading to other areas. • Re-painting every 5 years (regular practice for external façade).	Durability refer Fig. 3.6
Cleaning:	
• Periodic cleaning: Using high-foaming jet (without pressure) with neutral to mild alkaline. Avoid acidic, solvent and chlorinated chemicals.	

Table 7.27. Brick wall.

Maintenance	Cleaning Frequency
Cleaning:	
• If efflorescence forms, salt deposits to be brushed off. • Normal cleaning can be done using foaming jet with high-foaming alkaline (chlorinated or nonchlorinated). Avoid acidic chemical.	1 × /year

Table 7.28. Concrete wall.

Maintenance	Cleaning Frequency
Cleaning:	
• Regular cleaning using high-pressure foaming jet with foaming alkaline (chlorinated or nonchlorinated). Avoid acidic products.	1 × /year

Table 7.29. Natural stone floor and wall.

Material	Maintenance	Cleaning Frequency
Granite	Routine cleaning:	
	• Normal cleaning using high-foaming jet (without pressure) with neutral cleaner. Avoid acidic, high alkaline and chlorinated products. • For periodic cleaning of polished finished pieces and small areas, warm water applied with a clean, lint free cloth then wiped dry with a soft cloth is sufficient.	Residential: 2 × /week Commercial and institute: 3 × /day Offices and industrial: 2 × /day.
	Special cleaning:	
	• If necessary, cleaning, repointing and surface repairs in accordance with BS 6270: Part 1 or CP 67 Part 1: 1997	Special cleaning: 1 × /year
Marble	Routine cleaning:	
	• Warm water applied with a clean, lint free cloth, then wipe dry with a soft cloth. • If the marble has become soiled, a "soapless" cleaning solution is preferred; these include mild, phosphate-free cleaners, biodegradable liquid dish-soaps, general household neutral cleaners and soap powders. It has to be rinsed throughly. The use of proprietary stone leaner is also recommended, with method of application in conjunction with manufacturer instruction.	Residential: 2 × /week Commercial and institute: 3 × /day. Offices and industrial: 2 × /day. Special cleaning: 1 × /year

(*Continued*)

Table 7.29. (*Continued*)

Material	Maintenance	Cleaning Frequency
	• As a rule, do not use any cleaning solution which contains any form of acids, phosphorus, chlorine or scouring powders on polished marble, since this will make marble look dull (known as "orange peel") unless adhering to the cautions and/or formulas provided. • Any type of abrasive product should be avoided, in order to prevent scratches and loss of shine. Special cleaning: • If necessary, cleaning, repointing and surface repairs in accordance with BS 6270: Part 1 or CP 67 Part 1: 1997.	

Table 7.30. Ceramic tile floor and wall.

Material	Maintenance and Cleaning Method	Cleaning Frequency
Ceramic tile	Cleaning:	
	Regular mopping and buffing.	Residential: 2 × /week
	Hydro foaming jet with pressure to remove soil.	Commercial and institute: 3 × /day
	For heavier soiling, use a mild detergent solution, rinse well and wipe dry for more shine.	Offices and industrial: 2 × /day.
	Alternatively, using neutral to mild alkaline cleaner (chlorinated or nonchlorinate).	Special cleaning: 1 × /year

Table 7.31. Homogeneous tile floor and wall.

Material	Maintenance and Cleaning Method	Cleaning Frequency
Homogeneous tile	Cleaning: Regular mopping and buffing. Use plain water to remove soil. For heavier soiling, use a mild detergent solution, rinse well and wipe dry for more shine. Scrub with an electric floor washer or polisher-scrubber. Use a solution of 1/4 cup low-suds detergent, or 1–2 tablespoons of either washing soda or tri-sodium phosphate or commercial floor cleaning powder in 1 gallon water. Rinse well.	Residential: 2 × /week Commercial and institute: 3 × /day Offices and industrial: 2 × /day. Special cleaning: 1 × /year

7.6. Defect and Repair

The common defects of different materials are compiled in MM. Tables 7.32 and 7.33 illustrate examples of common defects for façades and wet areas, respectively.

Table 7.32. Common defect in materials used for façade.

Defect Cases	Causes	Remedy
Glass shattering 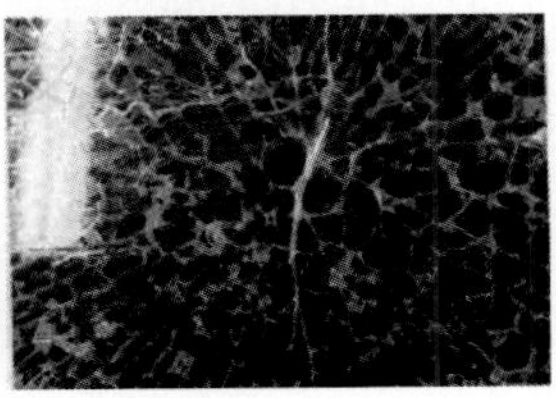	Inability of glass to accommodate movements. Impurities (nickel sulfide) present in tempered/toughened glass may expand on heating and contract on cooling to bring about sudden spontaneous shattering.	Replacement. Testing (i.e. impact test, boil test, impurity test, etc.) is necessary to prevent such defects.

(*Continued*)

Table 7.32. (*Continued*)

Defect Cases	Causes	Remedy
Fungus growth 	High moisture content or relative humidity is extreme; and when there is supply of organic material on which to feed on.	Treatment by fungicide. Depending on the level of staining the fungicide has to be left on the cladding for up to 7 days. All traces should be removed and rinsed.
Sealant deterioration 	Loss of solvent or plasticizers due to exposure to ultra-violet radiation and ozone. Sealant hardens and cracks.	Remove deteriorated sealant, clean and chaulk with sealant suitable for external conditions.
Sealant migration 	The migration of fluids (plasticizers, unreacted or unreactive polymers, antioxidants or other liquid formulation additives) out of sealant and into the porous or microporous substrate, such as stones, causing a dark band ("wet look") around the joint.	Select the right type of sealant for the substrate. Specify staining test. Polyurethane sealants are less prone to migration. Porous substrates such as stones must be properly sealed.

(*Continued*)

Table 7.32. (*Continued*)

Defect Cases	Causes	Remedy
Water penetration 	The source of moisture can be from condensation or from rain penetration through cracks or joints.	Repointing, water resistive surface coating and pressure grouting.
Efflorescence 	Water leaches out salt from cementitious substrate and upon evaporation crystallizes to form white deposits on the surface.	Efflorescence can be brushed off from the surface. The affected area can be washed with a solution of 10% phosphoric acid, allowed to stand for 10–15 min and then washed down with water.

Table 7.33. Common defect in wet area material.

Defect Cases	Causes	Repair
Tile staining 	Poor maintenance resulting in the accumulation of rust products.	Replace the corrode tap. Clean tile using suitable commercial tile cleaner. Tile cleaner should not be left to stand on the tile surface as it may contain acid that may etch the surface.

(*Continued*)

Table 7.33. (*Continued*)

Defect Cases	Causes	Repair
Grout deterioration 	May be due to harsh cleaning agent or low quality grout material, such as improper mixture/ proportion of grouting material	Scrap and re-grout.
Staining on ceiling boards 	Water leakage from above the false ceiling.	Check for water leakage or cracks. Repair leak and replace panel.
Pipe leakage 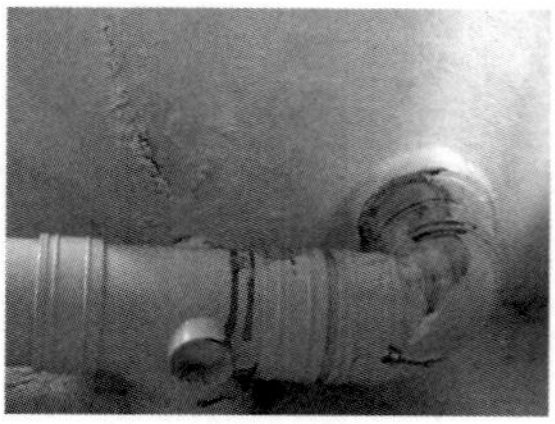	Leakage due to poor jointing between pipes.	Replace joints and affected parts.
Dampness or water mark 	Poor jointing or waterproofing. Cracks in grouting or on tile bedding material can lead to water seepage.	Torch and seal stone surface to prevent further water seepage. Remove stains with stone poultice. The poultice may need to be reapplied two or three times on stubborn stains.

References

[1] C. Hornbostel, *Construction Material-Types, Uses and Application*, John Wiley and Sons, New York, 1978.

[2] D. K. Doran, *Newnes Construction Materials Pocket Book*, Butterworth-Heinemann Ltd, Oxford, 1994.

[3] *Glass Guide*, edition 2000, Satint-Gobain Glass.

[4] Building and Construction Authority, *Good Industry Practices — Waterproofing for Internal Wet Area*, Conquas 21 Enhancement Series, Singapore, 2001.

[5] CITI training note, *Certificate in Waterproofing Supervision*, 2003.

[6] *BPG Building Fabric Component Life Manual*, Building Performance Group, London, 1999.

[7] C. Beall, *Thermal and Moisture Protection Manual: For Architects, Engineers and Contractors*, McGraw-Hill, New York, 1999.

[8] W. D. Callister, *Fundamentals of Materials Science and Engineering*, John Wiley and Sons, New York, 2001.

[9] J. I. Gersten and F. W. Smith, *The Physics and Chemistry of Materials*, John Wiley and Sons, Inc, New York, 2001.

[10] M. P. Groover, *Fundamentals of Modern Manufacturing: Materials, Processes and Systems*, Wiley, New York, 2002.

[11] *HAPM Component Life Manual*, E&FN Spon, London, 1992–1999.

[12] HMSO, *Housing Policy Review*, Vol. 3, 1986.

[13] M. Hollis and C. Gibson, *Surveying Buildings*, 2nd Edition, UK, 1986.

[14] J. M. Illston and P. L. J. Domone, *Construction Materials: Their Nature and Behaviour*, E&FN Spon, New York, 2001.

[15] B. Jahn, *Residential Construction Problem Solver*, McGraw-Hill, New York, 1998.

[16] T. Muneer, *Windows in Buildings-Thermal, Acoustic, Visual and Performance*, Architectural Press, Oxford, 2000.

[17] S. Kalpakjian and S. R. Schmid, *Manufacturing Engineering and Technology*, 4th Edition, Prentice Hall, New Jersey, 2001.

[18] K. Mathur and P. Kong Beng, *A Survey and Identification of Existing Building Defects and the Development of a Knowledge-Base*, NUS, Singapore, 1991.

[19] J. R. Meier and J. S. Russell, "Model process for implementing maintainability", *Journal of Construction Engineering and Management*, November/December, 440–450, 2000.

[20] B. A. Richardson, *Defects and Deterioration in Buildings*, E&FN Spon, New York, 2001.

[21] H. J. Rosen and T. Heineman, *Architectural Materials for Construction*, McGraw-Hill, New York, 1996.

[22] H. L. Simmons, *Repairing and Extending Finishes*, Van Nostrand Reinhold, New York, 1990.

[23] P. A. Stone, *Building Design Evaluation — Costs-in-Use*, Spons, London, 1980.

[24] G. D. Taylor, *Materials in Construction*, Pearson Education Ltd, London, 2000.

[25] H. R. Trechsel, *Manual on Moisture Control in Buildings*, ASTM, Philadelphia, 1994.

[26] S. Ledbetter and A. Keiller, *The Whole Life Performance of Façade,* Center for Window and Cladding Technology, University of Bath, UK, 2001.

[27] S. M. Levy, *Construction Building Envelope and Interior Finishes Databook*, McGraw-Hill, New York, 2000.

[28] A. S. Loh, *Design for Maintenance of External Façades in Commercial High Rise Buildings*, National University of Singapore, Singapore, 1989.

[29] M. S. Mamlouk, *Materials for Civil and Construction Engineers*, Addison-Wesley, Menlo Park, CA, 1999.

[30] J. R. Davis, *Alloy Digest Sourcebook: Stainless Steels,* ATSM International and Freund Publishing House Ltd, Ohio, 2000.

[31] A. K. Sinha, *Physical Metallurgy Handbook*, McGraw-Hill Inc, New York, 2003.

[32] R. E. Smallman and R. J. Bishop, *Modern Physical Metallurgy and Materials Engineering: Science, Process, Applications*, Butterworth-Heinemann, Boston, 1999.

[33] BCA, *Good Industry Practices — Ceramic Tiling*, Building and Construction Authority, Singapore, 1999.

[34] C. Beall, *Concrete and Masonry Databook*, McGraw-Hill, New York, 2002.

[35] A. G. King, *Ceramic Technology and Processing*, Noyes Publications: William Andrew Publishing, New York, 2002.

[36] W. H. Mosley, J. H. Bungey and R. Hulse, *Reinforced Concrete Design*, Macmillan, Basingstoke, Hants, 1999.

[37] E. G. Nawy, *Prestressed Concrete: A Fundamental Approach*, Prentice Hall, 2000.

[38] R. Novakovic and B. Korthaus, *Advanced Ceramics for Use in Highly Oxidizing and Corrosive Environments*, Enfield, N.H. Trans Tech Publications, Uetikon-Zuerich, Switzerland, 2001.

[39] R. Park and W. L. Gamble, *Reinforced Concrete Slabs*, Wiley, New York, 2000.

[40] P. H. Perkins, *Repair, Protection and Waterproofing of Concrete Structures*, Spon, London, 1997.

[41] N. Taly, *Design of Reinforced Masonry Structures*, McGraw-Hill, New York, 2001.

[42] R. F. Warner, *Concrete Structures*, Longman, Victoria, Australia, 1998.

[43] F. Bradley, *Natural Stone: A Guide to Selection*, Studio Marmo, Norton, New York, 1998.

[44] C. Hornbostel, *Construction Material — Types, Uses and Application*, John Wiley and Sons, New York, 1978.

[45] http://www.stonecaretechniques.com/11.htm. Retrieved June 2009.

[46] http://www.stoneinfo.com. Retrieved June 2009.

[47] http://www.unitedstone.com.sg. Retrieved June 2009.

[48] D. K. Doran, *Construction Material — Newnes Pocket Book*, Butterworth-Heinemann Ltd, Oxford, 1994.

[49] *Installation of Modular Stone Floor Tile*, Marble Institute of America, Columbus, Ohio, 2002

[50] F. Bradley, *Natural Stone: A Guide to Selection*, Studio Marmo, Norton, New York, 1998.

[51] E. Corbella, *The Architect's Handbook of Marble, Granite and Stone*, Van Nostrand Reinhold, New York, 1989.

[52] BCA, *Good Industry Practices — Marble and Granite Finishes*, Building and Construction Authority, Conquas Enhancement Series, Singapore, 2000.

[53] Marble Institute of America, *Installation of Modular Stone Floor Tile*, Columbus, Ohio, 2002.

[54] A. Compagno, *Intelligent Glass Façades: Material, Practice, Design*, Birkhauser, Switzerland, 1999.

[55] D. E. Czernik, *Gaskets: Design, Selection and Testing*, McGraw-Hill, New York, 1998.

[56] H. W. Krewinkel, *Glass Buildings: Material, Structure and Detail*, Birkhäuser Verlag, Boston, 1998.

[57] G. Sedlacek, *Glass for Structural Design*, Ernst and Sohn, 2001.

[58] *BPG Building Fabric Component Life Manual*, Building Performance Group, London, 1999.

[58] D. K. Doran, *Newnes Construction Materials Pocket Book*, Butterworth-Heinemann Ltd, Oxford, 1994.

[59] C. Hornbostel, *Construction Material — Types, Uses and Application*, John Wiley and Sons, New York, 1978.

[60] G. D. Taylor, *Material in Constructions*, Longman Group, London, 2000.

[61] T. Muneer, *Windows in Buildings — Thermal, Acoustic, Visual and Performance*, Architectural Press, Oxford, 2000.

[62] G. Sedlacek, *Glass for Structural Design*, Ernst & Sohn, 2001.

[63] C. Bray, *Dictionary of Glass: Materials and Techniques*, University of Pennsylvania Press, Philadelphia, 2001.

[64] A. Compagno, *Intelligent Glass Façades: Material, Practice, Design*, Birkhauser, Switzerland, 1999.

[65] Balkow *et al.*, *Glass Construction Manual*, Birkhauser, Publishers, Boston, 1999.

[66] M. Y. H. Bangash, *Prototype Building Structures: Analysis and Design*, Thomas Telford, London, 1999.

[67] H. W. Krewinkel, *Glass Buildings: Material, Structure and Detail*, Birkhäuser Verlag, Boston, 1998.

[68] M. P. Groover, *Fundamental of Modern Manufacturing, Material, Process and Systems*, John Wiley and Sons, Inc, New York, 2002.

[69] Alderson, Ann, *Glass in Buildings*, RIBA Publishing Centre for Accessible Environments, London, 2006.

[70] SWRI, *Sealants: The Professional's Guide,* Sealant, Waterproofing and Restoration Institute, Kansas City, 1995.

[71] M. Y. L. Chew and P. P. Tan, *Staining of Façades*, World Scientific, 2003.

[72] Building and Construction Authority, *Good Industry Practices — Waterproofing for Internal Wet Area*, Conquas 21 Enhancement Series, Singapore, 2001.

[73] M. T. Kubal, *Waterproofing the Building Envelope*, McGraw-Hill, New York, 1993.

[74] M. T. Kubal, *Construction Waterproofing Handbook*, McGraw-Hill, New York, 1999.

[75] Building and Construction Authority, *Good Industry Practices — Waterproofing for Internal Wet Area*, Conquas 21 Enhancement Series, Singapore, 2001.

[76] P. W. Pye, *Floor and Flooring*, Building Research Establishment, London, 1997.

[77] Singapore Standard CP 82: 1999, *Code of Practice for Waterproofing of Reinforced Concrete Buildings*, Singapore Productivity and Standard Board, Singapore, 1999.

[78] CITI training note; *Certificate in Waterproofing Supervision*, 2006.

[79] C. Beall, *Thermal and Moisture Protection Manual: For Architects, Engineers and Contractors*, McGraw-Hill, New York, 1999.

[80] M. T. Kubal, *Construction Waterproofing Handbook*, McGraw-Hill, New York, 1999.

[81] P. H. Perkins, *Repair, Protection and Waterproofing of Concrete Structures*, Spon, London, 1997.

[82] H. R. Trechsel, *Manual on Moisture Control in Buildings*, ASTM, Philadelphia, 1994.

[83] M. T. Kubal, *Waterproofing the Building Envelope*, McGraw-Hill, New York, 1993.

[84] M. P. Groover, *Fundamental of Modern Manufacturing, Material, Process and Systems*, John Wiley and Sons, Inc, New York, 2002.

[85] http://www.e-roofing.com/flat_roof_systems.html. Retrieved August 2000.

[86] www.xypex.com. Retrieved August 2000.

[87] www.gsfoamsystems.com. Retrieved August 2000.

[88] C. P. Hedlin, *Inspection and Maintenance of Flat Roofs*, Division of Building Research, National Research Council of Canada, Canadian Building Digest, 179, 1976.

CHAPTER 8
NONDESTRUCTIVE TESTS

8.1. General

A nondestructive test (NDT) is a test conducted to detect internal, surface or concealed defects in a material or component using techniques that do not cause any damage to the items being tested. A nondestructive tester usually comprises of the following:

(1) a search probe to a recognizable change in the medium caused by the test piece;
(2) a detector which has the ability to recognize the modification of the medium;
(3) a converter to change the detector's signal into an interpretable form;
(4) interpretation of the signal.

All NDTs are developed based on well-established physical principles to provide information on the quality of a material or component. Understanding the principles, conducting the test following the correct procedure and accurate interpretation of test results are crucial in the formation of correct judgment. Each test is devised for and used to evaluate the properties and characteristics of specific materials and components. The effectiveness of each test can only be achieved when they are applied for their intended purposes.

This chapter provides an introduction to the common portable NDTs and presents the basic workings and main features for each test. As there are numerous types and forms of portable NDTs commercially available, the tests discussed in this chapter are by no means exhaustive but only limited to those available in the laboratory where the author works.

The tests are classified into categories and discussed in sections as follows:

Category	Section
(a) Vibration	8.2
(b) Stress and strain	8.3
(c) Optical and mechanical	8.4
(d) Sonic and ultrasonic	8.5
(e) Thermal and infrared	8.6
(f) Electromagnetic, electronic and radiography	8.7

8.2. Vibration

8.2.1. *Accelerometer*

In addition as being an annoyance to building occupiers, vibration may be a serious concern to the structural integrity of a facility due to its possible adverse long-term effects. This is particularly crucial to weak structures and sensitive buildings such as historic buildings and monuments. Vibrations may also interfere with sensitive processes, such as those in hospital operating theatres, scientific research laboratories and high-tech industries [1].

The frequent internal and external sources of vibration are shown in Fig. 8.1.

Vibration may be (i) continuous (with magnitudes varying or remaining constant with time, e.g. machinery), (ii) impulsive (infrequent and occasional, e.g. heavy crane loading and unloading), (iii) intermittent (with the magnitude of each event being either constant or varying with time, e.g. trains, impact pile driving, jack hammers) [2].

A vibration measuring system comprises of:

- transducers, typically piezoelectric accelerometers;
- signal-conditioning equipment;
- a data recording and analysis system.

Accelerometers are inertia (resistance of an object to a change in its state of motion) measurement devices that convert mechanical motion to

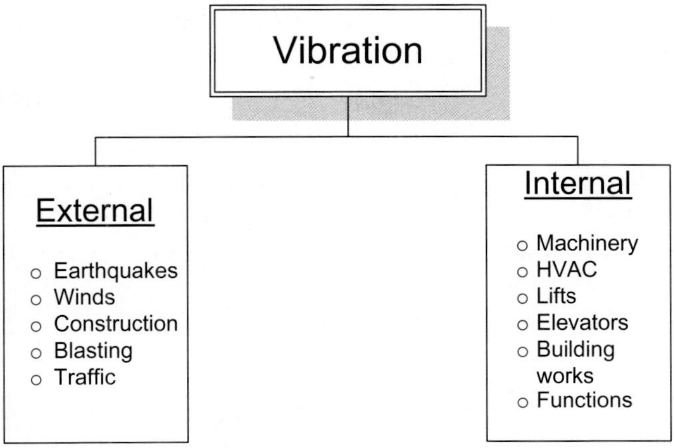

Fig. 8.1. External and internal sources of vibration.

an electrical signal. This signal is proportional to the vibration's acceleration using the piezoelectric principle. Piezoelectricity is the ability of some materials, e.g. certain ceramics to generate an electric potential in response to applied mechanical stress. The active element of an accelerometer is a piezoelectric material. A compression disk looks like a capacitor with the piezoceramic material sandwiched between two electrodes. A force applied perpendicular to the disk causes a charge production and a voltage at the electrodes.

Inertia measurement devices measure motion relative to a mass. This follows Newton's Third Law of Motion which states that a body acting on another will result in an equal action on the first. As an example, if the driver in a car is replaced with a mass suspended in an elastic mounting system, any acceleration of the car will cause movement of the mass relative to the car. The amount of displacement is proportional to the force causing the acceleration. The direction the mass moves is always opposite to the direction of the car's acceleration. When no acceleration is present, the mass will be at rest. When acceleration is present, the mass will lag in proportion to the acceleration force. In other words, the car moves but the mass wants to remain at rest.

Vibration is measured in all three axes, x-axis, y-axis and z-axis (Fig. 8.2).

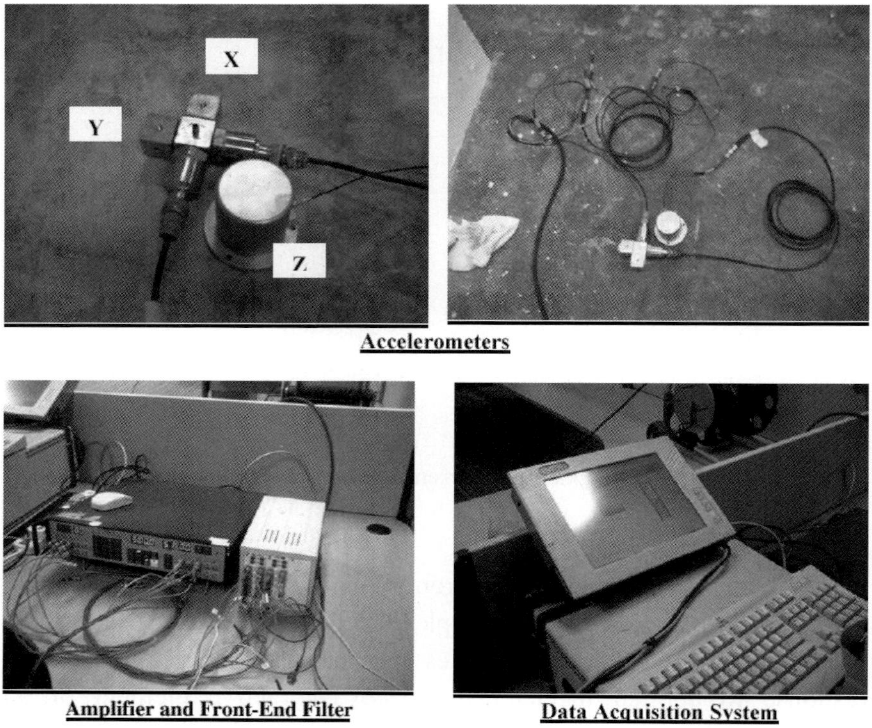

Fig. 8.2. A typical vibration measuring system.

8.2.2. *Tiltmeter*

A tiltmeter is a portable device used to monitor changes in the inclination of a structure [3]. Items in this monitoring system include a number of tilt plates, a portable tiltmeter and a readout unit (Fig. 8.3).

The portable tiltmeter uses a force-balanced servo-accelerometer to measure inclination. The accelerometer is housed in a rugged frame with machined surfaces that facilitate accurate positioning on the tilt plate. The bottom surface is used with horizontally-mounted tilt plates and the side surfaces are used with vertically-mounted tilt plates.

In the measurement, the tiltmeter is positioned onto the tilt plate and the displayed reading is noted. The tiltmeter is rotated 180° and a second reading is noted. The two readings are averaged and the changes in tilt are derived by comparing them with the initial reading.

(a) (b)

Fig. 8.3. (**a**) A tiltmeter with tilt plate on the wall and the digital reader. (**b**) Measurement using a tiltmeter.

8.3. Stress and Strain

Stress is defined as "force per area" or an object's internal resisting forces.

Stress normal to the plane is usually denoted as "normal stress" and can be expressed as

$$\sigma = F_n/A, \tag{8.1}$$

where

σ = normal stress ((Pa) N/m^2);
F_n = normal component force (N);
A = area (m^2).

Stress parallel to the plane is usually denoted as "shear stress" and can be expressed as

$$\tau = F_p/A, \tag{8.2}$$

where

τ = shear stress ((Pa) N/m^2, psi);
F_p = parallel component force (N, lb$_f$);
A = area (m^2, in^2).

Strain is defined as the displacement and deformation that occur due to stress.

Strain can be expressed as:

$$\varepsilon = dl/l_o, \tag{8.3}$$

where

dl = change of length (m, in);
l_o = initial length (m, in);
ε = unitless measure of engineering strain.

8.3.1. *Strain Gauges*

Strain may be compressive or tensile and is typically measured by strain gauges.

Figure 8.4 shows an example of a mechanical strain gauge used to measure the growth of a crack in a masonry foundation. The two halves of the device are rigidly attached to the opposite sides of the crack. The

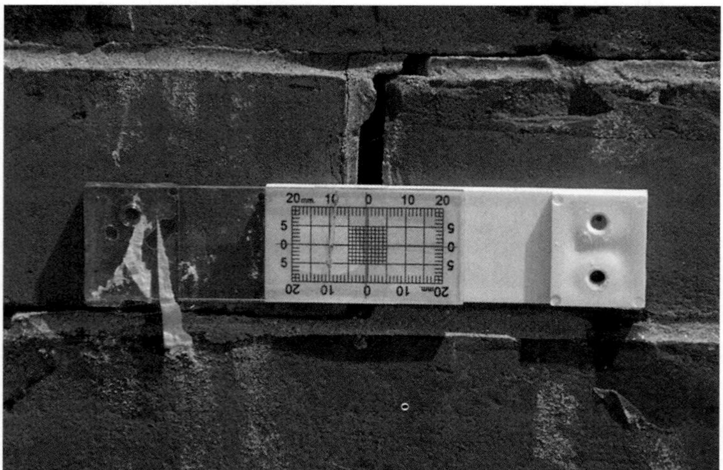

Fig. 8.4. A mechanical strain gauge installed on the Hudson–Athens Lighthouse to measure the growth of foundation cracks [4].

reference lines are on the transparent half and the grid is on the opaque white half. The device can monitor both horizontal and vertical movement (*X–Y* direction) over time. In this picture, the crack can be seen to have widened by approximately 0.3 mm (and no vertical movement) since the gauge was installed [4]. For more accurate measurements, compensation of temperature, moisture and relative humidity is required.

Another form of mechanical strain gauge is shown in Fig. 8.5. It consists of an invar main beam with two conical locating points with one of the locating points fixed in position and the other movable (Fig. 8.5a). The points sit in the pre-drilled stainless steel discs which are attached with adhesive to the specimen to be measured (Fig. 8.5b). The pivoting point's movements are measured by the strain gauge, which is attached to a base plate on the invar beam and the reading is translated and read from a meter (Fig. 8.5c). Thermal movement within the instrument is negligible.

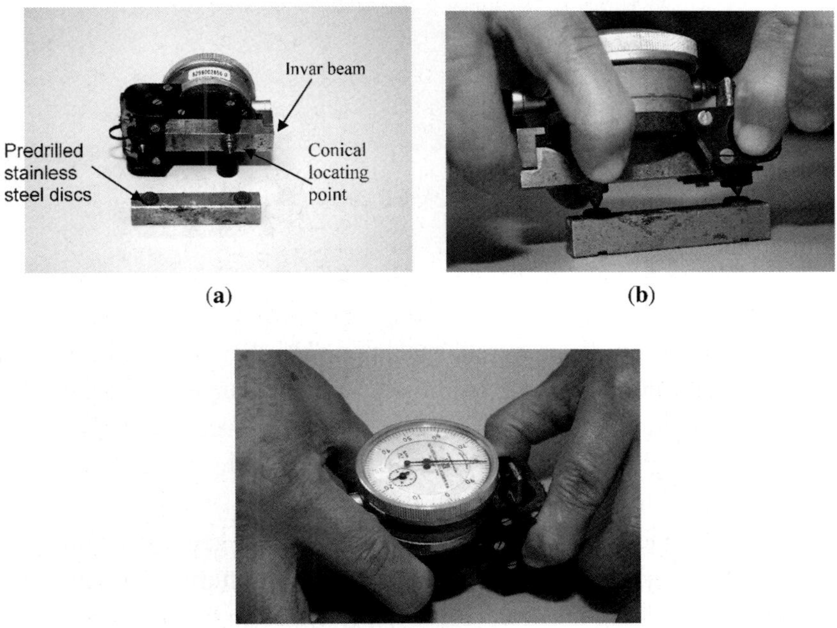

(a) (b) (c)

Fig. 8.5. (**a**) Invar main beam with two conical locating points. (**b**) Locating points sit in the pre-drilled stainless steel discs. (**c**) Meter showing reading translated from strain gauge.

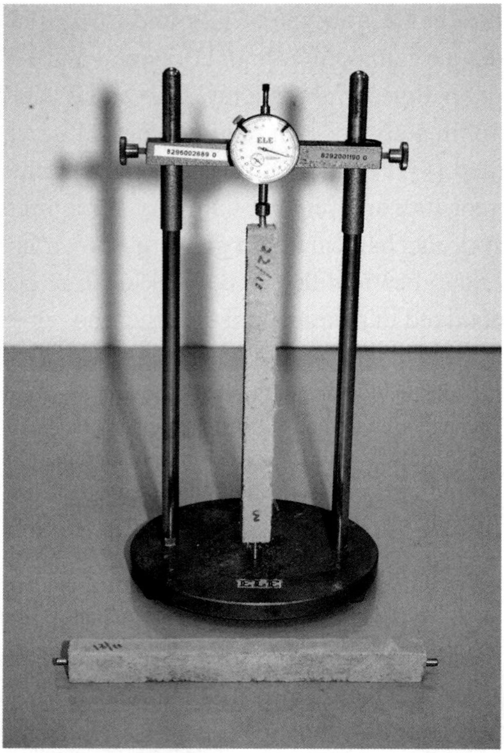

Fig. 8.6. A dial gauge.

Figure 8.6 shows the use of a dial gauge to accurately measure small graduation in length of a concrete specimen. The dial gauge consists of a graduated dial and needle to record the minor changes in length. The spring-loaded plunger of the gauge is inserted perpendicularly into a predrilled disc glued onto the concrete specimen to be measured and minor change in length is read from the meter.

When accurate measurements are needed, strain gauges of foil types are used. This type of strain gauge consists of a pattern of resistive foil mounted on a backing material (Fig. 8.7). It works on the principle that as the foil is subjected to stress, the resistance of the foil changes in a defined way [5].

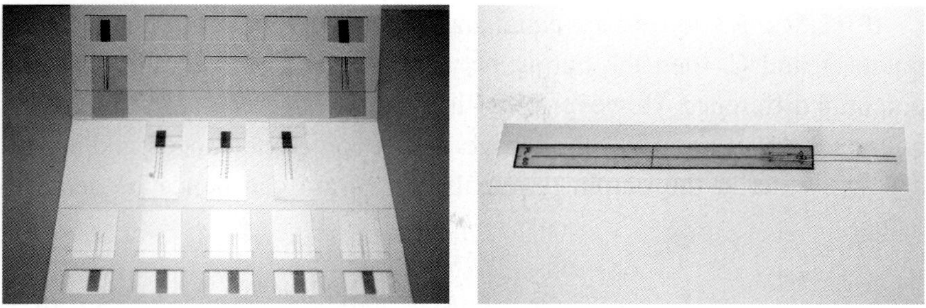

Fig. 8.7. A foil type strain gauge.

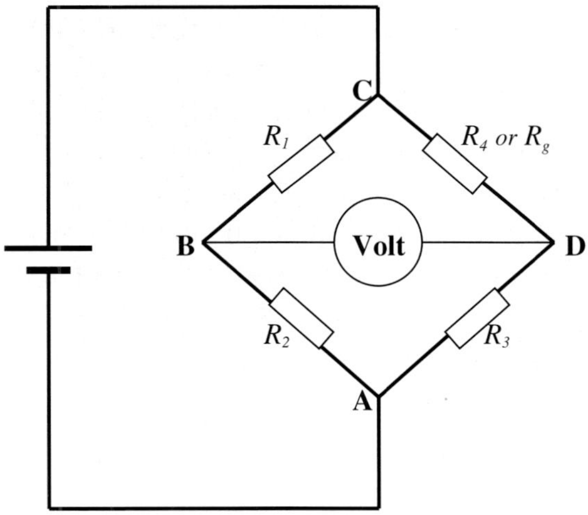

Fig. 8.8. Schematic diagram of a Wheatstone bridge circuit.

The change in resistance (R) is proportional to the strain sensitivity (S) of the wire's resistance or gauge factor (GF), which is given by:

$$\begin{aligned} \text{GF} &= (\Delta R/R)/(\Delta L/L) \\ &= (\Delta R/R)/\text{Strain}. \end{aligned} \tag{8.4}$$

Strain gage transducers usually employ four strain gage elements electrically connected to form a Wheatstone bridge circuit (Fig. 8.8) [5].

If R1, R2, R3 and R4 are equal and a voltage, VIN, is applied between points A and C, then the output between points B and D will show no potential difference. However, if R4 is changed to some value which does not equal R1, R2 and R3, the bridge will become unbalanced and a voltage will exist at the output terminals [6]. Strain can hence be deduced using:

$$\text{Strain} = (\Delta R_g/R_g)/\text{GF}. \tag{8.5}$$

8.3.2. *Load Cells*

The most direct way of measuring load or weight is by the use of a weight balancing mechanism. With the invention of Wheatstone bridge by Charles Wheatstone in 1843, load cells can be made with much smaller size and much higher accuracy (Fig. 8.9). Although there are many varieties of load cells, strain gauge based load cells are the most commonly used type. A strain gauge load cell is a transducer which converts force into a measurable electrical output. Except for laboratory works where precision mechanical balancing is needed (Fig. 8.10), strain gauge load cells dominate the weighing industry [7].

8.3.3. *Vibrating Wire Strain Gauge*

Different from the foil type strain gauge, it operates on the principle that a tensioned wire, when plucked, vibrates at a frequency that is proportional

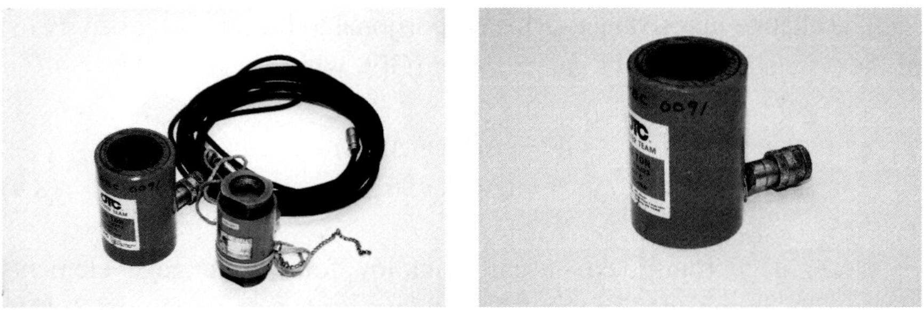

Fig. 8.9. Examples of load cells.

Fig. 8.10. Load mechanical balancing in a lab.

to the strain in the wire. Similar to a mechanical strain gauge, the device is positioned on both sides of the crack so that the wire is held in tension. Movement of the crack would result in a change in the tension of the wire which is picked up by using an electromagnet to pluck the wire and measure the frequency of vibration and deduce the strain (Fig. 8.11) [8, 9].

8.4. Optical and Mechanical

8.4.1. *Fiberscope*

A fiberscope or a flexible borescope is a flexible optical inspection device that consist of fiber optic bundles with an eyepiece at one end and a lens at the other.

A fiberscope usually comes in two bundles: (i) fiber optic light bundle which is for illuminating the object to be investigated and (ii) fiber optic image bundle which is to relay image to eyepiece [10]. It allows direct visual inspection for inaccessible part as observations can be made through a video monitor with high-resolution images of remote, difficult to reach areas. The flexible fibers allow it to be maneuver in hard to reach or inaccessible areas (Fig. 8.12).

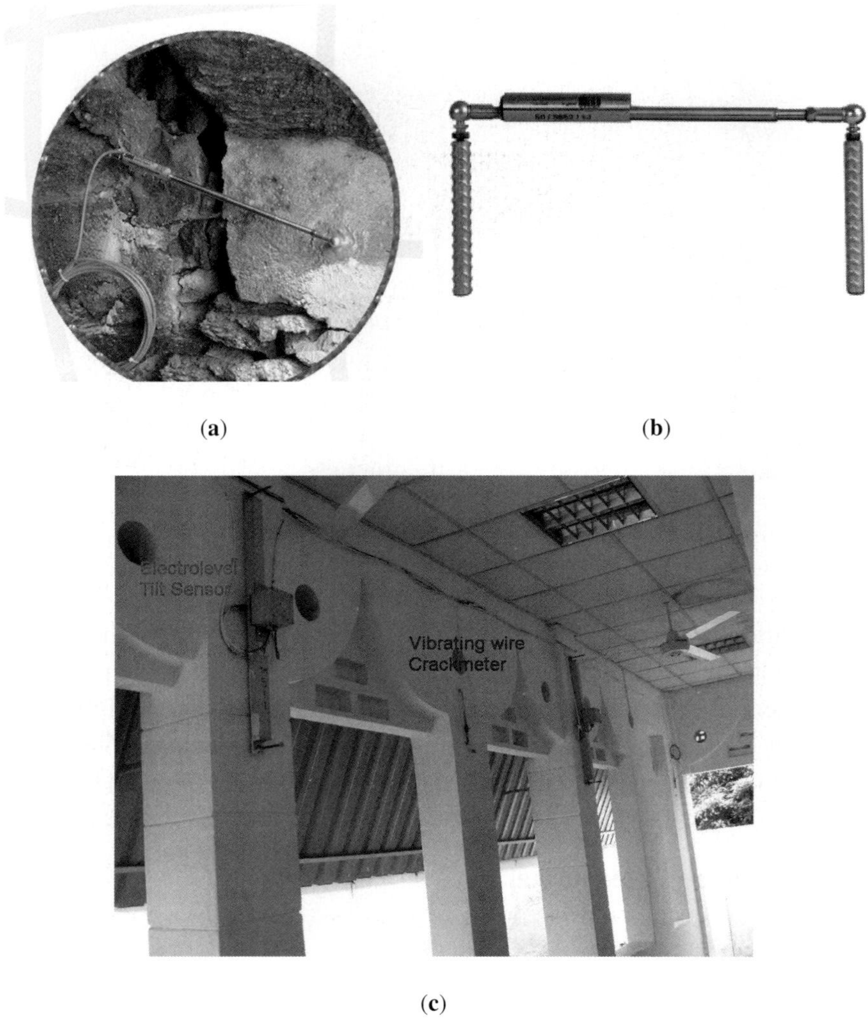

Fig. 8.11. Vibrating wire strain gauge (**a**) and (**b**) on a crack soil (*Courtesy* of Soil Instruments Ltd) and (**c**) on a building column.

It is often used to verify findings from other testing such as infrared thermography or impact-echo testing. Common applications include:

(1) Inspect connecting condition behind external cladding wall of a building.

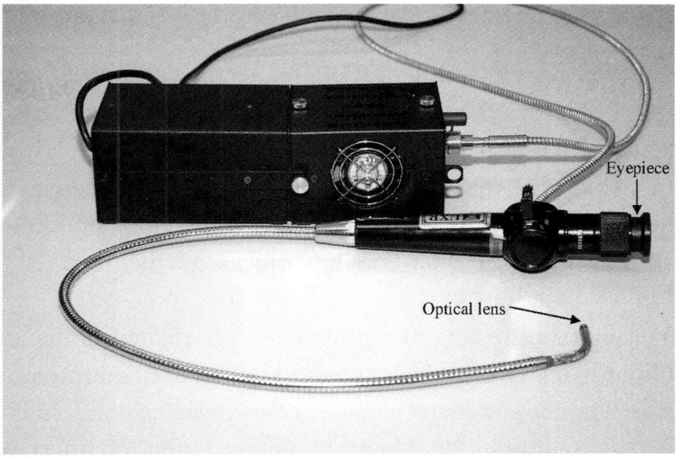

(**a**)

(**b**)

Fig. 8.12. (**a**) Fiberscope. (**b**) The use of fiberscope for inspecting service ducts.

(2) Inspect defects on concrete surface in a deep and narrow gap.
(3) Determine corrosion condition of steel tendon inside post-tensioning ducts of pre-stressed structural member.
(4) Investigate service pipes, ducts and other inaccessible areas.

8.4.2. *Glossmeter and Colormeter*

These are instruments to measure the optical properties of the surface of a material.

8.4.2.1. *Glossmeter*

A glossmeter measures specular reflection, or the capacity of a surface to reflect light. It is a scientific test internationally recognized to allow comparison of measurement values. As it is measuring the "gloss" of the surface of a material, the more direct light that is reflected, the more obvious will be the impression of gloss. Hence, smooth and polished surfaces reflect more light than rough surfaces as the light is diffusely scattered in all directions [11].

The measurement results of a glossmeter are related to the amount of reflected light from a black glass standard with a defined refractive index (Fig. 8.13a). The measurement value for this defined standard is equal to 100 gloss units (calibration). Materials with a higher or lower refractive index can have a measurement value above and below this value accordingly (Table 8.1) [12].

By using reflectometers, the reflected light of a surface is measured in an angle range. The light source is projected over the sample surface onto the opening of an aperture. A photoelectronic detector then measures the light passing through the aperture (Fig. 8.14). Common geometries and their performance indicators are shown in Table 8.1 [11].

8.4.2.2. *Color Meter/Spectrophotometer*

A color meter is a portable, hand-held spectrophotometer designed to measure, match and control the color of small, irregular surfaces as well as

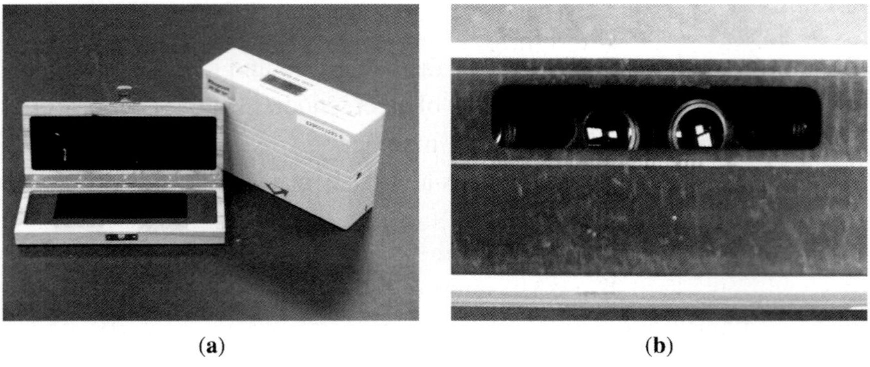

(a) (b)

Fig. 8.13. (**a**) Glossmeter and (**b**) underside of the glossmeter.

Table 8.1. Glossmeter with three geometries (20°, 60° and 85°) [11].

Gloss Range	60° Value	To Be Measured With
Semi gloss	10–70	60° geometry
High gloss	>70	20° geometry
Low gloss	<10	85° geometry

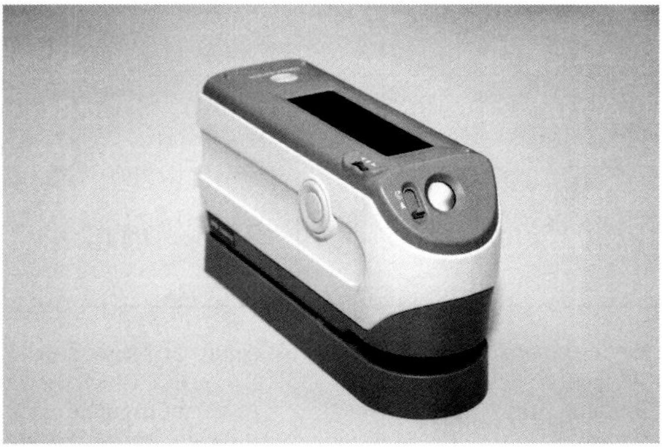

Fig. 8.14. Colormeter (*Courtesy*: Konica Minolta).

substrates with grain, texture or directional characteristics. It records colors of different wavelengths in terms of spectral reflectance and colorimetric values (Fig. 8.13).

8.4.3. *Liquid Penetrant Testing*

This method uses a colored liquid or a dye to detect surface-breaking flaws, such as cracks, laps and folds, on any nonabsorbent material's surface. It is based on the ability of the liquid or dye to be drawn into a "clean" surface-breaking flaw by capillary action [13]. After a period of time, excess surface penetrant is removed and a developer (dry white powder) is applied. The powder draws the penetrant from the flaw by reverse capillary action to reveal its presence (Figs. 8.15 and 8.16). As the colored indications are broader than the actual anomaly, it makes the anomaly more visible.

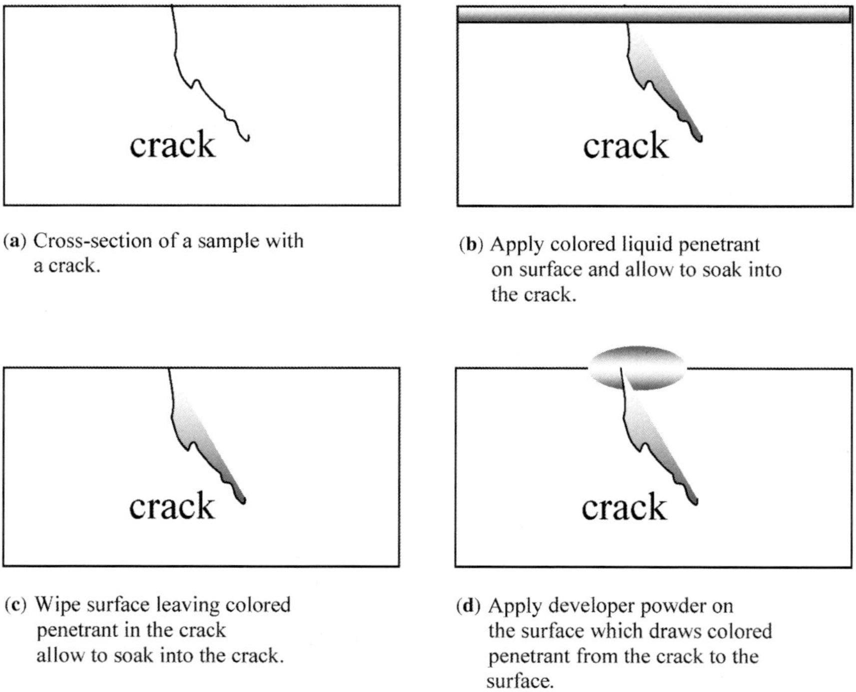

(**a**) Cross-section of a sample with a crack.

(**b**) Apply colored liquid penetrant on surface and allow to soak into the crack.

(**c**) Wipe surface leaving colored penetrant in the crack allow to soak into the crack.

(**d**) Apply developer powder on the surface which draws colored penetrant from the crack to the surface.

Fig. 8.15. Illustration of the working of liquid penetrant testing.

Fig. 8.16. Liquid penetrant test equipments.

8.4.4. *Elastic Recovery Tester*

The tester provides *in-situ* measurements of the performance of sealants based on elastic recovery and compressibility. The tester measures the elastic recovery property of sealants based on the principle of compression and relaxation mechanism, it also measures compressibility, i.e. the resistance to penetration. The tester is also equipped with an optical devise for examination at the tested spot throughout the testing process, specifically for monitoring of adhesion and cohesion performances (Fig. 8.17).

The fields of application are:

(1) quality check;
(2) maintenance and repair;
(3) assessment and diagnosis;
(4) quality control in sealant formulation, manufacturing as well as application on-site;
(5) prediction of service life;
(6) correlation of artificial and natural weathering.

8.4.5. *Rebound Hammer*

Rebound hammer test is an established test to estimate the compressive strength of concrete based on testing the hardness of the tested surface with an impact using a spring-loaded plunger (Fig. 8.18) [14–17].

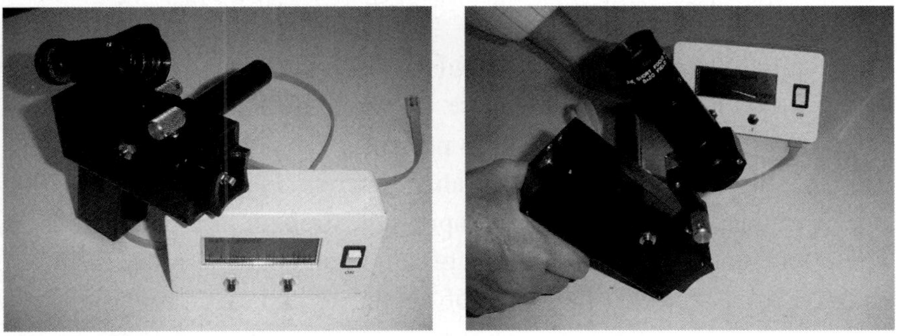

Fig. 8.17. Elastic recovery tester.

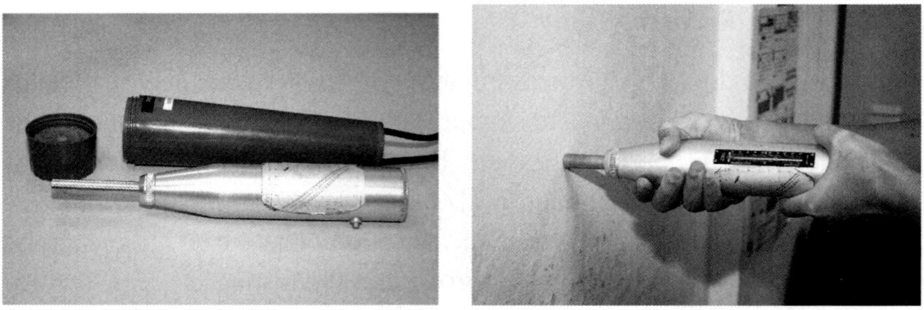

(**a**) (**b**)

Fig. 8.18. (**a**) Rebound hammer with the casing. (**b**) Rebound hammer test conducted on a concrete wall.

When the plunger of the rebound hammer is pressed against the concrete surface, the spring-loaded mass rebounds and the surface hardness of the concrete is recorded and compared to a graduated scale to relate to the compressive strength.

8.5. Sonic and Ultrasonic

8.5.1. *Ultrasonic Pulse Velocity Measurement*

This test is used to estimate the strength of concrete based on the measure of the time of travel of an ultrasonic pulse passing through the specimen. The tester consists of a pulse generator and a pulse receiver (Fig. 8.19). The time taken for the pulse to pass through the concrete is measured by electronic measuring circuits [18].

Usually, if large differences in pulse velocity are found within a structure for no apparent reason, there is strong reason to presume that defective or deteriorated concrete is present. High pulse velocity readings are generally indicative of good quality concrete. Fairly good correlation can be obtained between cube compressive strength and pulse velocity. These relations enable the strength of structural concrete to be predicted within ±20%, provided the types of aggregate and mix proportions are constant. Generally, the degree of damage is related to a reduction in pulse velocity. Cracks can also be detected.

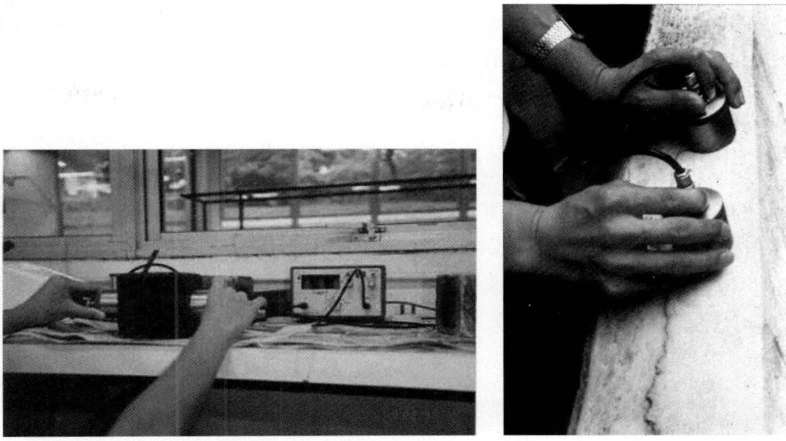

(**a**) Direct transmission (**b**) Indirect transmission

Fig. 8.19. Ultra sonic pulse velocity test.

Good acoustic coupling between the transducer face and the concrete surface is necessary to ensure accurate transit time measurement. This is done by ensuring that the concrete surface is free from dust and grit and covering the surface with a light or medium grease.

Direct transmission arrangement is usually employed for more accurate measurement where the two transducers are placed directly opposite to each other on the two sides of a concrete wall (Fig. 8.19a). However, examination of concrete by using diagonal paths and semi-direct arrangements, though not very accurate, are suitable for other purposes (Fig. 8.19b). During measurement, it is preferable to avoid path arrangement which contains steel reinforcing bars since pulse velocity measurements made in the vicinity of reinforcing steel may be high.

There is no unique relationship between the pulse velocity and strength as it is influenced by the concrete constituents and curing conditions. Nevertheless, if the mix proportions remain constant and only one type of coarse aggregate is used then it is possible to determine a specific relationship between pulse velocity and strength for *in-situ* concrete. Besides, moisture conditions also played a significant role in this testing.

Owing to these many factors, evaluation of the quality of concrete in structures is usually made on a comparative basis

8.5.2. *Impact-Echo Test*

Impact echo is a method for nondestructive evaluation of concrete and masonry based on the use of impact-generated compression waves that travel through the structure and are reflected by internal flaws and external surfaces (Fig. 8.20).

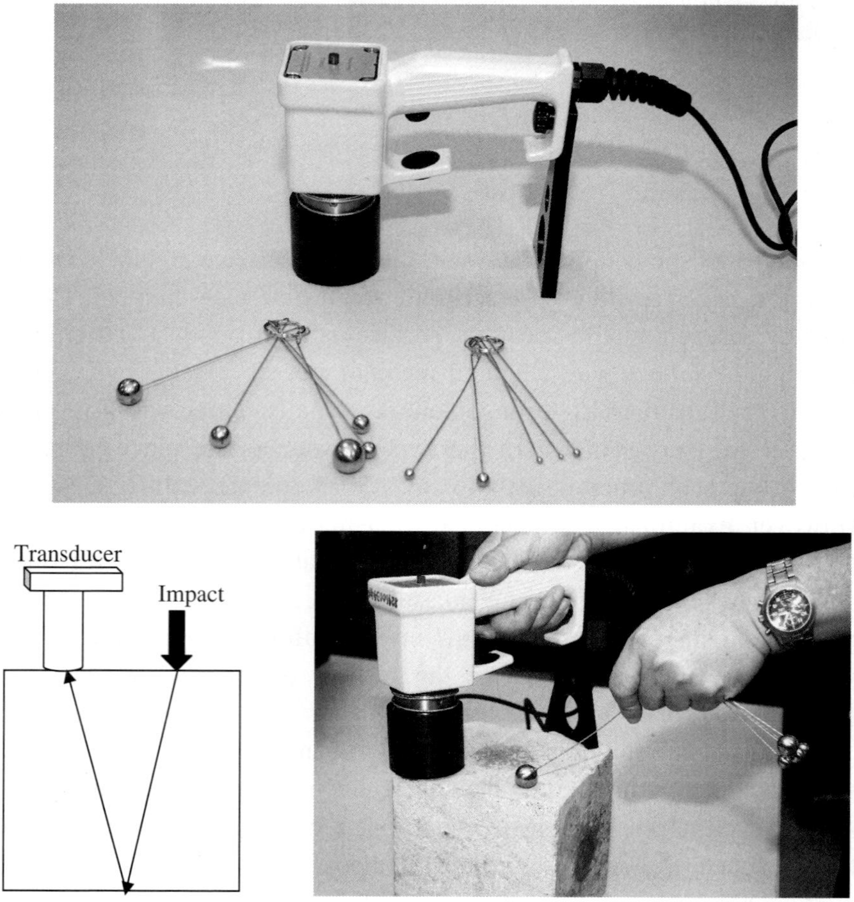

Fig. 8.20. Impact-echo tests.

It is used to determine:

— the location and extent of flaws such as cracks,
— delaminations,
— voids,
— honeycombing,
— voids in the subgrade directly beneath slabs and pavements.

In the test, a short-duration mechanical impact is given to the tested path, produced by tapping a small steel sphere against the tested surface. Low-frequency stress waves that generate propagate into the structure and are reflected by flaws and/or external surfaces. Surface displacements caused by reflections of these waves are recorded by a transducer, located adjacent to the impact. The displacement versus time signal is transformed into the frequency domain and plots of amplitude versus frequency (spectra) are obtained [19].

8.6. Thermal and Infrared

An infrared scanning camera is used to detect variations in infrared radiation output of a surface. Thermal gradients along the surface of a material arise because of the difference in surface temperature due to variation in moisture content, existence of cracks, delamination, electrical piping, etc. (Fig. 8.21). It is commonly used in building diagnostics of

(1) leakage — Moisture/electrical/air;
(2) crack and delamination of claddings [20];
(3) location of underground water/sewage pipes and defects;
(4) detection of energy loss in buildings;
(5) detection of reinforcement bars location.

Infrared thermography utilizes infrared detectors and optics to gather the infrared energy. It produces an output in the form of a thermal map known as thermogram.

The thermal image can be video taped or stored onto a storage disk to be analyzed later. Costly heat related problems caused by a component or

(**a**)

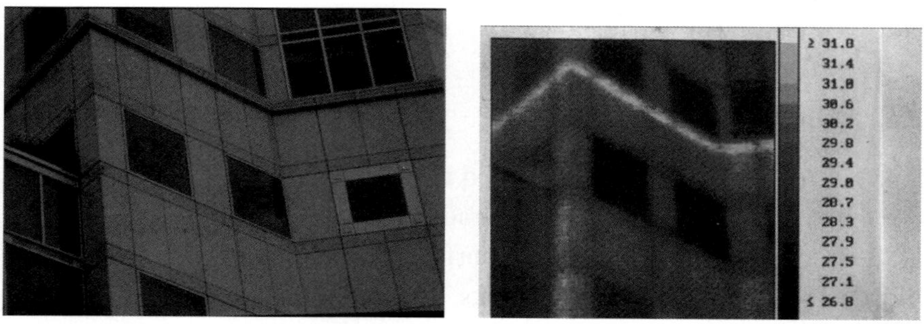

(**b**)

Fig. 8.21. (**a**) Infrared scanning camera. (**b**) Infrared thermographs.

material failure, poor design in building components and mechanical and electrical systems can be pinpointed.

Most modern thermal imaging instruments employ high-speed optics to scan the target point by point. They collect infrared radiation emanating from a target and direct it into a super cooled photo detector. The detector converts this incoming energy into a proportionate electric signal, which is then amplified. This amplified signal is then sent to a video processor and cathode ray tube (CRT) where it can be manipulated in a variety of ways for interpretive purposes. The image displayed on the face of the CRT is a temperature map in which variations in image density correspond to radiant energy differences in the original subject.

8.7. Electromagnetic, Electronics and Radiography

8.7.1. *Electromagnetic Cover Measurement*

The method is based on the basic principle that the presence of steel affects the magnetic field. An electromagnetic search probe is swept over the surface of the concrete under test. The presence of reinforcement within the range of the instrument is shown by movement of the indicator needle.

When the probe is moved until the deflection of the needle is at a maximum, the bar in question is then parallel to the alignment of the probe and directly beneath it. The needle indicates the cover on the appropriate scale for the diameter of the reinforcing bar [21].

Profometer and covermeter consist of electromagnetic cover devices used for determining the position and direction of reinforcement bars (Fig. 8.22).

Electromagnetic covermeters can be used for:

(1) quality control, to ensure correct location and cover to reinforcing bars after concrete placement;

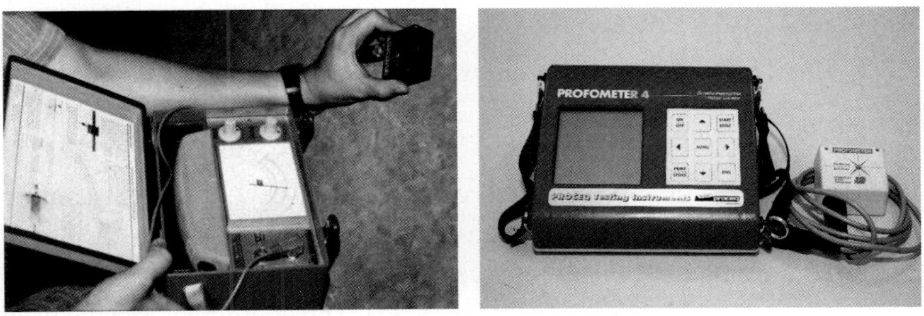

(a) (b)

Fig. 8.22. (**a**) Electromagnetic covermeter. (**b**) A profometer (*Courtesy*: Proceq).

(2) investigation of concrete members for which records are not available or need to be checked;
(3) location of reinforcement as a preliminary to some other form of testing in which reinforcement should be avoided or its nature taken into account, e.g. extraction of cores, ultrasonic pulse velocity measurement or "near to surface" methods;
(4) location of buried ferromagnetic objects other than reinforcement, e.g. water pipes, steel joints, lighting conduits.

8.7.2. *Leak Detection*

Leak detection is usually conducted using acoustical equipment which detects the sound or vibration induced by water as it escapes from pipes under pressure. The acoustical equipment usually comprises of listening devices such listening rods, aquaphones or sonoscopes and geophones or ground microphones. The listening devices are used to detect for leak sounds along the suspected location along a pipe.

Commercial leak detectors come with micro-processors which provide built-in analysis and correlation. Leak signals in the form of either sound or vibration at two points that bracket a suspected leak are detected. The position of the leak is determined using the built-in correlator based on the time shift between the leak signals calculated using the cross-correlation method [22]. The position of the leak is displayed on the screen of the detector. An example of a commercial leak detector is shown in Fig. 8.23.

8.7.3. *Environmental and Holidays Test for Paint*

8.7.3.1. *Environmental Paint Test*

It is essentially a dewmeter which computes the dew point temperature and hence the occurrence of condensation by measuring the surface temperature, air temperature and relative humidity at regular intervals and at different points on a paint surface (Fig. 8.24). The equipment may be used to check for condensation on the wall substrate before a paint job or to determine the moisture content of a painted wall before the next coat is applied to avoid adhesive failures.

Fig. 8.23. A leak detector (*Courtesy*: Gutermann International).

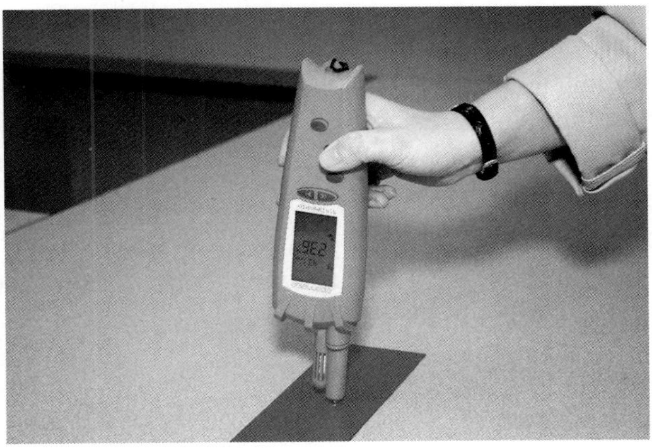

Fig. 8.24. Environmental paint test equipment.

The normal procedure in conducting the test is as follows:

(1) connect the probe to the dew meter;
(2) hold the dew meter in hand and point the probe to the surface to be measured;
(3) reading is taken off from the dewmeter.

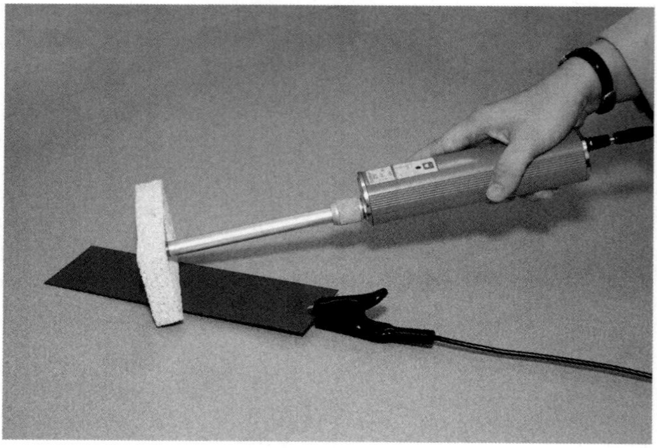

Fig. 8.25. A holiday or pinhole detector.

8.7.3.2. *Holiday Test*

A holiday or pinhole detector utilizing the wet sponge technique is suitable for locating pinholes through a nonconductive coating and a conductive substrate. A sponge, moistened with a wetting agent, is supplied with a low voltage. When the sponge is moved over a pinhole or holiday, liquid penetrates the flaw and flows to the substrate and completes the electrical circuit. The resulting current flow generates an audible and visual alarm in the detector (Fig. 8.25).

8.7.4. *Moisture Meter*

A moisture meter is a device designed to measure the moisture content of building materials such as roofing, insulation, drywall, plaster, wood, tile and fiberglass [23].

The common types of moisture meter are:

With pins — Use small probes that penetrate the material to be tested. It measures the amount of moisture on a material by its electrical conductivity. This is generally regarded as a more repeatable and accurate type of moisture measurement than the "search mode" (without pins).

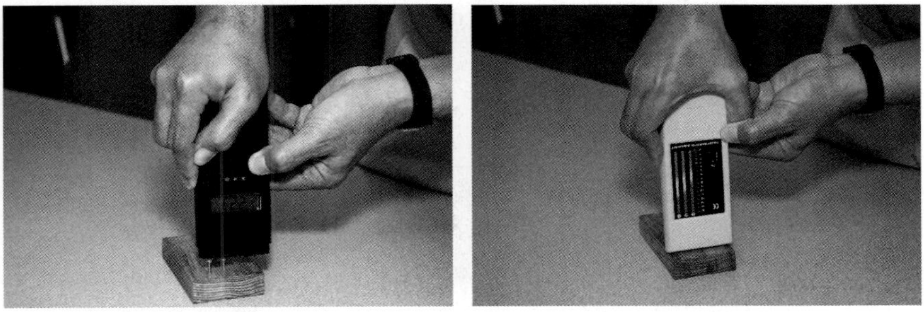

(**a**) With pins (**b**) Without pins

Fig. 8.26. Moisture meters.

Without pins — Low-frequency signals (electromagnetic waves) are transmitted into the material being tested to measure the change in impedance caused by the presence of moisture. The meter detects changes in the characteristics of the returned emissions, relates to the calibrated data and interprets into moisture content.

Figure 8.26 shows examples of moisture meters with and without pins.

8.7.5. *Ground Penetrating Radar* (*GPR*)

It is a high-frequency testing technique which generates data by the propagation and reflected energy of radar waves (electromagnetic waves).

The radar waves emitted by the antenna of the equipment are directed into the test material. The waves propagate through the material at a velocity defined by the electrical properties of the material. When a boundary of different electrical characteristics is encountered, part of the incident energy is reflected from the surface and the remainder travels across the boundary. The reflected (echo) wave is picked up by the receiver built into the antenna. The antenna is swept over the test surface and forms a continuous profile of the material condition (Fig. 8.27).

Antenna frequency is a major factor in depth penetration. The higher the frequency of the antenna, the shallower into the ground it will penetrate. A higher frequency antenna will also "see" smaller targets. Antenna choice is one of the most important factors in survey design (Table 8.2).

Table 8.2. Antenna frequency and depth penetration (*Courtesy*: GSSI).

Depth Range (Approximate)	Primary Antenna Choice	Secondary Antenna Choice	Appropriate Application
0–1.5 ft 0–0.5 m	1600 MHz	900 MHz	Structural concrete, roadways, bridge decks
0–3 ft 0–1 m	900 MHz	400 MHz	Concrete, shallow soils, archaeology
0–12 ft 0–9 m	400 MHz	200 MHz	Shallow geology, utilities, UST's, archaeology
0–25 ft 0–9 m	200 MHz	100 MHz	Geology, environmental, utility, archaeology
0–90 ft 0–30 m	100 MHz	Sub-Echo 40	Geologic profiling
Greater than 90 ft or 30 m	MLF (80, 40, 32, 20, 16 MHz)	20 m	Geologic profiling

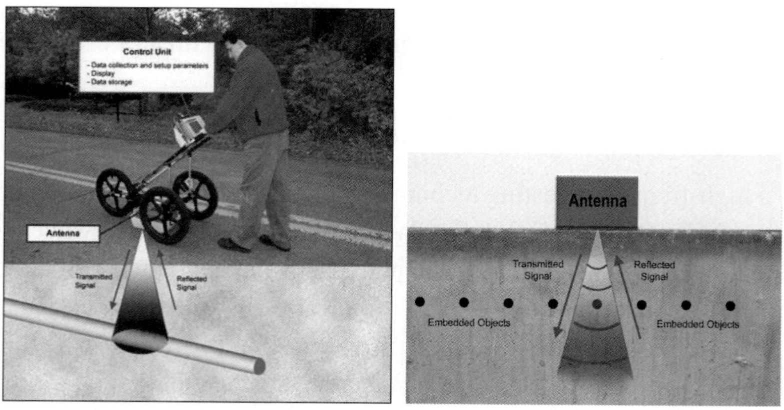

Fig. 8.27. Ground penetrating radar (*Courtesy*: GSSI).

Table 8.2 shows antenna frequency, approximate depth penetration and appropriate application [24].

Materials with a high dielectric are very conductive and thus attenuate the signal rapidly. Water saturation dramatically raises the dielectric of a material, so a survey area should be carefully inspected for signs of water penetration. Radar surveys should never be conducted through standing water, no matter how shallow. Depth penetration through a material with a high dielectric will not be very good.

Metals are considered to be a complete reflector and do not allow any amount of signal to pass through. Materials beneath a metal sheet, fine metal mesh, or pan decking will not be visible.

Radar energy is not emitted from the antenna in a straight line. It is emitted in a cone shape (Fig. 8.27). The two-way travel time for energy at the leading edge of the cone is longer than for energy directly beneath the antenna. This is because the leading edge of the cone represents the hypotenuse of a right triangle.

Because it takes longer for that energy to be received, it is recorded farther down in the profile. As the antenna is moved over a target, the distance between them decreases until the antenna is over the target and

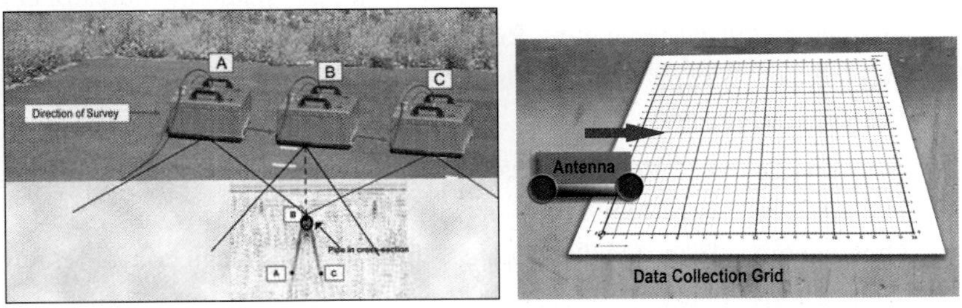

Fig. 8.28. Radar energy is emitted in a cone shape (*Courtesy*: GSSI).

Fig. 8.29. Ground penetrating radar in operation.

increases as the antenna is moved away. It is for this reason that a single target will appear in a data as a hyperbola, or inverted "U". The target is actually at the peak amplitude of the positive wavelet (Fig. 8.28). Figure 8.29 shows an example of GPR in operation.

References

[1] Institute for Research in Construction, National Research Council Canada, "*Traffic Vibrations in Buildings*", http://irc.nrc-cnrc.gc.ca/pubs/ctus/39_e.html. Retrieved May 2009.

[2] New South Wales Government, Department of Environment and Climate Change, "*Noise*", http://www.environment.nsw.gov.au/noise/index.htm. Retrieved May 2009.

[3] Slope Indicator, "*Portable Tiltmeter*", http://www.slopeindicator.com/instruments/tilt-portable.html. Retrieved May 2009.

[4] Wikipedia, "*Strain Gauge*", http://en.wikipedia.org/wiki/Strain_gauge. Retrieved May 2009.

[5] Sensorland.com, "*Strain Gauge*", http://www.sensorland.com/HowPage002.html. Retrieved May 2009.

[6] Omega.com, "*The Strain Gauge*", http://www.omega.com/Literature/Transactions/volume3/strain2.html. Retrieved May 2009.

[7] Honeywell, "*Load Cells*", http://content.honeywell.com/sensing/sensotec/loadcell.asp. Retrieved May 2009.

[8] Gage Technique, "*VW Embedment Strain Gauge*", http://www.gage-technique.demon.co.uk/instruments/sg-embedment.html. Retrieved May 2009.

[9] Soil Instrument, "*Vibrating Wire Crackmeter*", http://www.soil.co.uk/files/soil/J2_Vibrating%20Wire%20Crackmeter_Rev01.5.pdf. Retrieved May 2009.

[10] Fiberscope.net, "*General Information About Borescopes, Fiberscopes and Fiber Optic Scopes*", http://www.fiberscope.net/borescope_fiberscope_fiber_optic_scope.html#flex. Retrieved May 2009.

[11] Taber Industries, "*Glossmeter*", http://www.taberindustries.com/Products/Glossmeter/Taber_Industries_Glossmeters_Index.asp?ct=1&sc=14. Retrieved May 2009.

[12] Wikipedia, "*Glossmeter*", http://en.wikipedia.org/wiki/Glossmeter. Retrieved May 2009.

[13] TWI, "*Liquid Penetrant Inspection*", http://www.twi.co.uk/content/ksijm001.html. Retrieved May 2009.

[14] ASTM International, "*ASTM C803/C803M-03 — Standard Test Method for Penetration Resistance of Hardened Concrete*", January 2003.

[15] British Standards Institute, BS 1881–207 Testing Concrete Part 207: Recommendations for the Assessment of Concrete Strength by Near-to-Surface Tests (R), British Standards Institute, London, 1 January 1992.

[16] BS EN 12504–2:2001, "*Testing Concrete in Structures. Nondestructive Ttesting. Determination of Rebound Number*", 15/07/01.

[17] ASTM International, "*ASTM C 805/C 805M — Standard Test Method for Rebound Number of Hardened Concrete*", August 2008.

[18] BS EN 12504–4:2004, "*Testing Concrete. Determination of Ultrasonic Pulse Velocity*", 13/10/04.

[19] ASTM International, "*Standard Test Method for Measuring the P-Wave Speed and the Thickness of Concrete Plates Using the Impact-Echo Method*", 1 June 2004.

[20] ASTM International, "*ASTM D4788 — 03(2007) Standard Test Method for Detecting Delaminations in Bridge Decks Using Infrared Thermography*", 2007.

[21] British Standards Institute, *BS 1881: Part 204(1988). Recommendation on the Use of Electromagnetic Covermeters*, British Standards Institute, London.

[22] NRC, "*Leak Detection Methods for Plastic Water Distribution Pipes*", http://irc.nrc-cnrc.gc.ca/ui/bu/leakdetect_e.html. Retrieved June 2009.

[23] Wikipedia, "*Moisture Meter*", http://en.wikipedia.org/wiki/Moisture_meter. Retrieved May 2009.

[24] GSSI, "*Application — What is GPR?*", http://www.geophysical.com/WhatIsGPR.htm. Retrieved May 2009.

CHAPTER 9
MAINTAINABILITY GRADING SYSTEM

9.1. General

The global construction and real estate industries are facing shrinking maintenance budgets, rising construction and maintenance costs and increasing competitive global markets. It is of compelling logic to improve a building's level of maintainability right from the planning and design stages. The rationale of developing a maintainability benchmark is to enable the decision maker to select a more maintainable design option. Yardsticks like grading systems would assist in the selection of highly maintainable building designs.

The objective of grading systems is to promote buildings with high maintainability through a structured risk assessment protocol. It attempts to improve the performance of a building and reduce the life cycle cost (LCC) during its post occupancy period. The broad term "LCC" encompasses initial, future and demolition costs.

9.2. Grading System — State of the Art

Scale and scope of the entire available framework ranges from simple to chronologically complex; prescriptive to performance-based; grading of design-guideline to environmental, impact assessment. Most systems exhibit combination of all these approaches [1].

Mathematically such systems can be classified as:

— first generation: Nominal (pass–fail type certification);
— second generation: Simple additive;

— third generation: Weighed additive;
— others.

Major examples from each category are shown in Table 9.1.

9.2.1. *First Generation: Pass–Fail Systems*

Most of the examples are prescriptive certification programs for conventional building design — Judged against codes, standards or by-laws and with limited focus on energy and indoor environmental quality (IEQ). Few examples are listed in the following section.

9.2.1.1. *R-2000, Canada*

R-2000 [2] jointly developed by CHBA and NRC, started operating in 1981 to promote energy efficiency and reduction of greenhouse gas emissions of residential buildings (R-2000, n.d.). A performance-based energy target for space and water heating was established as per R-2000 standard and calculated for each house, taking into consideration the size, fuel type, lot orientation and climatic location. Upon fulfillment of mandates, certification is issued by the authority.

9.2.1.2. *P-Mark Certification System, Sweden*

The P-mark System [3] of 1989 by SP, Sweden primarily certifies of prefabricated detached houses, but can also be used at project level. It covers products (wooden windows, particleboards and nail plate trusses); quality system (crawl space, concrete slab foundations, basement walls, external walls, kitchen etc); or both against standard test methods or design rules. It involves (1) testing for ventilation, air-tightness of house and ventilation ducts and (2) inspection for HVAC performance, water-tightness of kitchen or toilet. In the long run, P-mark has resulted in a reduction in number of major faults [4].

Table 9.1. A summary of available grading systems.

Type	Year	Grading System	Developed By	Country
First generation	1981	R-2000	Canadian Home Builders' Association (CHBA), Natural Resources Canada (NRC)	Canada
	1989	P-mark	Swedish National Testing and Research Institute (SP)	Sweden
	1997	ELO and EM scheme	Danish Energy Authority	Denmark
	2001	Energy star	Environmental Protection Agency (EPA) and Department of Energy (DOE)	U.S.
Second generation	2000	Leadership in Energy and Environmental Design (LEED)	U.S. Green Building Council (USGBC)	U.S.
	2005	Green Mark	BCA	Singapore
Third generation	1990	Building Research Establishment Environmental Assessment Method (BREEAM)	Building Research Establishment (BRE) Ltd	U.K.
	1993	Building Environment Performance Assessment Criteria (BEPAC)	Environmental Research Group, University of British Columbia	Canada
	1996	Hong Kong Building Environmental Assessment Method (HK-BEAM)	HK-BEAM Society	Hong Kong
	2001	Housing Quality Assurance Law (HQAL)	Japan Government	Japan

(Continued)

Table 9.1. *(Continued)*

Type	Year	Grading System	Developed By	Country
	2002	Green Building Tool (GBTool)	International Initiative for a Sustainable Built Environment (iiSBC)	International
	2002	Global Environmental Method (GEM)	Global Alliance for Building Sustainability	UK
	2003	Green Star	Green Building Council of Australia (GBCA)	Australia
	2004	Green Globes	The Green Building Initiative (GBI)	U.S.
	2004	Go Green, Go Green Plus	Building Owners and Manufacturers Association (BOMA)	Canada
	2004	Maintainability Scoring System (MSS)	Department of Building, National University of Singapore (NUS)	Singapore
	2005	National Australian Built Environment Rating System (NABERS)	Department of the Environment and Heritage (DEH), Commercialized by Department of Energy, Utilities and Sustainability (DEUS)	Australia
Others	2004	Comprehensive Assessment System for Building Environmental Efficiency	Japan Sustainable Building Consortium (JSBC)	Japan

9.2.1.3. *ELO and EM Schemes, Denmark*

In 1997 Danish Energy Authority developed ELO and EM schemes as energy labeling program for existing large buildings (total floor area >1500 m^2) and small buildings, respectively, to cover various building types catering to residential, trade, private service and public uses. Both schemes consist of two parts, namely, an energy label and an energy plan. For energy label, three individual values for electricity, heat and water consumption are measured in a scale of A to M and compared with calculated values based on performance of similar buildings. In case of significant difference between these two values, recommendations are made as energy plan. Mandatory annual renewal of the label is exempted for two years if a project earns "A" for heating and electricity consumption.

9.2.1.4. *Energy Star, USA*

Energy Star [5] is a joint venture of U.S. EPA and DOE. Certified new homes must meet EPA guidelines and are at least 15% more energy efficient as per prescriptive and performance-based criteria set by 2006 International Energy Conservation Code (IECC). It includes effective insulation; high-performance windows; tight construction and ducts; efficient heating or cooling equipment; and Energy Star qualified lighting and appliances.

9.2.2. *Second Generation: Simple Additive Systems*

A set of second-generation, simple additive systems appeared to have gained popularity in a short while due to support from various government agencies. However, there is a very little scope for user modification to reflect regional differences or individual preferences, hence amendments are realized for such systems (ASMI, 2002b). For example, few logical drawbacks of LEED for new construction are rectified in LEED for home, which is under pilot stage. Similarly Green Mark has few more environmental proposals to complement its present form [6].

9.2.2.1. *LEED, USA*

In 2000, the United States Green Building Council (USGBC) proposed LEED and it presently includes various types of buildings, namely, new construction and major renovation (NC); existing buildings (EB); commercial interior (CI); core and shell (CS); homes (H); and neighborhood development (ND).

For LEED-NC [7], a total score of 69 is distributed among five major assessment areas namely, (1) sustainable sites; (2) water efficiency; (3) atmosphere; (4) materials and resources; (5) IEQ (Indoor Environmental Quality) and in addition a bonus for innovation and design. Among 69 points, seven are prerequisites or mandatory, 57 are core credits and additional five points are allocated for innovative design. LEED follows a simple additive system, where total score is calculated by adding each credit earned upon fulfilling one or more criteria or their alternative(s). The building is rated as Certified, Silver, Gold or Platinum based on a minimum score of 26, 33, 39 and 52, respectively, out of the total score of 69.

Unlike LEED-NC, LEED-H for homes can be customized for different climatic zones such as dry, wet or normal region, by allocating different points to (1) irrigation; (2) landscaping and surface water management; and (3) material durability. Critical issues are mandatory and provided with higher grades and intricate details. This pilot scheme is expected to undergo further refinements [7]. Though the total score for any building irrespective of its geographic location cannot exceed the maximum of 108 points, individual scoring of few criteria can be confusing, for example, a building in dry zone may earn 26 out of 24 for material and resources.

9.2.3. *Third Generation: Weighed Additive Systems*

Currently the highest-end tools recognize the different impacts of various parameters and therefore have adopted additive weighing systems. Determination of weightage mostly involves judgmental or conscious-based values due to inherent complexity and lack of objective basis. Expert opinion is sought to rank the parameters and then weighings are allocated by analyzing such data through various methods such as, analytic hierarchy process, statistical correlation, artificial neural networks, etc. Horvat and Fazio [8] classified the weighing into two major categories, namely, (1) pre-weighed credits such as HQAL or BREEAM and (2) weighing after scoring such as BEPAC or SBTool.

9.2.3.1. *BREEAM, UK*

BREEAM [9–12] was initiated by Building Research Establishment (BRE) of UK in 1990 and is applicable to residential houses (known as EcoHomes), industrial buildings, offices, prisons, retail facilities, schools and minor refurbishments. For any unconventional or multi-tenancy building, a "Bespoke BREEAM" is carried out with project specific assessment criteria and functions. Eight assessment categories that are evaluated include management, energy use, health and well-being, pollution, transport, land use, ecology, materials and water. EcoHomes cover all of them except management. Each category has sub-categories allocated with pre-weighed points that are either cumulative or otherwise, depending on performance against certain specified standards such as SAP 2005. The awarded credits add up to a final overall score, rated on a scale of Fail, Pass, Good, Very Good and Excellent. The benchmarking procedure and minimum requirement varies with building types and stages. For pre-weighed credits, integers (as in EcoHomes) can be more user-friendly than fractions (as in BREEAM office) or percentage scores of EcoHomes XB.

EcoHomes XB deals with minor refurbishments during routine planned maintenance as their effect cannot be captured solely by EcoHomes. It has a scoring system of 1–100 split into breaks of three without any rating. The smaller breaks reflect the effects of small works. A property once assessed, any further improvement can be monitored and the score is adjusted appropriately. The rating is limited to housing stocks as the management section for the stockholder's company policies is not applicable to individual ownership. Score for a whole development with different types of new houses is given by Eq. (9.1), while the same for existing buildings as mentioned in EcoHomes XB is derived by Eq. (9.2).

$$\text{Average score} = \frac{\sum_1^n [(\text{Individual score for house type } N) \times (\text{No. of units type } N)]}{\text{Total number of house units}}, \quad (9.1)$$

where,
n = the total number of house types,
N = each individual house type.
Note: Only identical homes can be classed as a "house type".

$$\text{Average Score} = \frac{\sum_1^n [(\text{Score for batch } N) \times (\text{No. of units type } N)]}{\text{Total number of units of all types that have been assessed}}, \quad (9.2)$$

where,
n = the total number of house types covered in all assessments to date,
N = number of dwellings of each individual house type.

In 1999, ECD (Energy Citation Database) Energy and Environment worked with TerraChoice to develop BREEAM Green Leaf eco-rating program for offices, multi-residential, industrial and recreational buildings. This question-based tool with percentage scoring is aimed to be more streamlined than BREEAM and easily harmonized with LEED-Canada.

9.2.3.2. *BEPAC, Canada*

BEPAC was developed in 1993 by the Environment Research Group at the University of British Columbia. The method is based on BREEAM and considers new or existing offices or commercial buildings for grading. Design and management criteria for both base building and tenancies are formulated as four modules, with five topic areas namely, (1) ozone layer protection; (2) environmental impact of energy use; (3) IEQ; (4) resource conservation; and (5) site and transportation. Although no hierarchy of importance is implied, the first two topic areas are observed to have profound implications [13]. For these two "performance-oriented" topics, points are awarded over a continuum for assessed performance. By contrast, rest of the topics are "feature-specific" and points are typically given if the feature is present. These criteria are ranked as essential, important and supplementary.

Each topic area constitutes of relevant criteria graded in a scale of 0–10 and few of them may have sub-criteria. Weighing on the points is based on its significance, its priority relative to other criteria within the same topic area, or the effort required to meet the criteria. The point scores are multiplied by the weighing factors to obtain individual credits for the five topic areas. Due to fundamental difference between the topic areas, the credits are not added but together they represent the building profile [8].

9.2.3.3. *Green Globes Rating System, USA*

This system was marketed by The Green Building Initiative (GBI) in 2004. It is based on BREEAM, UK and is the only grading system recognized by American National Standard Institute (ANSI). Its management tool includes an assessment protocol, rating system and guide for environment-friendly design of commercial and institutional buildings. Thirty-one parameters are grouped into seven categories namely, (1) project management — Policies and practices; (2) site; (3) energy; (4) water resources; (5) building materials and solid waste; (6) emissions and effluents; and (7) indoor environment. The checklist is in a questionnaire format with data to be filled in and points vary from 0 to 100 depending on significance and performance summing up to 1000. After obtaining a self assessment score of a minimum 35% in two stages of schematic design and construction document, a formal rating of one, two, three, or four Green Globes is awarded for a minimum total score of 35%, 55%, 70% and 85%, respectively [15].

9.2.3.4. *GEM, UK*

In 2002, Green Globes for existing buildings was introduced in the United Kingdom as the Global Environmental Method (GEM) under the control of GABS and RICS foundation. GEM's six sections namely, energy, water, resources, environmental management, indoor environment and emissions have 19-subsections summing up to 1020 points.

GEM weighing is the mean of weighing recommended by BREEAM, the Harvard Scale and the EPA. Comparing with buildings of similar age, type or geographical zone, two kinds of ratings are provided: (1) scores that give the percentage of possible points that have been awarded for implementing best practices; and (2) quintile ratings to show performance relative to other benchmarked buildings [16].

9.2.3.5. *Go Green and Go Green Plus, Canada*

In 2004, Green Globes for existing commercial buildings was adopted by the Building Owners and Manufacturers Association (BOMA) of Canada and presently known as Go Green which addresses 10 management and operation requirements under five environmental areas namely, (1) resource consumption; (2) waste reduction and recycling; (3) building materials; (4) interior environment; and (5) tenant awareness. A three-year valid certificate is issued upon meeting its "best practices" requirements [17].

Green Globes Plus is a more in-depth benchmarking tool that can be considered as superset of Go Green. Based on "yes/no" type questionnaire regarding 19 parameters grouped into six major criteria, such as energy, water, waste reduction, emission, indoor environment and environment management [18]. After individual assessment is done for each category, a minimum total score of 70% out of 1010 points earns a certification.

9.2.3.6. *HQAL, Japan*

Housing Quality Assurance Law (HQAL) of Japan Government came into effect in 2001 to improve quality and performance of residential buildings of various types namely, detached, apartment, pre-fabricated, built-on-site, timber, reinforced concrete or steel structure. Though it deals in detail with certification of manufacturers of wooden building material and also with dispute solving, HQAL deserves a special mention as it is the only program directly addressing durability in terms of reduced deterioration and enhanced service life [14]. Among four objectives, the third one sets "Housing Performance Indication Standards" based on recommendation of Japanese standards. These standards cover architectural design, construction process for foundation, structural framing process and interior finishing. The specific types of performance characteristics include: (1) structural performance; (2) fire safety; (3) durability; (4) ease of maintenance and management; (5) energy efficiency; (6) air quality; (7) ratio of exterior openings to total wall area; (8) noise transmission; and (9) barrier free design. A prescriptive and detailed "Judgment Standards for Defects" is proposed [19] against which dwelling units are ranked as 1 for fulfillment of requirement to 3 or 4 for better quality.

9.2.3.7. *MSS, Singapore*

Maintainability Scoring System (MSS) is developed by National University of Singapore to mitigate the higher LCC comprising of repair, replacement and cleaning cost due to harsher nature of tropical climate. A higher score indicates easier maintenance [20,21]. MSS defines building as combination of five major components namely, façade, basement, internal wet area, roof and M&E system. The system relies on a comprehensive database of "Material Manual" and "Defect Library" (http://www.hpbc.bdg.nus.edu.sg). Data collected from a detailed study of 450 buildings were modeled into MSS using back-propagation method of artificial neural network. Replacement and cleaning costs can be predicted for each sub-component based on the data fed in.

9.2.3.8. *Green Star Rating, Australia*

The Green Building Council of Australia (GBCA) developed the Green Star environmental rating system in 2003 for whole life cycle assessment of office buildings and extended the scope to pilot assessment for shopping centers, healthcare and educational facilities. Multi-unit residential buildings and convention facilities are scheduled to be launched by the end of 2007. The program rates a building in relation to eight environmental impact categories namely, management, IEQ, energy, transport, water, materials, land use and economy and pollution. Each category has various sub-categories and one or more points are allocated as per Green Star technical manual [22].

The single (overall) score of a project is determined by: (1) calculating each category score; (2) applying an environmental weighing to each category; (3) adding all weighed category scores together; and (4) adding any innovation points that may have been achieved. The calculation is as follows:

$$\text{Category score (\%)} = \text{No. of points achieved/No. of points available} \times 100\%, \tag{9.3}$$

$$\text{Weighed category score} = \frac{\sum \text{Category score (\%)} \times \text{Weighing factor (\%)}}{100}, \tag{9.4}$$

$$\text{Total score} = \text{Weighed category score} + \text{Points for innovation}. \tag{9.5}$$

For a particular project, nonapplicable credits are excluded from "points available". For example, use of recycled content of structural concrete is excluded for a refurbishment project. The category weighing was derived from variety of scientific and stakeholder inputs, as well as considering geographical location, to reflect issues of importance in each territory. For energy calculations, computer software tool Australian Building Greenhouse Rating (ABGR) is used to measure environmental impact of office building. ABGR evaluation for new and existing buildings in a scale of 1–5 is implemented through "commitment agreement" and "performance rating", respectively.

The maximum possible score for Green Star rating is 100 for the weighed categories with an additional 5 points for innovation. A grade of 1–6 stars is determined for overall minimum score of 10, 20, 30, 45, 60 and 75, respectively. GBCA certifies only the last three, i.e. 4 star or "Best Practice", 5 star or "Australian Excellence" and 6 star or "World Leader". The grading is simple and logical by excluding nonapplicable points, but unsuitable to use on a global basis as input for zones is denoted by Australian provinces rather than the climatic data.

9.2.3.9. *NABERS, Australia*

The National Australian Built Environment Rating System (NABERS) is a performance-based rating system in its developing stage designed by

DEUS of New South Wales, Australia for evaluation of existing buildings including home, office building or office tenancy during operation and is based on energy data of last 12 months [23]. Both existing ratings for office and the proposed program for home, utilize ABGR and Energy Smart Home Rating or ESHR [24] complemented by Green Star of GBCA and various software programs such as BASIX [25], NatHERS (Nationwide House Energy Rating Scheme) and FirstRate.

This benchmarking tool comprises of six environmental aspects namely, energy, transport, toxic materials, waste, indoor air quality (IAQ) and occupant satisfaction. Among those, the key impact categories identified by DEUS are energy use and greenhouse emissions, refrigerant use, water use, storm water runoff, storm water pollution, sewage outfall volume, landscape diversity, toxic material and IAQ. For each category relevance, realism and practicality are illustrated to measure them in a scale of 1–5 and finally combined into three scores as follows:

- Overall greenhouse score — Average of energy/greenhouse and refrigerant use.
- Overall water score — Average of water use, storm water runoff and sewage outfall volume.
- Site management score — Average of storm water pollution, landscape diversity, toxic materials, refrigerant ozone depletion and IAQ.

The final score is set at the lowest of these three "overall" scores, with a slight prioritization given to the first two. Refinement of benchmarks and simplification of overall scoring principle are major recommendations made in the first trial of NABERS.

9.2.3.10. *HK-BEAM, Hong Kong*

HK-BEAM or Hong Kong Building Environmental Assessment Method came into existence in 1996 through HK-BEAM Society. This BREEAM-based tool can be regarded as predecessor of SBTool and provides certification of new construction, major refurbishment and existing buildings of various types such as residential, commercial, industrial and institutional projects. Environmental issues are assessed in terms of local, global and indoor impacts grouped into five major areas including site aspect, material aspect, energy use, water use and IEQ. The final score depends on 72 parameters each carrying 1–10 performance-based credits and bonus points. Additionally, the framework includes 24 other parameters, five points for innovation and guidelines for annual energy use. Any nonstereotype building is judged on a case-by-case basis and any nonapplicable point, for example, inferior performance due to externality beyond the control of the client such as source of power supply, are excluded from "credits available".

Weighing of credits is based on expert opinion and other similar systems especially referring to the local standards. Few topics, for example, impact noise, for which local standards are not available, national or international codes are used. In case of conflicting rules, client can select either. For subjective issues, a checklist is used for equitable assessment. Alternative methods can be submitted along with method statement [26,27].

The award classification framework uses assessment benchmark as zero credit level. Overall assessment grade is based on % of applicable credits gained and it is mandatory to obtain a minimum % of credits for IEQ as it is given higher importance. For estate development, common features like site aspect, transport, water use undergo group assessment while IEQ and energy use are multiplied by ratio of floor area and summed up.

9.2.3.11. *GBTool 2005, International*

Based on BEPAC, an International Framework Committee (IFC) comprising of 13 countries proposed the Green Building Assessment (GBA) as an environmental performance assessment framework at international level. It resulted in the Green Building Challenge'98 (*GBC '98*) conference held in Vancouver, Canada [28] and Green Building Tool software (GBTool). GBC 2005 or SB05 is a continuation of the GBC '98 — 2002 presently involving 25 countries. *International Initiative for a Sustainable Built Environment* [29] is the official promoter of SBTool. Both new and renovation projects of residential, office, retail, hotel, motel, hospital, industrial, assembly, institutional and school can be graded along with a combination of any two, three or four different types of building occupancy.

In an Excel environment, GBTool is structured hierarchically in four levels, with the higher levels logically derived from the weighed aggregation of the lower ones as 1 goal, 7 issues, 29 categories and 109 criteria. Few criteria, for example, IEQ has sub-criteria. Issue and categories reflecting regional or city scale forms module-A. Its benchmarks and weights are fixed by national teams using diverse methods such as AHP or analytic hierarchy process [30], Pearson's product–moment correlation [31], etc. Criteria and sub-criteria contained in module-B are project specific and used as self-assessment as set by module-A. In both cases, default values can be used. Nonapplicable parameters such as HVAC in a naturally ventilated building, weigh zero. Weighings at the sub-criteria and criteria level are fixed and are evenly weighed, unless in exceptional cases. Pre-defined weighings of category are selected as per Table 9.5. The values range between –2 and +5, where,

- −1 = unsatisfactory,
- 0 = minimum acceptable performance,
- 5 = considerably advanced of current practice,
- 1–4 = intermediate performance levels and
- 2 = normal default or null value (except for mandatory parameters ranging from 3 to 5).

All issues have individual score of 100 and totals up to 100%. These values are auto-generated within SBTool in any of the following ways [32]:

- Explicitly defined target value: difference between "best" performance target and the benchmark divided by 5.
- Fixed interval: The subsequent values for scores are calculated as difference with the benchmarked value.
- Percentage scale: For criteria measured as % of resource saved or used compared to fixed % value of benchmark. The scale intervals can be set to determine the full range.

9.2.4. *Other Examples*

9.2.4.1. *CASBEE, Japan*

CASBEE (Comprehensive Assessment System for Building Environmental Efficiency), developed by Japan Sustainable Building Consortium (JSBC) differs from other assessment system by BEE (Building Environmental Efficiency) — A quotient to measure environmental performance of building. For both residential and nonresidential buildings, four grading tools for building life cycle are (1) Tool-0: CASBEE-PD for pre-design; (2) Tool-1: CASBEE-NC for new construction; (3) Tool-2: CASBEE-EB for existing building; and (4) Tool-3: CASBEE-RN for renovation [33].

The basic structure of CASBEE has four major aspects with 80 items re-grouped into (1) Q or Quality: the environmental quality and performance of the building and (2) L or Loadings: the building's environmental loading with three sub-groups each. A scale of 1–5 is used with 3 as standard grade or ordinary level for each assessment item. Grades are tabulated on score sheet, weighed using pre-defined co-efficient for each to get overall scores $S_{Q1\sim Q3}$ or $S_{LR1\sim LR3}$. Next overall scores S_Q and S_{LR} are calculated to compute BEE as:

$$\text{BEE} = \frac{Q\text{ (Building environmental quality and performance)}}{L\text{ (Building environmental loadings)}} = \frac{25\times(S_Q-1)}{25\times(5-S_{LR})}, \tag{9.6}$$

$$\text{Building complex score} = \Sigma\text{ (Score for each building type} \times \text{Corresponding floor ratio)}. \tag{9.7}$$

The same logic applies for a building with multiple use. Weighing coefficient for each assessment field for different kinds of building are based on the outcome of a questionnaire survey of designers, building owners and operators, related officials and others users. The responses were analyzed by analytic hierarchy process (AHP). For example, Level 1 is awarded for noise level >50 dB for office, hospital, hotel, apartment but for school a much lower value of 45 dB is recommended. Similarly, multiple scoring is provided for points with variations such as the criteria to reflect special regional characteristics. Examples of such customizations are regional versions developed in Nagoya and Osaka [34].

CASBEE calculations are done in excel environment and graphical representation in assessment sheet includes BEE, radar chart and bar graph. For BEE, points are plotted with Q values on the Y axis and L values on the X axis to determine the BEE position, in a scale of class C (poor), class B^-, class B^+, class A and class S (excellent), in order of increasing BEE value.

9.3. Maintainability and Management

In maintenance management, maintainability is a critical decision making process which involves the optimum selection of building elements that require minimum cleaning, repair and replacement. In general, this is not a straightforward task, given (1) the large number of materials and variety of systems and designs, (2) variation of performance and (3) difference in the risk of failures associated with different systems and components.

Building maintainability is hence a composition of many factors. It could be modeled in various forms, one of which is LCC [35]. In such case, optimum maintainability, i.e. the optimum LCC is set at the minimum level of risk. Defining the minimum level of risk involves a series of decision-making process known in the construction industry as "risk management".

The most common method for contemporary risk management, the "five-tiered" model might be used for maintainability management. This might be expressed as [36]:

(1) risk identification,
(2) risk estimation,
(3) risk evaluation,
(4) risk response and
(5) risk monitoring.

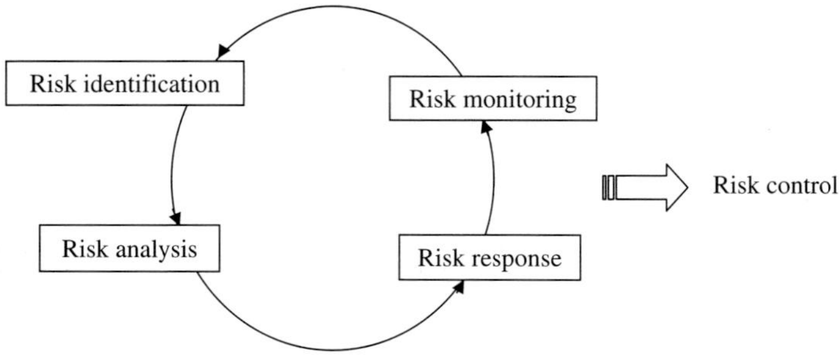

Fig. 9.1. Risk management cycle.

Of the five steps, risk estimation and evaluation are closely related and thus are collectively referred to as risk analysis. Similarly, risk response and risk monitoring could also be combined into a single broader category such as risk control (Fig. 9.1).

9.3.1. *Risk Identification and Primary Risk Factors*

The number of risk factors that describe a typical risk environment of maintainability could often run into hundreds. However, not all of them are significant when considering the local conditions of the user. For example, risk of snowfall in affecting the maintainability of façades is unlikely to be experienced by buildings located along the tropical belt. Hence, in identifying risks to maintainability, historical maintenance data from a large pool of buildings would be most useful.

Risk factors that determine the level of LCC and hence maintainability of housing stocks were reported [37–41]. The significance of quality of constructional components at the time of completion, technical aspects of the aging process, maintenance methods and practices and level maintenance as preliminary factors of LCC was identified.

The following sections look at the following risk factors

(1) design (water-tightness, stress accommodation, water flow control, accessibility, downtime and material performance) risk,
(2) construction (quality) risk,
(3) maintenance (quality) risk,

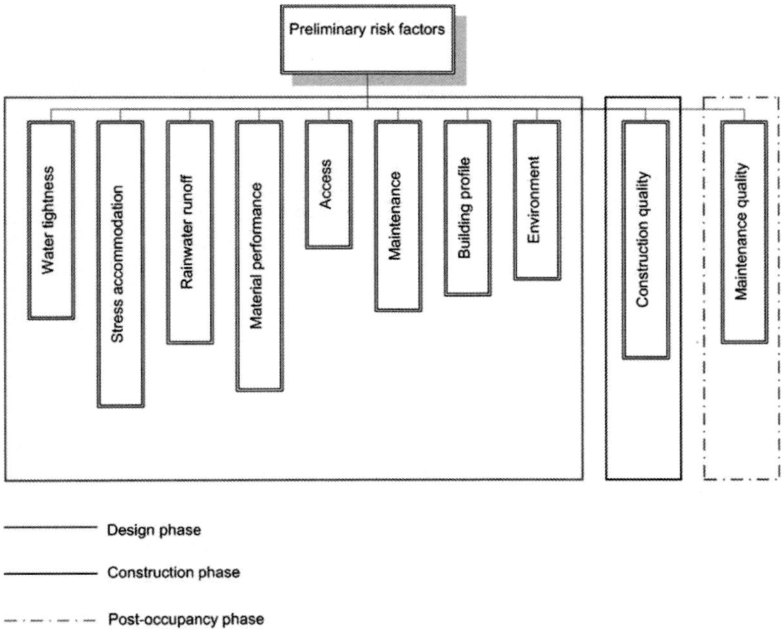

Fig. 9.2. Façade preliminary risk factors.

(4) risk from environment and usage, and
(5) risk from building's profile.

Façade as an element is used as an example in the following sections. Figure 9.2 illustrates primary risk factors related to façade.

From a detailed survey [35], 22 risk factors that were significant to a façade's maintainability were identified. These factors were further grouped into 14 clusters (Table 9.2).

Details of the risk factors of other components are described in the Maintainability Website (http://www.hpbc.bdg.nus.edu.sg/) under the topic "Maintainability Scoring System".

9.3.2. *Risk Analysis*

Once significant risk factors have been identified, they could be analyzed in depth. Such risk analysis was concerned with the severity of each risk attribute to LCC. Over the past years, several formal techniques have

Table 9.2. Façade risk factors.

Risk Clusters	Risk Factors	Elements
Wall selection	1. First line defense	1. Type of finishes and joints
	2. Second line defense	2. Type of wall system
Wall detail	3. Shape	3. Proper detail to drain away the water and prevent staining
	4. Grid	4. Size, detailing
	5. Joint	5. Type, spacing, size and detailing and exposure
Wall maintainability	6. Accessibility	6. Building shape, provision for access and coverage
	7. Flexibility	7. Wall complexity, removability (panel), down time and access system
Window selection	8. Fixing method	8. Exposure
	9. Panel detail	9. Panel, frame and drainage system
Window detail	10. Fixing details	10. Protection detail, drainage detail and louver detail
Window maintainability	11. Accessibility for cleaning	11. Method of opening
Material selection	12. Durability	12. Lifespan
Material detail	13. Performance	13. Resistance to tropical conditions and interface compatibility
Material maintainability	14. Cleanability	14. Cleaning requirement such as methods, frequencies, etc.
	15. Inspection	15. Maintenance requirements such as inspections, frequencies, etc.
Construction	16. Workmanship quality	16. Construction quality assessment score (CONQUAS)
Maintenance practice	17. Method	17. Maintenance inspections, equipments and chemical used for cleaning
	18. Efficiency	18. Frequencies of inspection and cleaning cycles
Environmental	19. Topography	19. Pollution level
	20. Orientation	20. Parks, expressways, major roads, light traffic roads and cost line
Building age	21. Age 1–30 years	
Building height	22. Five groups such that height less than 5, 5–12, 13–40, 41–60 and more than 60 stories	

been developed to facilitate risk analysis. They ranged from simple techniques such as elementary risk analysis to more sophisticated techniques that involved modern concepts such as application of expert systems and neural networks.

To carry out a risk analysis, it is necessary to estimate and quantify each of the risk. Probabilistic techniques and discrete interval methods were few of the techniques that could be used. In practice, various indices that were associated with parameters describing risk factors were specified in their own dimension and units. For instance, risk of water penetration was well-described by depth of penetration while accessibility was considered as area of coverage. However, from an analytical point of view, using a homogeneous set of inputs was much more computationally feasible than a set of heterogeneous inputs.

9.4. Maintainability Prediction and Integration

Maintainability is a design parameter and hence should be addressed right from the design phase. In previous chapters, the major elements of a facility is discussed in the form of risk analysis and subsequently converted to sets of checklist. The relevant chapters and tables for each of the elements are summarized in Table 9.3.

This section discusses work conducted in NUS of which the elements of a facility were graded in terms of cause and criticality. Guidelines in

Table 9.3. Summary of checklists for elements.

	Element	Chapter	Table
Civil and Architecture	Wet area	2	2.7, 9.6
	Façade	3	3.3, 9.5
	Basement	4	4.3, 9.4
	Roof	5	5.1, 9.7
Mechanical and Electrical	Sanitary-Plumbing	6	4.1, 9.8
	HVAC	6	4.2, 9.9
	Elevators and Lifts	6	4.3, 9.10
	Electrical	6	4.4, 9.11
	Fire protection	6	4.5, 9.12

the form of maintainability checklist were developed. The relevant factors, their significance and relative weights were evaluated. The result may be used, with relevant adjustment to the different needs of the user, as a benchmark for good design, construction and maintenance practices and provides guidelines for optimum selection of maintenance strategies.

The work focused on major building systems and services maintained by central facility management. The elements or subsystems were grouped under two main systems, civil- architectural (C&A — Basement, façade, wet area and roof) and mechanical–electrical (M&E — Sanitary-plumbing, HVAC, elevator, electrical and fire protection). The work was conducted in three phases:

- defect analysis;
- benchmarking for design, construction and maintenance;
- integration of subsystems based on maintainability parameters.

9.4.1. *Defect Analysis*

As per FMECA rules, the nine subsystems were divided into components. List of associated defects, their causes and effects were noted from literature, site investigation, discussion with building professionals and facility managers (FM) and past maintenance records. The causes were divided into: (1) design/specification; (2) construction/installation; (3) maintenance; and (4) external factors, e.g. age, vandalism, etc. Criticality was expressed through defect frequency and severity given by impact on (1) economy; (2) system performance; and (3) health, safety and comfort. Severity Index (Sv) and Criticality Index (Cr) were proposed as:

$$\mathrm{Sv} = \frac{\overline{X}_{EC} + \overline{X}_{SP} + \overline{X}_{HS}}{3} \quad \text{and} \quad \mathrm{Cr} = \frac{\overline{X}_{FR}}{5} \times \frac{\mathrm{Sv}}{5},$$

where $\overline{X}_{EC}$, $\overline{X}_{SP}$, $\overline{X}_{HS}$ and $\overline{X}_{FR}$ are the mean rating (out of 5) or weighted average for economic impact, system performance impact, health–safety–comfort impact and defect frequency, respectively.

The grading scale for frequency and severity depends on the situation. Here a five-point scale was selected to suit the five-point Likert scale of questionnaire. Such structured data collection strategy ensures unbiased and unambiguous response. For frequency, 1 = very rare (once in 10 years. or rarer); 2 = rare (once in 5 years); 3 = occasional (yearly); 4 = frequent (quarterly); 5 = very frequent (monthly or more frequent). For severity, 1 = negligible; 2 = slight; 3 = moderate; 4 = serious; 5 = very serious, hazardous or fatal. In a questionnaire survey, experienced FMs graded the defects in terms of frequency and three severity parameters. A critical defect has ≥1 of these four factors found significant in statistical *t*-test with 95% confidence interval. The null hypothesis ($H_{0:}\ \mu \leq \mu_0$) was tested against the alternative hypothesis (H_1: $\mu > \mu_0$), where μ is the population mean and μ_0(=3) is critical rating above which a factor was called significant.

9.4.2. *Benchmarking for Design, Construction and Maintenance*

For each subsystem, guidelines to achieve higher maintainability were developed in the form of defect-mitigating checklist factors based on literature and DL (Defect Library). These factors were grouped under design–construction–maintenance aspects and further sub-grouped under components of the subsystem. Factors' relative weights (RW) are their ability to mitigate both critical and noncritical defects and scores was set as 5 = full compliance with the requirements; 1 = violation of the same and 0 = factors not applicable. Weighted sum of the MS of each factors indicate MS for individual subsystem. The ideal score ≈ 100%. Using 80–20 rule, RW of *i*-th factor RW_i is:

$$\text{RW}_\text{i} = W_i \Big/ \sum\nolimits_{i=1}^{N} W_i \quad \text{for} \quad W_i = \left(0.8\ \times \sum\nolimits_{p=i}^{P} \text{Cr}_p\right) + \left(0.2 \times \sum\nolimits_{q=i}^{Q} \text{Cr}_q\right),$$

where W_i = weight of a factor, N = no. of factors for a subsystem, P, Q = no. of critical and noncritical defects respectively, Cr = criticality index for a defect. This generic formula is applicable to any component or any building phase. In that case RWs of only relevant factors are to be renormalized to obtain a new set of RWs.

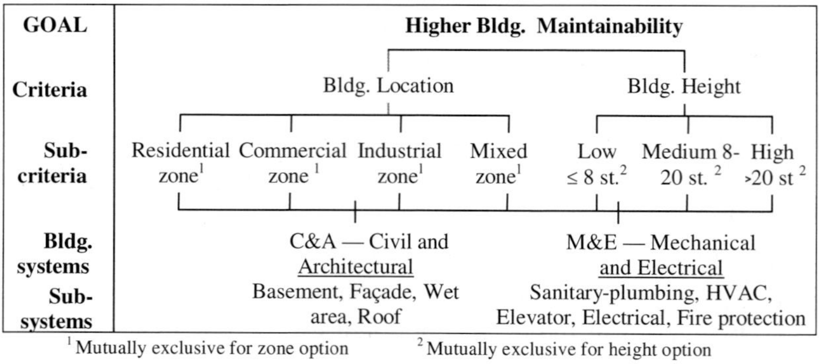

Fig. 9.3. The AHP model to integrate building subsystems into COMASS.

9.4.3. *Integration of Subsystems Based on Maintainability Parameters*

Figure 9.3 shows the AHP model for integration of nine subsystems in terms of maintainability. As per problem definition, higher building maintainability was goal and subsystems were at lowest strata. Expert FMs as respondents considered technical viability of subsystems (objective parameter) and business profile (subjective parameter) rely on height and zone, respectively. Hence these two were criteria and their different options (mutually exclusive) were sub-criteria. From individual responses, group decision for RW of subsystems was derived as per standard method. The numeric values were matched to grading logic and the MS of the building was yielded from weighted sum of MS of subsystems.

Tables 9.4–9.12 show the maintainability factors for various elements with their relative weights for reference only. Users should use the relevant weights according to their own needs for the best performance.

9.4.4. *Overview of Scoring of All Subsystems*

For the entire building a total of 731 guidelines were proposed (Table 9.13). Design related guidelines were undoubtedly the highest in number. Details of the guideline is tabulated in the Maintainability Website (http://www.hpbc.bdg.nus.edu.sg/).

Table 9.4. Maintainability factors for basement and their RWs.

Component		Maintainability Factors	Asso. Critical Defects	Asso. Noncritical Defects	Wt.	RW
Design phase						
W/P system	a1	Usage	A3, A5		0.454	0.021
	a2	Application feasibility	A3, A5		0.454	0.021
	a3	Soil condn.	A3, A5		0.454	0.021
Concrete	a4	System/construction method selection	A20	A12	0.302	0.014
	a5	Design (mix and rebar)	A3, A4. A5	A1, A111, A17	1.117	0.052
	a6	Joint details		A1, A12, A17	0.333	0.016
	a7	Pipe penetrations	A19		0.289	0.014
W/P membrane (in Type A system)	a8	Location	A3		0.258	0.012
	a9	Application	A3		0.258	0.012
	a10	Material selection	A3		0.258	0.012
	a11	Material properties	A3, A4	A8	0.622	0.029
	a12	Joint details		A1	0.283	0.013
	a13	Pipe penetrations	A19		0.289	0.014
Waterstop (in Type B system)	a14	Shape	A20	A1	0.571	0.027
	a15	Material	A2	A1	0.571	0.027
	a16	Detailing	A2	A1	0.571	0.027
Cavity (in Type C system)	a17	Cavity wall design		A6, A11	0.065	0.003
	a18	Masonry block and mortar selection		A6	0.033	0.002
	a19	Cavity floor selection	A20		0.278	0.013
	a20	Cavity floor design	A20	A6	0.311	0.015

(*Continued*)

Table 9.4. (*Continued*)

Component		Maintainability Factors	Asso. Critical Defects	Asso. Noncritical Defects	Wt.	RW
Drainage system	a21	Pump sump	A20		0.278	0.013
	a22	Catchments	A20		0.278	0.013
Flooring	a23	Screed	A20	A17	0.278	0.013
	a24	Additives	A16		0.229	0.011
	a25	Finishing	A16		0.229	0.011
Wall finishes	a26	Paint selection		A8, A10	0.230	0.011
	a27	Tile selection		A12, A13	0.061	0.003
Ancillary facilities	a28	Coordination among professions		A15	0.244	0.011
					Subtotal	0.450
Construction phase						
RCC	a29	Excavation and formwork	A20	A12	0.278	0.013
	a30	Rebar laying		A1	0.283	0.013
	a31	Material quality	A20	A11, A17	0.057	0.003
	a32	Casting and curing	A3, A4, A5	A1, A12, A17	1.110	0.052
	a33	Check and test		A17	0.025	0.001
W/P install	a34	Substrate and material quality	A3		0.258	0.012
	a35	Application	A3, A19	A1	0.830	0.039
	a36	Inspect and test	A3	A1	0.541	0.025
	a37	Protection	A3		0.258	0.012

(*Continued*)

Table 9.4. (*Continued*)

Component		Maintainability Factors	Asso. Critical Defects	Asso. Noncritical Defects	Wt.	RW
Waterstop	a38	Installation	A2, A4	A1	0.895	0.042
Cavity	a39	Material quality and condition		A12	0.024	0.001
	a40	Laying	A20		0.278	0.013
	a41	Finishing		A14	0.032	0.002
Flooring	a42	Screed	A20		0.278	0.013
	a43	Joints		A17	0.025	0.001
	a44	Checking	A20	A17	0.303	0.014
Paint and plaster	a45	Substrate and material quality		A8, A9	0.219	0.010
	a46	Application	A7		0.212	0.010
Tiling	a47	Screed and tile prep.		A12, A14	0.056	0.003
	a48	Tiling		A13, A14	0.070	0.003
	a49	Protection		A13	0.037	0.002
					Subtotal	0.285
Maintenance phase						
General	a50	Inspection	A4, A5, A15, A19	A2, 18	1.569	0.074
Drainage system	a51	Check	A20		0.278	0.013
	a52	Clean	A20		0.278	0.013
Flooring	a53	Clean	A20		0.360	0.017

(*Continued*)

Table 9.4. (*Continued*)

Component		**Maintainability Factors**	**Asso. Critical Defects**	**Asso. Noncritical Defects**	**Wt.**	**RW**
Tiles	a54	Clean	A16, A20	A17	0.419	0.020
Painted wall	a55	Clean		A10, A18	0.266	0.012
External factors	a56	Soil permeability	A1, A3, A5		0.541	0.025
	a57	Aggressive chemical	A1, A3		0.541	0.025
	a58	Depth	A1, A3		0.541	0.025
	a59	Building age	A1, A3, A4		0.865	0.041
					Subtotal	0.265
					Total	**1.000**

Table 9.5. Maintainability factors for façade and their RWs.

Component	Maintainability Factors	Critical Defects Mitigated	Noncritical Defects Mitigated	Wt.	RW
Design phase					
System selection	b1 Water resistance	B3	B4, B9, B14, B25, B29, B30	0.351	0.016
	b2 Complexity	B49		0.192	0.009
	b3 Removeability	B49		0.192	0.009
	b4 Accessibility		B48	0.060	0.003
	b5 Availability	B49		0.192	0.009
Finishes	b6 Water resistance: Traditional wall	B3	B9, B14, B25, B30	0.298	0.014
	b7 -Do-: Curtain wall/cladding	B43	B40	0.261	0.012
	b8 Cleanability	B5, B13, B21, B43, B49	B10, B24, B31	1.250	0.059
	b9 Inspect. freq.	B49		0.192	0.009
Material	b10 Masonry block	B3, B5	B1, B2, B10, B14, B25	0.560	0.026
	b11 Plaster	B13		0.241	0.011
	b12 Stone		B27 ,B28, B29, B30, B31	0.330	0.015
	b13 Tiles		B23, B24, B26	0.097	0.005
	b14 Paint	B21	B16, B18, B19, B20	0.318	0.015
	b15 Coating		B34, B39	0.069	0.003
	b16 Metal	B33	B32, B35, B36, B38, B41, B42	0.338	0.016
	b17 Glass	B43, B44		0.413	0.019

(*Continued*)

Table 9.5. *(Continued)*

Component	Maintainability Factors		Critical Defects Mitigated	Noncritical Defects Mitigated	Wt.	RW
Wall joint	b18	Water resistance	B3	B9, B14, B25, B30	0.298	0.014
	b19	Grid size		B6, B22, B27, B35	0.074	0.003
	b20	Exposure		B6, B10	0.075	0.004
	b21	Drainage	B43	B40	0.261	0.012
Expansion joint	b22	Location	B3	B1, B11, B27, B28, B35, B36	0.332	0.016
	b23	Spacing		B22, B23, B27, B28, B35, B36	0.127	0.006
	b24	Rigidity	B44	B35, B36, B45	0.273	0.013
Sealant	b25	Type		B6, B42	0.065	0.003
	b26	Geometry	B44	B6, B42	0.243	0.011
	b27	Backer rod	B44	B6, B42	0.243	0.011
	b28	Exposure		B42, B45	0.066	0.003
Building shape	b29	Plan	B13, B21, B31, B49	B48	0.980	0.046
	b30	Massing	B49	B48	0.252	0.012
	b31	Regularity	B5, B13, B21, B31, B33, B43	B9, B10, B24	1.457	0.068
	b32	Projection	B49	B48	0.252	0.012
Access system	b33	Type	B49		0.192	0.009
	b34	Coverage	B49		0.192	0.009
	b35	Number	B49		0.192	0.009

(Continued)

Table 9.5. (*Continued*)

Component	Maintainability Factors	Critical Defects Mitigated	Noncritical Defects Mitigated	Wt.	RW
Window	b36 Position	B46		0.213	0.010
	b37 Panel detail	B46		0.213	0.010
	b38 Louvre detail	B5, B13, B21, B33	B9, B10	0.915	0.043
	b39 Shading	B5, B13, B21, B33, B46	B9, B10	1.128	0.053
	b40 Drainage detail	B46		0.213	0.010
	b41 Accessibility	B47		0.219	0.010
Ancillary	b42 Structural	B7, B8, B21	B1, B2, B10, B22, B23, B35	0.640	0.030
				Subtotal	0.677
Construction phase					
Brickwork	b43 Material quality	B3, B5	B1, B2	0.463	0.022
	b44 Laying		B4	0.036	0.002
	b45 Finishing	B3	B4	0.233	0.011
Block masonry	b46 Material quality		B1, B2	0.035	0.002
	b47 Laying		B4	0.036	0.002
	b48 Finishing	B3		0.197	0.009
PC	b49 Preparation	B7		0.030	0.001
	b50 Erection and jointing	B8	B10	0.173	0.008

(*Continued*)

Table 9.5. *(Continued)*

Component	Maintainability Factors	Critical Defects Mitigated	Noncritical Defects Mitigated	Wt.	RW
Plastering	b51 Substrate prep.	B12, B15		0.061	0.003
	b52 Material quality		B11, B12, B14, B15	0.123	0.006
	b53 Application and curing	B13	B11, B12	0.309	0.014
Painting	b54 Substrate prep.		B16, B17	0.069	0.003
	b55 Material quality		B16	0.022	0.001
	b56 Application		B17	0.047	0.002
Tiling (ceramic and stone)	b57 Base prep.		B22, B23, B28	0.049	0.002
	b58 Tile prep.		B23, B26	0.046	0.002
	b59 Tiling		B23, B27, B28, B30	0.079	0.004
	b60 Grouting		B24	0.051	0.002
	b61 Inspection		B22, B23, B24, B26	0.112	0.005
	b62 Protection		B24, B26	0.073	0.003
Metal	b63 Material quality and preparation		B32, B34, B35, B37, B38, B40	0.159	0.007
	b64 Setting fixings	B44	B32, B37, B40	0.275	0.013
	b65 Panel erection	B44	B32, B34, B35, B36, B40, B41	0.318	0.015
Sealant application	b66 Substrate prep.	B44	B6, B42, B45	0.274	0.013
	b67 Material quality		B6, B42, B45	0.096	0.005
	b68 Application	B43	B6, B42, B45	0.331	0.016
				Subtotal	0.173

(Continued)

Table 9.5. (*Continued*)

Component	Maintainability Factors	Critical Defects Mitigated	Noncritical Defects Mitigated	Wt.	RW
Maintenance phase					
Clean	b69 Masonry	B5		0.230	0.011
	b70 Conc. and plaster	B8, B13	B9, B10, B14	0.466	0.022
	b71 Painted	B21	B17	0.278	0.013
	b72 Stone cladding	B31		0.257	0.012
	b73 Tile cladding		B23, B24, B25, B26, B30	0.146	0.007
	b74 Glass curtain	B43	B47	0.454	0.021
	b75 Metal cladding	B33	B32, B38, B39	0.274	0.013
External	b76 Building age	B8	B10, B20, B22, B35	0.219	0.010
	b77 Exposure	B5, B7, B8, B31	B2, B10, B19, B20, B39, B42, B45	0.676	0.032
	b78 Height		B49	0.192	0.009
				Subtotal	0.149
				Total	**1.000**

Table 9.6. Maintainability factors for wet area and their RWs.

Component		Maintainability Factors	Asso. Critical Defects	Asso. Noncritical Defects	Wt.	RW
Design phase						
Floor	c1	Zoning	C2	C1, C11	0.163	0.010
	c2	Gradient		C18	0.024	0.002
	c3	Concrete slab	C2, C4	C1, C3, C5, C6, C11	0.350	0.022
	c4	Screed		C1, C5, C6, C11, C18	0.118	0.008
Water proofing	c5	Material select	C2, C4	C1, C11	0.290	0.019
	c6	Material property	C2, C4	C1, C11	0.290	0.019
	c7	Application feasibility	C2, C4	C11	0.259	0.017
	c8	Joint details		C1, C11	0.055	0.004
	c9	Penetration details	C4, C15, C16	C11	0.547	0.035
	c10	Fixture detailing	C2	C11	0.131	0.008
Plumbing	c11	Wet wall or layout	C2	C11, C13	0.428	0.028
	c12	Penetration plan	C2, C4, C15, C16	C11	0.655	0.042
	c13	No of fl./wall penetration	C4, C15	C11	0.340	0.022
	c14	Pipe accessibility	C4, C15	C11, C12, C17	0.385	0.025
	c15	Wall accessibility	C13	C7	0.313	0.020
Fixture and fittings	c16	Gen. quality		C19	0.238	0.015
	c17	Basin selection	C13		0.296	0.019
	c18	Basin layout	C13		0.296	0.019
	c19	WC and urinal selection	C13		0.296	0.019
	c20	WC and urinal layout	C13		0.296	0.019

(Continued)

Table 9.6. (*Continued*)

Component		Maintainability Factors	Asso. Critical Defects	Asso. Noncritical Defects	Wt.	RW
	c21	Bath selection	C13	C18	0.320	0.021
	c22	Shower layout	C13		0.296	0.019
	c23	Bathtub layout		C18	0.024	0.002
	c24	Piping material selection	C19	C12, C17	0.280	0.018
Tiles on floor	c25	Movement joint location		C5, C6, C8	0.076	0.005
and wall	c26	Movement jt. detail		C5, C6	0.040	0.003
	c27	Tiles selection		C5, C6, C7, C8	0.092	0.006
	c28	Bedding material		C6, C8	0.039	0.003
	c29	Grout selection		C6, C7, C9	0.059	0.004
Paints	c30	Paint property	C10, C13	C11	0.432	0.028
	c31	Water resistance	C13	C11	0.320	0.021
Ancillary facilities	c32	Coordination among professions	C13	C14, C16	0.549	0.035
					Subtotal	0.533
Construction phase						
Floor	c33	Slab casting	C2, C4	C1, C3, C6, C11,C12	0.352	0.023
	c34	Embedding of services	C4, C15, C16	C11	0.547	0.035
	c35	Laying the slope		C18	0.024	0.002
Water proofing	c36	Substrate condition	C2, C4	C1, C11	0.290	0.019
	c37	Quality of material	C2, C4	C1, C11	0.290	0.019
	c38	Application	C2, C4, C15, C16	C1, C11	0.686	0.044
	c39	Inspection and testing	C2, C4	C1, C11	0.290	0.019
	c40	Protection	C4, C15, C16	C11	0.547	0.035

(*Continued*)

Table 9.6. (*Continued*)

Component		Maintainability Factors	Asso. Critical Defects	Asso. Noncritical Defects	Wt.	RW
Tiles on floor and wall	c41	Base (substrate and screed) preparation		C6, C8	0.059	0.004
	c42	Preparation of tiles		C6, C8, C9	0.079	0.005
	c43	Tiling		C5, C6, C8, C9	0.096	0.006
	c44	Grouting		C6, C7, C9	0.059	0.004
	c45	Inspect and check		C5, C6, C7, C9	0.076	0.005
	c46	Protection		C5, C6	0.040	0.003
Paints	c47	Substrate prep.	C10	C11	0.135	0.009
	c48	Material quality	C10	C11	0.135	0.009
	c49	Application	C10	C12	0.131	0.008
Fix. and fittings	c50	Fixing and connection	C15, C16	C11, C14	0.282	0.018
					Subtotal	0.265
Maintenance phase						
General	c51	Inspection	C4, C15	C1, C11, C12, C17	0.416	0.027
Tile finish	c52	Cleaning	C13		0.296	0.019
Marble	c53	Cleaning	C13		0.296	0.019
Granite	c54	Cleaning	C13		0.296	0.019
Paints	c55	Cleaning	C13		0.296	0.019
Fix fittings	c56	Cleaning	C19		0.238	0.015
External factors	c57	Building age	C2, C4, C16	C1, C11	0.497	0.032
	c58	level of usage	C19	C7	0.254	0.016
	c59	Vandalism	C13, C19	C7	0.550	0.035
					Subtotal	0.202
					Total	**1.000**

Table 9.7. Maintainability factors for roof and their RWs.

Component	Maintainability Factors		Asso. Critical Defects	Asso. Noncritical Defects	Wt.	RW
Design						
Roofing sys. deck	d1	Selection	D3	D11, D15	0.230	0.014
	d2	Struct. conc.		D1	0.025	0.001
	d3	Bearing capacity	D3, D5, D21	D11	0.679	0.040
	d4	Deflection	D5, D21	D6, D11	0.538	0.032
	d5	Movement jt.	D3	D2, D4, D11, D16	0.280	0.017
	d6	Vapor barrier		D6, D12	0.060	0.004
	d7	Parapet	D5	D4	0.235	0.014
	d8	Pipe and equip.	D3, D25		0.323	0.019
WP membrane	d9	Selection		D6, D10, D11, D12, D13, D14	0.152	0.009
LAM	d10	Jt. detail	D3	D4, D14	0.224	0.013
	d11	Penetration	D3, D22, D25	D4, D14	0.602	0.036
Preformed	d12	Jt. detail/flashing	D3	D2, D4, D14	0.248	0.015
	d13	Penetration	D3, D22, D25	D4, D14	0.602	0.036
	d14	Venting		D6	0.038	0.002
Insulation	d15	Properties		D4, D11	0.045	0.003
	d16	Material	D21	D5	0.477	0.028
Tiles/panels	d17	Properties		D16, D17	0.050	0.003
Sealant	d18	Type		D18	0.043	0.003
	d19	Joint geometry		D18, D20	0.064	0.004
	d20	Bk. rod detail		D18, D20	0.064	0.004

(Continued)

Table 9.7. (*Continued*)

Component	Maintainability Factors		Asso. Critical Defects	Asso. Noncritical Defects	Wt.	RW
Roof drainage	d21	Design rate	D22		0.235	0.014
	d22	Effective runoff	D22		0.235	0.014
	d23	Angle of slope	D22		0.235	0.014
	d24	Uniformity of slope	D4, D21	D5	0.499	0.030
	d25	Ease of construction	D21	D5	0.477	0.028
Outlet	d26	Size and no.	D22		0.235	0.014
	d27	Location	D21	D5	0.477	0.028
	d28	Protection	D22		0.235	0.014
RWDP	d29	RWDP size and material	D22		0.235	0.014
	d30	Slope	D22		0.235	0.014
	d31	Access	D22		0.235	0.014
	d32	Fixing	D22		0.235	0.014
	d33	Jointing	D22	D24	0.268	0.016
	d34	Termination	D23		0.098	0.006
Ancillary facility	d35	Coordination	D3, D27	D14, D24, D26	0.387	0.023
					Subtotal	0.553
Construction phase						
Deck	d36	Concreting	D4	D1	0.047	0.003
	d37	Precast unit		D1	0.025	0.001
	d38	Propping		D1	0.025	0.001

(*Continued*)

Table 9.7. (*Continued*)

Component	Maintainability Factors		Asso. Critical Defects	Asso. Noncritical Defects	Wt.	RW
	d39	Const. joint	D3	D2	0.204	0.012
	d40	Curing		D1, D6	0.063	0.004
	d41	Protection	D2	D3, D9, D11	0.259	0.015
	d42	Finished surface	D2, D4, D21	D3, D5, D6, D9, D11, D12	0.817	0.049
WP-LAM	d43	Storage and prep.		D6, D9, D10, D11	0.117	0.007
	d44	LAM apply		D6, D9, D12, D14, D15	0.141	0.008
	d45	Membrane application	D21	D6, D9, D10, D13, D14, D15	0.429	0.025
	d46	Joint and penetration	D4, D21	D3, D5, D11, D14, D24	0.757	0.045
	d47	Protection		D9, D14	0.054	0.003
	d48	Testing	D21	D5	0.477	0.028
Insulation	d49	Layout		D16	0.031	0.002
Tiles/panels	d50	Base	D21	D17	0.282	0.017
	d51	Tiling		D17	0.019	0.001
Sealant	d52	Substrate		D18	0.043	0.003
	d53	Quality check		D18	0.043	0.003
	d54	Application		D18, D20	0.064	0.004
Drainage	d55	Jointing	D21, D22	D5, D24	0.745	0.044
	d56	Testing	D21, D22	D5, D24	0.745	0.044
					Subtotal	0.320

(*Continued*)

Table 9.7. *(Continued)*

Component	Maintainability Factors		Asso. Critical Defects	Asso. Noncritical Defects	Wt.	RW
Maintenance phase						
Roof surface	d57	Check	D2, D19	D3, D7, D8, D20	0.425	0.025
	d58	Clean	D19	D5	0.375	0.022
Drainage	d59	Check	D4	D12	0.257	0.015
	d60	Clean		D12	0.235	0.014
Details	d61	Check	D2	D3, D5	0.417	0.025
External factors	d62	Usage		D7, D8, D9, D15	0.097	0.006
	d63	Building age	D2	D3, D18, D20	0.268	0.016
	d64	Exposure		D15, D18	0.071	0.004
					Subtotal	0.127
					Total	**1.000**

Table 9.8. Maintainability factors for sanitary-plumbing system and their RWs.

Component	Maintainability Factors		Critical Defects Mitigated	Noncritical Defects Mitigated	Wt.	RW
Design phase						
Cold water supply piping	e1	Size	E5	E1	0.366	0.009
	e2	Material	E2, E3, E4, E5		0.957	0.025
	e3	Strength		E1	0.064	0.002
	e4	Layout	E3, E4, E5, E6, E14	E7	1.142	0.030
	e5	Safe distance		E1	0.064	0.002
	e6	Cross connection	E3		0.144	0.004
	e7	Protection (u. gr.)		E1	0.064	0.002
	e8	Support (u. gr.)		E1	0.064	0.002
	e9	Protection (ovr. gr.)	E2		0.266	0.007
	e10	Support (ovr. gr.)	E6		0.237	0.006
	e11	Access	E3	E7	0.203	0.005
	e12	Noise prevention	E6		0.237	0.006
	e13	Joints	E3, E4, E5		0.690	0.018
	e14	Penetration	E2	E9	0.320	0.008
	e15	Cleaning facility	E3, E5		0.446	0.012
	e16	Isolation valve	E3	E7	0.203	0.005
	e17	Air valve	E5, E6, E14		0.695	0.018
Hot water supply	e18	Piping design	E11		0.282	0.007
	e19	Material	E12		0.202	0.005
	e20	Temp. control	E3, E11, E12		0.628	0.016

(Continued)

Table 9.8. *(Continued)*

Component	Maintainability Factors		Critical Defects Mitigated	Noncritical Defects Mitigated	Wt.	RW
Storage tank	e21	Capacity		E10	0.029	0.001
	e22	Location	E3		0.144	0.004
	e23	Material	E3, E8	E9	0.523	0.014
	e24	Access	E3	E10	0.173	0.004
	e25	Circulation	E3	E9, E10	0.226	0.006
	e26	Ventilation	E3		0.144	0.004
Pumps	e27	Capacity	E5, E13, E15		0.868	0.022
	e28	Material		E16	0.063	0.002
	e29	Housing	E13		0.302	0.008
Sanitary appliances	e30	Material	E18, E19, E21		0.955	0.025
	e31	Fixing	E19, E21		0.613	0.016
	e32	Supply	E3, E18		0.486	0.013
	e33	Discharge	E19		0.364	0.009
	e34	Floor waste and trap	E19, E20		0.712	0.018
	e35	Access	E19		0.364	0.009
Sanitary pipe	e36	System select and venting	E18, E24, E25		0.735	0.019
	e37	Pipe design E18	E23, E24, E25		0.691	0.018
	e38	layout	E24	E26	0.253	0.007
	e39	Access	E24	E26	0.253	0.007
	e40	Joints	E23		0.298	0.008

(Continued)

Table 9.8. (*Continued*)

Component		Maintainability Factors	Critical Defects Mitigated	Noncritical Defects Mitigated	Wt.	RW
Sewage ejector and solid diverter tank	e41	Pit design	E27, E29		0.478	0.012
	e42	Location	E29		0.257	0.007
	e43	Access	E27		0.221	0.006
	e44	Services	E27		0.221	0.006
	e45	Pump and piping	E29		0.257	0.007
Sewer drain	e46	Material	E3	E31	0.207	0.005
	e47	Jointing	E3, E30		0.340	0.009
	e48	Location	E3		0.144	0.004
	e49	Layout	E30		0.196	0.005
					Design subtotal	0.460
Construction phase						
Water supply	e50	Storage and handling	E2, E11, E12		0.728	0.019
	e51	Laying	E3, E5, E6, E14	E7	0.898	0.023
	e52	Joint and penetration	E3, E4, E5	E9	0.744	0.019
	e53	Insulation	E2, E12		0.468	0.012
	e54	Backfill		E1	0.064	0.002
	e55	Testing	E4, E5	E1	0.610	0.016

(*Continued*)

Table 9.8. *(Continued)*

Component	Maintainability Factors		Critical Defects Mitigated	Noncritical Defects Mitigated	Wt.	RW
Storage tank	e56	Achieve water-tightness	E3, E8	E9	0.523	0.014
	e57	Device installation	E3	E10	0.173	0.004
	e58	Commissioning	E3, E8	E9	0.523	0.014
	e59	disinfection	E3		0.144	0.004
Pumps	e60	Mounting	E13, E14		0.458	0.012
	e61	Testing	E13, E15		0.565	0.015
Sanitary appliance	e62	Storage	E18		0.342	0.009
	e63	Fixing	E19, E20, E21		0.961	0.025
	e64	Testing	E19, E20		0.712	0.018
	e65	Protection	E18, E20		0.690	0.018
Sanitary pipe	e66	Storage and handling	E23, E24		0.515	0.013
	e67	Fixing	E24	E26	0.253	0.007
	e68	Jointing	E23, E24	E26	0.551	0.014
	e69	Protection	E23, E24		0.515	0.013
	e70	Test	E23		0.298	0.008
Swg. eject and solid divert	e71	Installation	E29		0.257	0.007
	e72	Testing	E29		0.257	0.007
Sewer	e73	Laying	E30	E31	0.259	0.007
	e74	Testing	E3, E30		0.340	0.009
					Subtotal	0.306

(Continued)

Table 9.8. *(Continued)*

Component	Maintainability Factors		Critical Defects Mitigated	Noncritical Defects Mitigated	Wt.	RW
Maintenance phase						
Cold water supply	e75	Check	E2, E3, E4, E5		0.957	0.025
	e76	Clean	E3, E5		0.446	0.012
Hot water supply	e77	Check	E3, E11		0.426	0.011
	e78	Clean	E11, E12		0.484	0.013
Storage	e79	Check	E3, E8		0.470	0.012
	e80	Clean	E3, E8	E10	0.499	0.013
Pumps	e81	Check	E13		0.302	0.008
	e82	Clean	E13, E15	E16	0.628	0.016
	e83	Test	E5	E17	0.327	0.008
Sanitary appliances	e84	Check	E19, E21, E24, E27	E22	1.103	0.028
	e85	Clean	E18, E20		0.690	0.018
	e86	Replace	E20, E27		0.569	0.015
Sanitary pipe	e87	Check	E23, E24		0.515	0.013
	e88	Clean	E24	E26	0.253	0.007
Swg. eject and	e89	Check	E29		0.257	0.007
solid divert	E90	clean	E27		0.221	0.006
Sewer	e91	Check		E31	0.063	0.002
	e92	Clean	E30		0.196	0.005
Ext. factor	e93	Building age	E2, E12, E28	E22	0.659	0.017
					Subtotal	0.234
					Total	1.000

Table 9.9. Maintainability factors for HVAC system and their RWs.

Component	Maintainability Factors		Asso. Critical Defects	Asso. Noncrit. Defects	Wt.	RW
Design phase						
AHU room	f1	Location	F38		0.098	0.004
	f2	Enclosure	F38	F3, F11	0.160	0.006
	f3	Access	F4, F10, F13		0.972	0.038
	f4	Mounting	F2, F38		0.317	0.012
	f5	Isolator	F2, F38		0.317	0.012
	f6	Services	F13		0.412	0.016
Air intake	f7	Location	F10	F37	0.345	0.013
	f8	Protection		F31, F37	0.054	0.002
	f9	Fan and motor	F1	F7	0.359	0.014
Drain pan	f10	Material	F5	F3	0.147	0.006
	f11	Drainage	F8	F31, F37	0.252	0.010
	f12	Drain off pipe	F8	F9	0.229	0.009
Filler	f13	Selection	F10		0.304	0.012
	f14	Layout	F10		0.304	0.012
	f15	Pressure gauge	F10		0.304	0.012
Cooling coil	f16	Material		F3	0.043	0.002
	f17	Coil row	F4	F7	0.292	0.011
FCU	f18	Location	F10	F37	0.345	0.013
	f19	Housing		F3	0.043	0.002
	f20	Access	F13, F39		0.528	0.020

(*Continued*)

Table 9.9. (*Continued*)

Component	Maintainability Factors		Asso. Critical Defects	Asso. Noncrit. Defects	Wt.	RW
	f21	Isolator	F2		0.219	0.008
	f22	Cooling coil	F4	F7	0.292	0.011
	f23	Drain pan material	F5	F3	0.147	0.006
	f24	Drainage	F8		0.198	0.008
	f25	Drain off pipe	F8	F9	0.229	0.009
	f26	Filter type	F10		0.304	0.012
Chiller	f27	Capacity	F19	F18, F27	0.173	0.007
	f28	Refrigerant	F17		0.170	0.007
	f29	Compressor	F19	F35	0.150	0.006
	f30	Condenser	F14	F18	0.359	0.014
	f31	Piping	F22		0.229	0.009
	f32	Purge unit	F17	F18	0.192	0.007
	f33	Mounting	F19		0.110	0.004
	f34	Thermal insulation	F22		0.229	0.009
	f35	Leak detector	F16, F17		0.289	0.011
	f36	Plant room	F19, F20, F23		0.475	0.018
Cooling tower	f37	Design	F24		0.347	0.013
	f38	Type	F14, F18		0.359	0.014
	f39	Material		F25	0.024	0.001
	f40	Location	F39	F27	0.157	0.006

(*Continued*)

Table 9.9. (*Continued*)

Component	Maintainability Factors		Asso. Critical Defects	Asso. Noncrit. Defects	Wt.	RW
	f41	Access		F26	0.028	0.001
	f42	Vibe/noise reduction	F28		0.067	0.003
	f43	Legionella prevention		F26	0.028	0.001
Air distribution	f44	Duct material	F30		0.128	0.005
	f45	Insulation		F29	0.046	0.002
	f46	Return air	F39		0.116	0.004
	f47	Air velocity		F33	0.033	0.001
	f48	Access	F32		0.369	0.014
	f49	Wet exhaust		F31	0.013	0.001
	f50	Terminal location	F34	F29	0.253	0.010
	f51	Terminal type	F34, F39		0.323	0.012
	f52	VAV box access	F39		0.116	0.004
	f53	Damper selection		F33	0.033	0.001
	f54	Controller location	F36		0.248	0.010
					Sub-total	0.475
Construction phase						
AHU/FCU	f55	Housing		F11	0.019	0.001
	f56	Filter	F4, F10, F32	F6	0.977	0.038
	f57	Fan and motor	F1, F2, F38		0.640	0.025

(*Continued*)

Table 9.9. (*Continued*)

Component	Maintainability Factors		Asso. Critical Defects	Asso. Noncrit. Defects	Wt.	RW
	f58	Cooling coil		F6	0.048	0.002
	f59	Drain pan and pipe	F5, F8	F3	0.346	0.013
Chiller	f60	General	F38	F15	0.139	0.005
	f61	Motor and controller	F16, F17		0.289	0.011
	f62	Fluid system	F21, F23	F35	0.406	0.016
Air distribution	f63	Duct quality	F28	F25	0.091	0.004
	f64	Fixing	F30, F32	F29	0.543	0.021
	f65	Testing	F30	F31	0.142	0.005
Cooling tower	f66	Installation		F33	0.033	0.001
					Subtotal	0.142
Maintenance phase						
AHU/FCU housing	f67	Clean	F13		0.412	0.016
	f68	Check	F12, F38	F3, F11	0.312	0.012
Cooling coil	f69	Clean	F4	F3, F6	0.346	0.013
	f70	Check	F4	F6	0.304	0.012
Drain pan	f71	Clean	F8		0.198	0.008
	f72	Check	F8	F9	0.229	0.009
Air intake and fan	f73	Clean	F2, F10, F32		0.893	0.035
	f74	Check	F1, F2, F10, F32		1.216	0.047

(*Continued*)

Table 9.9. (*Continued*)

Component	Maintainability Factors		Asso. Critical Defects	Asso. Noncrit. Defects	Wt.	RW
Chiller	f75	Check	F16, F20, F22		0.471	0.018
Compressor	f76	Check and test	F19, F21, F38	F35	0.373	0.014
	f77	Clean	F21		0.126	0.005
Condenser	f78	Check		F15	0.041	0.002
	f79	Clean	F14	F15, F18	0.497	0.019
	f80	Test	F19		0.110	0.004
Evaporator	f81	Check		F18	0.021	0.001
	f82	Clean		F3	0.043	0.002
Refrigerant	f83	Check	F16, F17		0.289	0.011
	f84	Purge	F17	F18	0.192	0.007
Control	f85	Check		F33	0.033	0.001
	f86	Clean		F35	0.039	0.002
Cooling tower	f87	Check	F28	F25	0.091	0.004
	f88	Clean and lubricate	F14, F24, F28, F39	F18, F25	0.913	0.035
	f89	Water treatment	F24, F39	F26	0.491	0.019
Air distribution	f90	Check	F34, F36, F39	F29, F33	0.689	0.027
	f91	Clean	F32, F34, F39		0.692	0.027
Ext. factor	f92	Building age	F2, F5, F12, F19, F23	F3	0.870	0.034
					Subtotal	0.383
					Total	**1.000**

Table 9.10. Maintainability factors for elevator and their RWs.

Component	Maintainability Factors		Asso. Critical Defects	Asso. Noncritical Defects	Wt.	RW
Design phase						
General	g1	Goal setting		G1, G4, G59	0.127	0.006
	g2	Performance selection	G6	G1	0.279	0.013
Machine room	g3	Location		G4	0.041	0.002
	g4	Size	G9	G8	0.306	0.014
	g5	Structure	G9	G11	0.243	0.011
	g6	Access	G9		0.217	0.010
	g7	Services	G10	G11	0.161	0.007
Machines in general	g8	Placing	G15	G18, G21	0.240	0.011
	g9	Load calculation		G16, G18	0.093	0.004
	g10	Material for parts	G15	G18, G21	0.240	0.011
	g11	Sheave dimension		G21	0.027	0.001
	g12	Sheave geometry		G21, G41	0.055	0.002
Roping system	g13	Load calculation	G3		0.193	0.009
	g14	Roping method	G3	G40	0.237	0.011
	g15	No. of strand	G22	G16, G40	0.198	0.009
	g16	Rope type	G22	G16, G40, G41	0.226	0.010
	g17	Rope lay	G22	G16	0.154	0.007

(Continued)

Table 9.10. (*Continued*)

Component	Maintainability Factors		Asso. Critical Defects	Asso. Noncritical Defects	Wt.	RW
Car	g18	Size and shape	G1, G4		0.092	0.004
	g19	Enclosure		G38	0.031	0.001
	g20	Internal finish	G31	G35	0.227	0.010
	g21	COP button	G6, G32, G42, G43		0.984	0.044
	g22	Services	G30, G31	G33	0.372	0.017
	g23	Car top	G34		0.170	0.008
	g24	Car bottom		G26	0.022	0.001
Hoistway	g25	Space design		G8, G37, G38	0.081	0.004
	g26	Landing opening		G8, G38	0.045	0.002
	g27	Material		G38, G39	0.075	0.003
	g28	Guiderail		G37	0.036	0.002
	g29	Services		G38	0.031	0.001
	g30	Counter weight		G4	0.041	0.002
	g31	Safety gear type		G4	0.041	0.002
	g32	Safety gear location		G25	0.016	0.001
Lobby/car door	g33	Material		G50, G51	0.090	0.004
	g34	Strength		G50, G51	0.090	0.004
	g35	Locks	G7, G52		0.239	0.011
	g36	Operation		G46	0.035	0.002
Landing	g37	Material selection	G54		0.324	0.015
	g38	Fire protection	G55	G56	0.245	0.011

(*Continued*)

Table 9.10. *(Continued)*

Component	Maintainability Factors		Asso. Critical Defects	Asso. Noncritical Defects	Wt.	RW
Pit	g39	Size	G28, G57	G8	0.450	0.020
	g40	Structure	G57	G8	0.327	0.015
	g41	Access	G28, G57		0.436	0.020
	g42	Services	G58	G8	0.147	0.007
	g43	Partitions		G8	0.014	0.001
Ancillary facilities	g44	Coordination	G10, G19, G24, G31	G12, G38, G39	0.704	0.032
					Subtotal	0.379
Construction phase						
Structure	g45	Quality	G57	G11, G37	0.374	0.017
	g46	Protection	G57		0.313	0.014
	g47	Landing opening	G3		0.193	0.009
Equipment	g48	Controller	G6, G14	G2, G17, G46	0.409	0.018
	g49	Traction machine and motor	G15	G18	0.093	0.004
	g50	Brake assembly	G19	G4, G5, G20	0.246	0.011
	g51	Lift car	G27	G23	0.107	0.005
	g52	Rope		G4, G40	0.085	0.004
	g53	Doors		G46, G47, G48	0.087	0.004
	g54	Cable		G26, G29, G59	0.076	0.003
					Subtotal	0.089

(Continued)

Table 9.10. (*Continued*)

Component		Maintainability Factors	Asso. Critical Defects	Asso. Noncritical Defects	Wt.	RW
Maintenance phase						
Machine room	g55	Cleaning	G9		0.217	0.010
	g56	Wkg. condn.	G10	G8	0.150	0.007
	g57	Check and record	G9	G11	0.243	0.011
Controller	g58	Clean	G14	G1, G13	0.155	0.007
	g59	Check and record	G6, G7, G14, G42, G49, G52	G2, G12, G13, G46, G47	1.140	0.051
Traction machine and motor	g60	Clean and lubricate	G15, G18	G4, G17	0.286	0.013
	g61	Check and record	G15	G1	0.222	0.010
Brake assembly	g62	Clean and lubricate	G19	G5, G20	0.204	0.009
	g63	Check and record		G5, G20	0.089	0.004
Governor machine	g64	Clean and lubricate	G22	G21	0.129	0.006
	g65	Check and record	G22		0.102	0.005
Lift car body	g66	Clean and lubricate	G27, G28	G25	0.219	0.010
	g67	Check	G3, G27	G23, G29	0.319	0.014
	g68	Safety	G7, G24	G34	0.295	0.013
Car interior	g69	Clean	G6		0.229	0.010
	g70	Checking	G30, G32, G42, G43	G33	0.942	0.042
Hoistway	g71	Clean	G36	G8, G38, G39	0.191	0.009
	g72	Check		G37	0.036	0.002

(*Continued*)

Table 9.10. (*Continued*)

Component	Maintainability Factors		Asso. Critical Defects	Asso. Noncritical Defects	Wt.	RW
Roping system	g73	Lubricate		G16, G37, G41	0.115	0.005
	g74	Check and record		G40	0.044	0.002
Lobby/car door	g75	Clean	G43	G44, G46, G47	0.383	0.017
	g76	Check	G7, G42, G43, G49, G52, G53	G44, G45, G46	1.201	0.054
Lobby	g77	Clean	G55		0.209	0.009
	g78	Check	G55	G56	0.245	0.011
Pit	g79	Clean and lubricate	G57	G8, G59	0.363	0.016
	g80	Check and record	G55, G58	G4, G59	0.599	0.027
	g81	Safety		G8	0.392	0.018
External factors	g82	Age of the building	G19, G22	G2, G18, G40, G46	0.363	0.016
	g83	Level of usage	G6, G30, G32, G42, G43	G46, G50	1.257	0.057
	g84	Vandalism	G6, G7, G30, G32, G42, G43, G53	G47, G50	1.453	0.066
					Subtotal	0.532
					Total	**1.000**

Table 9.11. Maintainability factors for electrical system and their RWs.

Component	Maintainability Factors		Asso. Critical Defect	Asso. Noncritical Defect	Wt.	RW
Design phase						
System in general	h1	Capacity	H2		0.191	0.005
	h2	Phase	H2		0.191	0.005
	h3	Location	H12		0.186	0.005
	h4	Label/notice	H12	H6, H18	0.247	0.006
Transformer	h5	Selection	H2, H3, H10, H11, H13, H21	H8, H22	1.193	0.029
	h6	Mounting		H8	0.032	0.001
	h7	Location		H8	0.032	0.001
	h8	Access	H11, H12		0.326	0.008
	h9	Room design	H11, H12	H8	0.358	0.009
	h10	Noise reduction		H8	0.032	0.001
Cable and wiring	h11	Material	HI, H13, H16		0.829	0.020
	h12	Size	H1, H2, H5, H13, H16		1.299	0.031
	h13	Cable selection	H1, H2, H13, H16	H14	1.053	0.026
	h14	Insulation	H1, H2, H13, H16		1.020	0.025
	h15	Joints	H15		0.189	0.005
	h16	Thermal effect	H1, H2, H4, H13, H16, H27	H18	1.427	0.035
	h17	Ph. Protection		H6, H14	0.066	0.002
	h18	Laying method	H5	H14	0.313	0.008
	h19	Buswasy select	H1, H2, H5, H13, H16		1.299	0.031
	h20	Location	H13	H14	0.357	0.009
	h21	Layout	H1, H2, H4, H13, H16	H18	1.216	0.029

(*Continued*)

Table 9.11. *(Continued)*

Component	Maintainability Factors		Asso. Critical Defect	Asso. Noncritical Defect	Wt.	RW
Connector	h22	Jn./pull box		H6, H14, H17	0.101	0.002
	h23	Splice	H5, H15		0.469	0.011
	h24	Termination	H5, H15		0.469	0.011
	h25	Gen. spec.	H19	H18, H23	0.272	0.007
	h26	Housing	H3, H4, H19, H20	H18, H23	0.899	0.022
	h27	Switchboard		H18	0.029	0.001
	h28	Panelboard spec.	H19		0.210	0.005
	h29	Panelboard nos.	H3		0.181	0.004
	h30	E. rm./closet	H4	H18	0.197	0.005
	h31	Outlet box	H20		0.279	0.007
	h32	Switch		H23	0.033	0.001
	h33	Receptacle	H20	H22	0.305	0.007
	h34	Short-circuit	H1, H13, H16, H27	H7, H24, H31	1.136	0.028
	h35	Ground fault	H4, H16	H18	0.488	0.012
Lighting	h36	Aesthetics	H28		0.211	0.005
	h37	Flexibility	H28		0.211	0.005
	h38	Maintainability	H28	H30	0.245	0.006
	h39	Lamp selection	H28	H29	0.237	0.006
	h40	Ballast selection		H32	0.041	0.001

(Continued)

Table 9.11. *(Continued)*

Component	Maintainability Factors		Asso. Critical Defect	Asso. Noncritical Defect	Wt.	RW
Standby and emergency power	h41	Source selection	H33, H35, H36		0.460	0.011
	h42	Housing	H33, H36		0.348	0.008
	h43	Generator selection	H33, H35, H36		0.460	0.011
	h44	Control	H36		0.144	0.004
	h45	Battery selection	H36		0.144	0.004
	h46	Battery discharge	H36		0.144	0.004
Grounding	h47	Shape	H38		0.125	0.003
	h48	Conductor	H38	H37	0.153	0.004
	h49	Electrode	H38	H37	0.153	0.004
	h50	Location	H38	H39	0.254	0.006
	h51	Networking	H38		0.125	0.003
	h52	Access	H38	H39	0.254	0.006
LPS	h53	Shape	H38		0.125	0.003
	h54	Material	H38	H37	0.153	0.004
	h55	Joint	H38	H37	0.153	0.004
	h56	Fastener	H38	H37	0.153	0.004
Ancillary facility	h57	Coordination	H15, H28	H39	0.529	0.013
					Subtotal	0.527

(Continued)

Table 9.11. *(Continued)*

Component	Maintainability Factors		Asso. Critical Defect	Asso. Noncritical Defect	Wt.	RW
Construction phase						
Transformer	h58	Inspection	H11, H12	H8	0.358	0.009
	h59	Elec. test	H2, H3		0.371	0.009
Cable	h60	Prep.	H1, H13, H16	H14	0.719	0.017
	h61	Pulling	H15, H16	H14	0.513	0.012
	h62	Jointing	H1, H13, H15, H16	H14, H17	1.087	0.026
	h63	Protection	H5, H15	H14	0.502	0.012
	h64	Testing	H2		0.191	0.005
Busway	h65	Prep.	H1, H13, H16	H14	0.862	0.021
	h66	Install	H1, H13, H15, H16		1.018	0.025
	h67	Testing	H2		0.191	0.005
Dist. and prot. device	h68	Wiring	H2, H4, H25	H18	0.721	0.017
	h69	Connection	H4, H5, H25	H14, H17, H18	0.878	0.021
	h70	Testing	H2, H16	H14, H17	0.550	0.013
Lighting	h71	Fixing		H32	0.041	0.001
	h72	Testing		H29	0.034	0.001
Standby and emergency power	h73	Gen. install	H33, H36		0.348	0.008
	h74	Generator commissioning	H33, H36		0.348	0.008
	h75	Battery install	H36		0.144	0.004

(Continued)

Table 9.11. (*Continued*)

Component	Maintainability Factors		Asso. Critical Defect	Asso. Noncritical Defect	Wt.	RW
Grounding	h76	Prep.	H38	H37	0.153	0.004
	h77	Install	H38	H39	0.254	0.006
	h78	Test	H38		0.125	0.003
LPS	h79	Install	H38	H37	0.153	0.004
	h80	Test	H38		0.125	0.003
					Subtotal	0.235
Maintenance phase						
Transformer	h81	Clean	H8	H11, H12	0.358	0.009
	h82	Check	H8, H10	H11	0.288	0.007
	h83	Test		H11	0.140	0.003
Wiring	h84	Clean	H15, H16	H14	0.513	0.012
	h85	Check		H14, H17	0.069	0.002
Distr. equipment	h86	Clean		H18	0.029	0.001
	h87	Check		H14, H17	0.069	0.002
	h88	Test		H24	0.029	0.001
Protective equipment	h89	General	H3, H25		0.514	0.012
	h90	CB	H1, H3, H13, H16, H25		1.343	0.033
	h91	Fuse	H1, H3, H13, H16		1.010	0.024
	h92	GFCI	H13, H16, H25		0.948	0.023

(*Continued*)

Table 9.11. (*Continued*)

Component	Maintainability Factors		Asso. Critical Defect	Asso. Noncritical Defect	Wt.	RW
Lighting	h93	Clean	H28, H35		0.323	0.008
	h94	Check	H27, H28	H30, H31, H32	0.531	0.013
	h95	Replace	H27, H28	H31	0.456	0.011
Standby power	h96	Check	H36	H34	0.142	0.003
Generator	h97	Clean	H36	H34	0.142	0.003
	h98	Check	H33		0.203	0.005
	h99	Test	H35		0.113	0.003
Battery	h100	Clean	H36		0.144	0.004
	h101	Check	H36		0.144	0.004
	h102	Calibrate and test	H36		0.144	0.004
Grounding and LPS	h103	Check	H38	H37	0.153	0.004
	h104	Test	H38	H39	0.254	0.006
Ext. factors	h105	Building age	H4, H5, H12, H25, H38	H18, H30	1.155	0.028
	h106	Use		H22, H23	0.058	0.001
	h107	Vandalism	H12, H39	H14, H18, H22, H23, H39	0.560	0.014
					Subtotal	0.238
					Total	**1.000**

Table 9.12. Maintainability factors for fire protection system and their RWs.

Component		Maintainability Factors	Asso. Critical Defect	Asso. Noncritical Defect	Wt.	RW
Design phase						
Detector	j1	Performance	J1		0.251	0.015
	j2	Applicability	J1		0.251	0.015
	j3	Mounting	J1		0.251	0.015
Alarm	j4	Selection		J3	0.027	0.002
	j5	Location		J3	0.027	0.002
	j6	Connectivity		J4	0.046	0.003
Alarm panel	j7	Location		J4	0.046	0.003
	j8	Legibility		J3, J4	0.027	0.002
	j9	Indicator		J3	0.073	0.004
	j10	Safety		J4	0.046	0.003
	j11	Connectivity		J4	0.046	0.003
	j12	Power supply		J2, J3, J4	0.106	0.006
Hydrant	j13	Location		J4, J6	0.083	0.005
	j14	Security		J6		0.002
	j15	Material		J6, J8	J8	0.004
	j16	Supply		J8		0.002
Hosereel	j17	Cabinet location	J11		0.120	0.007
	j18	Cabinet spec.	J11	J12	0.161	0.009
	j19	Hose	J10		0.194	0.011

(Continued)

Table 9.12. (*Continued*)

Component	Maintainability Factors		Asso. Critical Defect	Asso. Noncritical Defect	Wt.	RW
Sprinkler	j20	Selection	J14		0.209	0.012
	j21	Location	J15		0.152	0.009
	j22	Material	J16, J17	J20, J21	0.403	0.024
	j23	Support		J18	0.032	0.002
	j24	Water quality	J16	J21, J25	0.264	0.016
	j25	Pump location	J19		0.135	0.008
Portable fire extinguisher	j26	Selection	J29		0.182	0.011
	j27	Location	J26		0.234	0.014
	j28	Quality	J26, J28	J30	0.464	0.027
Fire door	j29	Selection		J32	0.048	0.003
	j30	Operation	J31		0.221	0.013
	j31	Detailing	J31	J32	0.269	0.016
	j32	Marking		J32	0.048	0.003
Escape lighting	j33	Power supply	J34		0.175	0.010
	j34	System selection	J34		0.175	0.010
	j35	Lamp selection	J34		0.175	0.010
	j36	Battery discharge	J34		0.175	0.010
	j37	Battery location	J34		0.175	0.010
	j38	Charger	J34		0.175	0.010
Smoke ctrl.	j39	Safety and back-up	J36		0.116	0.007
					Sub-total	0.337

(*Continued*)

Table 9.12. (*Continued*)

Component	Maintainability Factors		Asso. Critical Defect	Asso. Noncritical Defect	Wt.	RW
Construction phase						
Detector and alarm	j40	Wiring		J2, J4	0.079	0.005
	j41	Mounting	J1		0.251	0.015
	j42	Testing	J1		0.251	0.015
Hydrant	j43	Storage		J6	0.037	0.002
	j44	Install		J5, J6	0.082	0.005
	j45	Testing	J7	J6, J8	0.266	0.016
Hosereel	j46	Base work	J11		0.120	0.007
	j47	Commissioning	J10, J11		0.314	0.018
Sprinkler	j48	Mounting	J15		0.152	0.009
	j49	Support	J15	J18	0.184	0.011
	j50	Supply network	J17, J19	J21, J24	0.345	0.020
Portable extinguisher	j51	Mounting		J27, J30	0.070	0.004
	j52	Protection	J26, J28		0.427	0.025
Fire door	j53	Storage and handling		J31	0.048	0.003
	j54	Fixing	J31		0.221	0.013
	j55	Checking	J31		0.221	0.013
Light	j56	Fixing	J34		0.175	0.010
	j57	Marking	J34		0.175	0.010
	j58	Testing	J34		0.175	0.010

(*Continued*)

Table 9.12. *(Continued)*

Component	Maintainability Factors		Asso. Critical Defect	Asso. Noncritical Defect	Wt.	RW
Smoke ctrl.	j59	Install	J36		0.116	0.007
					Sub-total	0.218
Maintenance phase						
Detector and alarm	j60	Clean	J1	J4	0.297	0.017
	j61	Test	J1	J4	0.297	0.017
	j62	Check and record		J2, J3, J4	0.106	0.006
Hydrant	j63	Clean		J6, J8	0.072	0.004
	j64	Check		J6, J9	0.076	0.004
	j65	Test and record		J6	0.037	0.002
Hosereel	j66	Clean	J11		0.120	0.007
	j67	Check	J10, J10		0.314	0.018
	j68	Test	J10		0.194	0.011
Sprinkler	j69	Clean	J14, J15		0.361	0.021
	j70	Check	J14, J15		0.361	0.021
Supply	j71	Clean	J16, J17		0.345	0.020
	j72	Check	J16, J19	J20, J23	0.396	0.023
	j73	Test	J19		0.135	0.008
Valves	j74	Clean		J21, J24, J25	0.106	0.006
	j75	Check		J21, J22, J23, J24, J25	0.149	0.009
	j76	Test		J21, J24	0.080	0.005

(Continued)

Table 9.12. *(Continued)*

Component	**Maintainability Factors**		**Asso. Critical Defect**	**Asso. Noncritical Defect**	**Wt.**	**RW**
Portable extinguisher	j77	Check	J26, J28, J29	J27, J30	0.679	0.040
	j78	Replace/recharge	J28, J29		0.374	0.022
	j79	Test and record	J28		0.193	0.011
Fire door	j80	Clean	J31, J33	J32	0.448	0.026
	j81	Check and record	J31	J32	0.269	0.016
	j82	Lubricate	J33		0.178	0.010
	j83	Repair and replace		J32	0.048	0.003
Lighting	j84	Clean	J34		0.175	0.010
	j85	Check and record	J34		0.175	0.010
	j86	Calibrate and test	J34		0.175	0.010
Smoke ctrl.	j87	Check and test	J36	J35	0.151	0.009
External factors	j88	Building age	J10, J33	J18, J21	0.439	0.026
	j89	Level of usage	J31	J32	0.269	0.016
	j90	Vandalism	J26, J28	J13, J22, J24	0.548	0.032
					Subtotal	0.445
					Total	**1.000**

Table 9.13. No. of maintainability factors.

System	Subsystem	Total	Design No.	Design %	Construction No.	Construction %	Maintenance No.	Maintenance %
C&A	Basement	63	30	47.62	24	38.10	9	14.29
	Façade	79	42	53.16	26	32.91	11	13.92
	Wet area	59	32	54.24	18	30.51	9	15.25
	Roof	64	35	54.69	21	32.81	8	12.50
	Avg.	66.25	34.75	52.43	22.25	33.58	9.25	13.99
M&E	San-plumb.	93	49	52.69	25	26.88	19	20.43
	HVAC	92	54	58.70	12	13.04	26	28.26
	Elevator	84	44	52.38	10	11.90	30	35.71
	Electrical	107	57	53.27	23	21.50	27	25.23
	Fire prot.	90	39	43.33	20	22.22	30	33.33
	Avg.	93.20	48.60	52.07	18.00	19.11	26.40	28.59
Entire bldg.		731	382	52.26	179	24.49	169	23.12

References

[1] J. A. Todd, D. Crawley, S. Geissler and G. Lindsey, "Comparative assessment of environmental performance tools and the role of the green building challenge", *Building Research and Information*, **29**(5), 324–335, 2001.

[2] *R-2000.* (n.d.), http://r2000.chba.ca/. Retrieved 13 October 2006.

[3] SP Technical Research Institute of Sweden, "The P-Mark", http://www.sp.se/EN/UNITS/CERTIFICATION/PRODUCT/P_MARK/Sidor/default.aspx, May 2009.

[4] R. Anneling, "The P-mark system for prefabricated houses in Sweden", *Swedish National Testing and Research Institute CADDET Newsletter No.* 1, 1998.

[5] EPA, "Energy Star", http://www.energystar.gov/, May 2009.

[6] L. T. Sing, "Green Mark for building: Promoting environmental friendly building", Presentation at *SEI Professional Sharing Series*, Singapore, July 2005.

[7] U.S. Green Building Council, "Rating system for pilot demonstration of LEED for homes program", Version 1.72, http://www.usgbc.org/FileHandling/show_general_file.asp?DocumentID=855. Retrieved 19 October 2006.

[8] M. Horvat and P. Fazio, "Comparative review of existing certification programs and performance assessment tools for residential buildings", *Architectural Science Review*, **48**(1), 69–80, 2005.

[9] Building Research Establishment Ltd, "BREEAM Offices 2006: Pre-assessment estimator: Design and procurement", http://www.breeam.org/pdf/OFF_DP_Pre-assessmentEstimator2006Rev00.pdf. Retrieved 6 November 2006.

[10] Building Research Establishment Ltd, "BREEAM Offices 2006: Pre-assessment estimator: Management and operation", http://www.breeam.org/pdf/OFF_MO_Pre-assessmentEstimator2006Rev00.pdf. Retrieved 6 November 2006.

[11] Building Research Establishment Ltd, "EcoHomes 2006 — The environmental rating for homes", http://www.breeam.org/pdf/EcoHomes2006Guidance_v1_2_April2006.pdf. Retrieved 6 November 2006.

[12] Building Research Establishment Ltd, "EcoHomes XB: The environmental rating for existing housing", http://www.breeam.org/ecohomes.html. Retrieved 6 November 2006.

[13] International Energy Agency, "IEA Annex 31: Energy related environmental impact of buildings", http://www.iisbe.org/annex31/index.html. Retrieved 14 October 2006.

[14] M. Horvat and P. Fazio, "Comparative review of existing certification programs and performance assessment tools for residential buildings", *Architectural Science Review*, **48**(1), 69–80, 2005.

[15] *Green Globes v.1 Rating System.* (n.d.), http://www.thegbi.com/greenglobes/questionnaire.asp. Retrieved 15 October 2006.

[16] *Global Environmental Method UK.* (n.d.), http://www.greenglobes.com/existing/homeuk.asp. Retrieved 14 October 2006.

[17] Building Owners and Manufacturers Association, Canada. "Go Green: Overview and application process", http://www.bomagogreen.com/docs/Overview_Process.pdf. Retrieved 15 October 2006.

[18] Building Owners and Manufacturers Association, Canada. "Go Green plus: Program guide", http://www.bomagogreen.com/docs/GoGreenPlus_ProgramGuide_20060512.pdf. Retrieved 15 October 2006.

[19] I. Eastin and P. Boardman, "Regulatory change in the Japanese residential construction industry", *CINTRAFOR Newsletter*, **16**(2), 2000, http://forests.org/archive/asia/jahosect.htm. Retrieved 28 September 2006.

[20] M. Y. L. Chew and N. De Silva, "Factorial method for performance assessment of building façades", *Journal of Construction Engineering and Management*, **130**(4), 525–533, 2004.

[21] M. Y. L. Chew, N. D. D. S. Egodage and S. S. Tan, "A neural network approach to assessing the building façade maintainability in the tropics", *Construction Management and Economics*", **22**(6), 581–594, 2003.

[22] *Green Star* (n.d.), http://www.gbcaus.org/. Retrieved 9 October 2006.

[23] Department of Energy, Utilities and Sustainability, NSW, "NABERS office buildings trial", http://www.nabers.com.au/nabersdocuments/report_nov_8.pdf. Retrieved 26 September 2006.

[24] Energy Smart Home (n.d.), http://www.energysmarthome.com.au/. Retrieved 5 November 2006.

[25] *Building Sustainability Index* (n.d.), http://www.basix.nsw.gov.au/. Retrieved 10 October 2006.

[26] HK-BEAM Society, *HK-BEAM 5/04: Existing buildings*, http://www.hk-beam.org.hk/fileLibrary/5-04%20(full)%2021feb05.pdf. Retrieved 27 October 2006.

[27] HK-BEAM Society, *HK-BEAM 4/04: New buildings*, http://www.hk-beam.org.hk/fileLibrary/_4-04%20New%20Buildings%20(Full%20Version).pdf. Retrieved 27 October 2006.

[28] Natural Resources Canada, "Conference proceedings", *Green Building Challenge'98*, Vancouver, Canada, 1998.

[29] International Initiative for a Sustainable Built Environment (iiSBE), (n.d.), *GBC '98 – 2005 process overview*, http://www.iisbe.org/iisbe/gbc2k5/gbc2k5-start.htm. Retrieved 23 August 2006.

[30] K. F. Chang, C. M. Chiang and P. C. Chou, "Adapting aspects of GBTool 2005 — Searching for suitability in Taiwan", *Building and Environment*, **42**(1), 310–316, 2007.

[31] W. L. Lee and J. Burnett, "Customisation of GBTool in Hong Kong", *Building and Environment*, **41**(12), 1831–1846, 2006.

[32] R. J. Cole and N. Larsson, "Green building challenge 2002: GBTool user manual", http://www.iisbe.org/iisbe/gbc2k5/gbc2k5-start.htm. Retrieved 23 August 2006.

[33] Japan Sustainable Building Consortium, *CASBEE-NC* (*for New construction*) technical manual, http://www.ibec.or.jp/CASBEE/english/download.htm. Retrieved 30 September 2006.

[34] Institute for Building Environment and Energy Conservation, *CASBEE for new construction* Tool-1, 2004 Edition, Japan, 2004.

[35] M. Y. L. Chew, "Maintainability of buildings under tropical conditions", R-296-000-046-112/490, National University of Singapore, March 2004.

[36] British Standard (1996) BS8444: Risk Management: Part 3 — Guide to Risk Analysis of Technical Systems, British Standard Institution, London.

[37] W. V. Honstede, "Research into quality of housing stock in the Netherlands", L. K. Quah (ed.), *Building Maintenance and Modernisation Worldwide*, **2**, 659–668, Longman, Singapore, 1990.

[38] L. M. Gambardella and M. Moroni, "Expert systems application to building pathology diagnosis: Methodology", *Proc. 2nd Euro. Conf. on Application of Artificial Intelligence and Robotics to Bldg., Arch. and Civil Engg.*, Liege, Belgium.

[39] Olubodun, F, *An Empirical Approach to the Evaluation of Factors in Local Authority Housing*, Salford, Manchester.

[40] S. Assaf, A. Al-Hammad and M. Al-Shihah, "The effect of faulty construction on building maintenance", *Building Research and Information*, **23**, 175–181, 1995.

[41] A. Al-Hammad and S. Assaf, "The effect of faulty design on building mainenance", *Journal of Quality in Maintenance*, **3**(1), 29–39, 1997.

INDEX